Autodesk Simulation Mechanical 2016 for Designers

(3rd Edition)

CADCIM Technologies
525 St. Andrews Drive
Schererville, IN 46375, USA
(www.cadcim.com)

Contributing Author
Sham Tickoo
Professor
Department of Mechanical Engineering Technology
Purdue University Calumet
Hammond, Indiana, USA

CADCIM Technologies

Autodesk Simulation Mechanical 2016 for Designers, 3rd Edition
Sham Tickoo

CADCIM Technologies
525 St Andrews Drive
Schererville, Indiana 46375, USA
www.cadcim.com

ISBN 978-1-936646-98-2

NOTICE TO THE READER

www.cadcim.com

DEDICATION

*To teachers, who make it possible to disseminate knowledge
to enlighten the young and curious minds
of our future generations*

*To students, who are dedicated to learning new technologies
and making the world a better place to live in*

SPECIAL RECOGNITION

*A special thanks to Mr. Denis Cadu and the ADN team of Autodesk Inc.
for their valuable support and professional guidance to
procure the software for writing this textbook*

THANKS

*To the faculty and students of the MET department of
Purdue University Calumet for their cooperation*

To employees of CADCIM Technologies for their valuable help

Online Training Program Offered by CADCIM Technologies

CADCIM Technologies provides effective and affordable virtual online training on various software packages including Computer Aided Design, Manufacturing and Engineering (CAD/CAM/CAE), computer programming languages, animation, architecture, and GIS. The training is delivered 'live' via Internet at any time, any place, and at any pace to individuals as well as to the students of colleges, universities, and CAD/CAM/CAE training centers. The main features of this program are:

Training for Students and Companies in a Classroom Setting

Highly experienced instructors and qualified engineers at CADCIM Technologies conduct the classes under the guidance of Prof. Sham Tickoo of Purdue University Calumet, USA. This team has authored several textbooks that are rated "one of the best" in their categories and are used in various colleges, universities, and training centers in North America, Europe, and in other parts of the world.

Training for Individuals

CADCIM Technologies with its cost effective and time saving initiative strives to deliver the training in the comfort of your home or work place, thereby relieving you from the hassles of traveling to training centers.

Training Offered on Software Packages

CADCIM provides basic and advanced training on the following software packages:

CAD/CAM/CAE: CATIA, Pro/ENGINEER Wildfire, Creo Parametric, Creo Direct, SOLIDWORKS, Autodesk Inventor, Solid Edge, NX, AutoCAD, AutoCAD LT, AutoCAD Plant 3D, Customizing AutoCAD, AutoCAD MEP, AutoCAD Plant 3D, Autodesk Simulation Mechanical, EdgeCAM, and ANSYS

Architecture and GIS: Autodesk Revit Architecture, AutoCAD Civil 3D, Autodesk Revit Structure, AutoCAD Map 3D, Revit MEP, Navisworks, Primavera Project Planner, and Bentley STAAD Pro

Animation and Styling: Autodesk 3ds Max, , Autodesk Maya, Autodesk Alias, Foundry NukeX, and MAXON CINEMA 4D

Computer Programming: C++, VB.NET, Oracle, AJAX, and Java

*For more information, please visit the following link: **http://www.cadcim.com***

Note
If you are a faculty member, you can access the teaching resources by clicking on the following link: ***http://www.cadcim.com/Registration.aspx***. The student resources are available at ***http://www.cadcim.com***. We also provide Live **Virtual Online Training** on various software packages. For more information, write us at ***sales@cadcim.com***.

Table of Contents

Chapter 2: Introduction to Autodesk Simulation Mechanical

Chapter 3: Importing and Exporting Geometry

Chapter 4: Creating and Modifying a Geometry

Chapter 5: Meshing-I

Chapter 6: Meshing-II

Chapter 7: Working with Joints and Contacts

Chapter 8: Defining Materials and Boundary Conditions

Chapter 9: Performing Analysis and Viewing Results

Chapter 10: Advanced Structural Analysis

- **Real-world Mechanical Engineering Projects as Tutorials**
  The author has used about 30 real-world mechanical engineering projects as tutorials in this book. This will enable the readers to relate the tutorials to the real-world models in the mechanical engineering industry. In addition, there are about 25 exercises based on the real-world mechanical engineering projects.

- **Tips and Notes**
  Additional information related to various topics is provided to the users in the form of tips and notes.

- **Learning Objectives**
  The first page of every chapter summarizes the topics that are covered in the chapter.

- **Self-Evaluation Test, Review Questions, and Exercises**
  The chapters with Self-Evaluation Test that enables the users to assess their knowledge of the chapter. The answers to Self-Evaluation Test are given at the end of the chapter. Also, Review Questions and Exercises are given at the end of the chapters and they can be used by the Instructors as test questions and exercises.

- **Heavily Illustrated Text**
  The text in this textbook is heavily illustrated with the help of around 400 line diagrams and 500 screen captures.

Symbols Used in the Textbook

Note

The author has provided additional information to the users about the topic being discussed in the form of Notes.

Tip

Special information and techniques are provided in the form of Tips that will increase the efficiency of the users.

Formatting Conventions Used in the Textbook

Please refer to the following list for the formatting conventions used in this textbook.

- Names of tools, buttons, options, panels, tabs, and Ribbon are written in boldface.

  Example: The **Generate 3D Mesh** tool, the **OK** button, the **Mesh** panel, the **Analysis** tab, and so on.

- Names of dialog boxes, drop-downs, drop-down lists, list boxes, areas, edit boxes, check boxes, and radio buttons are written in boldface.

  Example: The **Model Mesh Settings** dialog box, the **Beam and Truss** drop-down in the **Stress** panel of the **Results Contours** tab, the **Type** drop-down list in the **Model Mesh Settings** dialog box, the **Size** edit box in the **Model Mesh Settings** dialog box, the **Do not dismiss dialog after bolt generation** check box in the **Generate Bolted Connection** dialog box, the **Axial Force** radio button in the **Generate Bolted Connection** dialog box, and so on.

- Values entered in edit boxes are written in boldface.

  Example: Enter **5** in the **Bolt diameter** edit box.

- Names and paths of the files are written in italics.

  Example: *C:\Autodesk Simulation Mechanical\c04\ Tut03*, and so on

- The methods of invoking a tool/option from the **Ribbon**, **Quick Access Toolbar**, **Application Menu**, are enclosed in a shaded box.

  Ribbon: Analysis > Run Simulation
  Quick Access Toolbar: New
  Application Menu: New

Naming Conventions Used in the Textbook

Tool

If you click on an item in a panel of the **Ribbon** and a command is invoked to create/edit an object or perform some action, then that item is termed as **tool**.

For example:
To Edit: **Divide** tool, **Intersect** tool, **Trim** tool
To Mesh: **Generate 3D Mesh** tool
To Analysis: **Run Simulation** tool
Action: **Zoom All** tool, **Pan** tool, **Copy Object** tool

If you click on an item in a panel of the **Ribbon** and a dialog box is invoked wherein you can set the properties to create/edit an object then that item is also termed as **tool**, refer to Figure 1.

For example:
To Create: **Line** tool, **Joint** tool, **Bolt** tool
To Mesh: **Generate 3D Mesh** tool

Student Resources

• **Technical Support**

You can get online technical support by contacting *techsupport@cadcim.com*.

• **Part Files**

The part files used in illustrations and tutorials are available for free download.

If you face any problem in accessing these files, please contact the publisher at ***sales@cadcim.com*** or the author at ***stickoo@purduecal.edu*** or ***tickoo525@gmail.com***.

Stay Connected

You can now stay connected with us through Facebook and Twitter to get the latest information about our textbooks, videos, and teaching/learning resources. To stay informed of such updates, follow us on Facebook (***www.facebook.com/cadcim***) and Twitter (***@cadcimtech***). You can also subscribe to our You Tube channel (***www.youtube.com/cadcimtech***) to get the information about our latest video tutorials.

Chapter 1

Introduction to FEA

Learning Objectives

After completing this chapter, you will be able to:
- *Understand basic concepts and the general working of FEA*
- *Understand advantages and limitations of FEA*
- *Understand the type of analysis*
- *Understand important terms and definitions used in FEA*
- *Understand theories of failure in FEA*

Stiffness Matrix

The stiffness matrix represents the resistance offered by a body to withstand the load applied. In the previous equation, the following part represents the stiffness matrix (K):

$$\begin{bmatrix} K_1 + K_2 & -K_2 \\ -K_2 & K_2 \end{bmatrix}$$

This matrix is relatively simple because it comprises only one pair of spring, but it turns complex when the number of springs increases.

Degree of Freedom

The Degree of freedom is defined as the ability of a node to translate or transmit the load. In the previous example, you are only concerned with the displacement and forces. By making one endpoint fixed, one degree of freedom for displacement is removed. So, now the model has two degrees of freedom. The number of degrees of freedom in a model determines the number of equations required to solve the mathematical model.

Boundary Conditions

The boundary conditions are used to eliminate the unknowns in a system. A set of equations that is solvable is meaningless without the input. In the previous example, the boundary condition was $X_0 = 0$, and the input forces were F1 and F2. In either ways, the displacements could have been specified in place of forces as boundary conditions and the mathematical model could have been solved for the forces. In other words, the boundary conditions help you reduce or eliminate unknowns in the system.

The FEA technique needs the finite element model (FEM) for its final solution as it does not use the solid model. FEM consists of nodes, keypoints, elements, material properties, loading, and boundary conditions.

Nodes, Elements, and Element Types

Before proceeding further, you must be familiar with commonly used terms such as nodes, elements, and element types. These terms are discussed next.

Nodes

An independent entity in space is called a node. Nodes are similar to the points in geometry and represent the corner points of an element. You can change the shape of an element by moving the nodes in space. The shape of a node is shown in Figure 1-3.

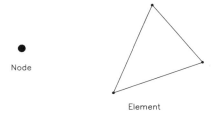

Figure 1-3 *A node and an element*

Elements

An element is an entity into which the system under study is divided. The shape (area, length, and volume) of an element is specified by nodes. Figure 1-3 shows a triangular shaped element.

Element Types

The following are the basic types of the elements:

Point Element

A point element is in the form of a point and therefore has only one node.

1D Element

A 1D element has the shape of a line or curve, therefore a minimum of two nodes are required to define it. There can be higher order elements that have additional nodes (at the middle of the edge of the element). The element that does not have a node at the middle of the edge of the element is called a linear element. The elements with node at the mid of the edges are called quadratic or second order elements. Figure 1-4 shows some line elements.

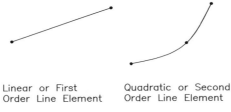

Linear or First   Quadratic or Second
Order Line Element   Order Line Element

Figure 1-4 *The 1D elements*

2D Element

An 2D element has the shape of a quadrilateral or a triangle, therefore it requires a minimum of three or four nodes to define it. Some of the 2D elements are shown in Figure 1-5.

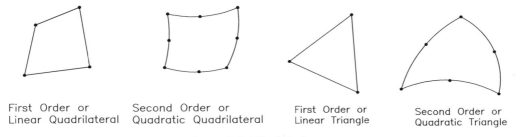

First Order or   Second Order or   First Order or   Second Order or
Linear Quadrilateral   Quadratic Quadrilateral Linear Triangle   Quadratic Triangle

Figure 1-5 *The 2D elements*

 Note

In this chapter, only the basic introduction of element types has been covered.

3D Element

A 3D element has the shape of a hexahedron (8 nodes), wedge (6 nodes), tetrahedron (4 nodes), or a pyramid (5 nodes). Some of the 3D elements are shown in Figure 1-6.

First order Linear Second order   First order Linear   Second order
Hexahedran   Quadratic Hexahedran Wedge   Quadratic Wedge

Figure 1-6 *The 3D elements*

Areas for Application of FEA

FEA is a very important tool for designing. It is used in the following areas:

1. Structural strength design
2. Structural interaction with fluid flows
3. Shock analysis
4. Acoustics
5. Thermal analysis
6. Vibrations
7. Crash simulations
8. Fluid flows
9. Electrical analysis
10. Mass diffusion
11. Buckling problems
12. Dynamic analysis
13. Electromagnetic analysis
14. Coupled analysis, and so on.

General Procedure of Conducting Finite Element Analysis

To conduct the finite element analysis, you need to follow certain steps. These steps are given next.

1. Set the type of analysis to be carried out.
2. Create or import the model.
3. Define the element type.
4. Divide the given problem into nodes and elements (generate a mesh).
5. Apply material properties and boundary conditions.
6. Solve the unknown quantities at nodes.
7. Interpret the results.

FEA through Autodesk Simulation Mechanical

In Autodesk Simulation Mechanical, the general process of finite element analysis is divided into three main phases, namely preprocessor, solution, and postprocessor, refer to Figure 1-7.

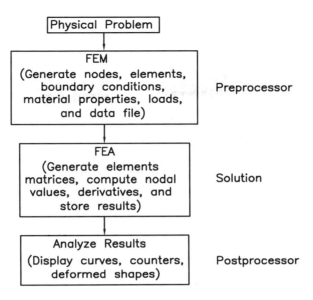

Figure 1-7 *FEA through Autodesk Simulation Mechanical*

Preprocessor

The preprocessor is a program that processes the input data to produce the output that is used as input in the subsequent phase (solution). Following are the input data that need to be given to the preprocessor:

1. Analysis type (structural or thermal, static or dynamic, and linear or nonlinear)
2. Element type
3. Real constants
4. Material properties
5. Geometric model
6. Meshed model
7. Loadings and boundary conditions

The input data will be preprocessed for the output data and preprocessor will generate the data files automatically with the help of users. These data files will be used by the subsequent phase (solution), refer to Figure 1-7.

Solution

Solution phase is completely automatic. The FEA software generates the element matrices, computes nodal values and derivatives, and stores the resulting data in files. These files are further used by the subsequent phase (postprocessor) to review and analyze the results through graphic display and tabular listings, refer to Figure 1-7.

Postprocessor

The output of the solution phase (result data files) is in numerical form and consists of nodal values of the field variable and its derivatives. For example, in structural analysis, the output is nodal displacement and stress in the elements. The postprocessor processes the result data and displays it in graphical form to check or analyze the result. The graphical output gives detailed information about the resultant data. The postprocessor phase is automatic and generates the graphical output in the form specified by the user, refer to Figure 1-7.

Effective Utilization of FEA

Some prerequisites for effective utilization of FEA from the perspective of engineers and FEA software are discussed next.

Engineers

An engineer who wants to work with this tool must have sound knowledge of Strength of Materials (for structural analysis), Heat Transfer, Thermodynamics (for thermal analysis), and a good analytical/designing skill. Besides this, the engineer must have a fair knowledge of advantages and limitations of the FEA software being used.

Software

The FEA software should be selected based on the following considerations:

1. Analysis type to be performed.
2. Flexibility and accuracy of the tool.
3. Hardware configuration of your system.

Nowadays, the CAE / FEA software can simulate the performance of most of the systems. In other words, anything that can be converted into a mathematical equation can be simulated using the FEA techniques. Usually, the most popular principle of GIGO (Garbage In Garbage Out) applies to FEA. Therefore, you should be very careful while giving/accepting the inputs for analysis. The careful planning is the key to a successful analysis.

Advantages and Limitations of FEA Software

Following are the advantages and limitations of FEA software:

Advantages

1. It reduces the amount of prototype testing; thereby saving the cost and time involved in performing design testing.
2. It gives graphical representation of the result of analysis.
3. The finite element modeling and analysis are performed in the preprocessor phase and the solution phase, which if done manually, will consume a lot of time and in some cases, may be impossible to carry out.
4. Variables such as stress, temperature can be measured at any desired point in the model.
5. It helps optimize the design.
6. It is used to simulate the designs that are not suitable for prototype testing such as surgical implants (artificial knees).
7. It helps you create more reliable, high quality, and competitive designs.

Limitations

1. It provides approximate solutions.
2. FEA packages are costly.

3. Qualified personnel are required to perform the analysis.
4. The results give solutions but not remedies.
5. Features such as bolts and welded joints cannot be accommodated in the model. This may lead to approximation and errors in the result obtained.
6. For more accurate result, more computer space and time are required.

KEY ASSUMPTIONS IN FEA

There are four basic assumptions that affect the quality of the solution and must be considered before carrying out finite element analysis. These assumptions are not comprehensive, but cover a wide variety of situations applicable to the problem. Make sure to use only those assumptions that apply to the analysis under consideration.

Assumptions Related to Geometry

1. When the displacement is small, the linear solution can be consider.
2. Stress behavior outside the area of interest is not important so the geometric simplifications in those areas will not affect the outcome.
3. Only internal fillets in the area of interest will be included in the solution.
4. Local behavior at the corners, joints, and intersection of geometries is of primary interest therefore no special modeling of these areas is required.
5. Decorative external features will be assumed insignificant for the stiffness and performance of the part, and will be omitted from the model.
6. The variation in mass due to the suppressed features is negligible.

Assumptions Related to Material Properties

1. Material properties will remain in the linear region and nonlinear behavior of the material property cannot be accepted. For example, it is understood that either the stress levels exceeding the yield point or the excessive displacement will cause a component failure.
2. Material properties are not affected by the load rate.
3. The component is free from surface imperfections that can produce stress risers.
4. All simulations will assume room temperature unless specified otherwise.
5. The effect of relative humidity or water absorption on the material used will be neglected.
6. No compensation will be made to account for the effect of chemicals, corrosives, wears or other factors that may have an impact on the long term structural integrity.

Assumptions Related to Boundary Conditions

1. Displacements will be small so that the magnitude, orientation, and distribution of the load remains constant throughout the process of deformation.
2. Frictional loss in the system is considered to be negligible.
3. All interfacing components will be assumed rigid.
4. The portion of the structure being studied is assumed a part separate from the rest of the system. As a result, the reaction or input from the adjacent features is neglected.

Assumptions Related to Fasteners

1. Residual stresses due to fabrication, preloading on bolts, welding, or other manufacturing or assembly processes will be neglected.
2. All the welds between the components will be considered ideal and continuous.

3. The failure of fasteners will not be considered.
4. Loads on the threaded portion of the parts is supposed to be evenly distributed among the engaged threads.
5. Stiffness of bearings, radially or axially, will be considered infinite or rigid.

TYPES OF ANALYSIS

The following types of analysis can be performed using FEA software:

1. Structural analysis
2. Thermal analysis
3. Fluid flow analysis
4. Electromagnetic field analysis
5. Coupled field analysis

Structural Analysis

In structural analysis, first the nodal degrees of freedom (displacement) are calculated and then the stress, strains, and reaction forces are calculated from the nodal displacements. The classification of the structural analysis is shown in Figure 1-8.

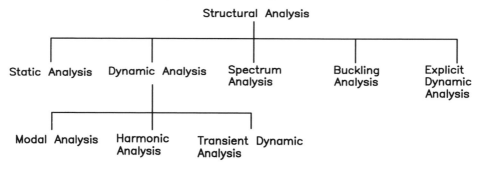

Figure 1-8 *Types of structural analysis*

Static Analysis

In static analysis, the load or field conditions do not vary with respect to time and therefore, it is assumed that the load or field conditions are applied gradually not suddenly. The system under analysis can be linear or nonlinear. Inertia and damping effects are ignored in static structural analysis. In static structural analysis, the following matrices are solved:

$$[K] \times [X] = [F]$$

Where,

         K = Stiffness Matrix
         X = Displacement Matrix
         F = Load Matrix

The above equation is called as the force balance equation for a linear system. If the elements of matrix [K] are a function of [X], the system is known as a nonlinear system. Nonlinear systems include large deformation, plasticity, creep, and so on. The loadings that can be applied in a static analysis include:

1. Externally applied forces and pressures
2. Steady-state inertial forces (such as gravity or rotational velocity)
3. Imposed (non-zero) displacements
4. Temperatures (for thermal strain)
5. Fluences (for nuclear swelling)

The outputs that can be expected from a FEA software are given next.

1. Displacements
2. Strains
3. Stresses
4. Reaction forces

Dynamic Analysis

In dynamic analysis, the load or field conditions do vary with time. The assumption here is that the load or field conditions are applied suddenly. The system can be linear or nonlinear. The dynamic load includes oscillating loads, impacts, collisions, and random loads. The dynamic analysis is classified into the following three main categories:

Modal Analysis

It is used to calculate the natural frequency and mode shape of a structure.

Harmonic Analysis

It is used to calculate the response of the structure to harmonically time varying loads.

Transient Dynamic Analysis

It is used to calculate the response of the structure to arbitrary time varying loads.

In dynamic analysis, the following matrices are solved:

For the system without any external load:

$$[M] \times \text{Double Derivative of } [X] + [K] \times [X] = 0$$

where

M = Mass Matrix
K = Stiffness Matrix
X = Displacement Matrix

For the system with external load:

$$[M] \times \text{Double Derivative of } [X] + [K] \times [X] = [F]$$

where

K = Stiffness Matrix
X = Displacement Matrix
F = Load Matrix

The above equations are called as force balance equations for a dynamic system. By solving the above set of equations, you will be able to extract the natural frequencies of a system. The load types applied in the dynamic analysis are the same as that in the static analysis. The outputs that can be expected from a software are:

1. Natural frequencies
2. Mode shapes
3. Displacements
4. Strains
5. Stresses
6. Reaction forces

All the outputs mentioned above can be obtained with respect to time.

Spectrum Analysis
This is an extension of the modal analysis and is used to calculate the stress and strain due to the response of the spectrum (random vibrations).

Buckling Analysis
This type of analysis is used to calculate the buckling load and the buckling mode shape. The slender structures and the structures with slender part loaded in the axial direction buckle under relatively small loads. For such structures, the buckling load becomes a critical design factor.

Explicit Dynamic Analysis
This type of structural analysis is used to calculate fast solutions for large deformation dynamics and complex contact problems.

Thermal Analysis
Thermal analysis is used to determine the temperature distribution and related thermal quantities such as:

1. Thermal distribution
2. Amount of heat loss or gain
3. Thermal gradients
4. Thermal fluxes

All the primary heat transfer modes such as conduction, convection, and radiation can be simulated. You can perform two types of thermal analysis, steady-state and transient.

Steady State Thermal Analysis
In this analysis, the system is studied under steady thermal loads with respect to time.

Transient Thermal Analysis
In this analysis, the system is studied under varying thermal loads with respect to time.

Fluid Flow Analysis

This analysis is used to determine the flow distribution and temperature of a fluid. It simulate the laminar and turbulent flow, compressible and electronic packaging, automotive design, and so on. The outputs that can be expected from the fluid flow analysis are:

1. Velocities
2. Pressures
3. Temperatures
4. Film coefficients

Electromagnetic Field Analysis

This type of analysis is used to determine the magnetic fields in electromagnetic devices. The types of electromagnetic analyses are:

1. Static analysis
2. Harmonic analysis
3. Transient analysis

Coupled Field Analysis

This type of analysis considers the mutual interaction between two or more fields. It is impossible to solve the fields separately because they are interdependent. Therefore, you need a program that can solve both the physical problems by combining them.

For example, if a component is exposed to heat, you may first require to study the thermal characteristics of the component and then the effect of the thermal heating on the structural stability.

Alternatively, if a component is bent into different shapes using one of the metal forming processes and then subjected to heating, the thermal characteristics of the component will depend on the new shape of the component and therefore the shape of the component has to be predicted through structural simulations first. This is called as coupled field analysis.

IMPORTANT TERMS AND DEFINITIONS

Some of the important terms and definitions used in FEA are discussed next.

Strength

When a material is subjected to an external load, deformation occurs and the resistance offered by the material against this deformation is by the virtue of its strength.

Load

The external force acting on a body is called the load.

Stress

The force of resistance offered by a body per unit area against the deformation is called stress. The stress is induced in the body while the load is applied on it. Stress is calculated as load per unit area.

$$p = F/A$$

where

p = Stress in N/mm^2
F = Applied Force in Newton
A = Cross-Sectional Area in mm^2

The material can undergo various types of stresses which are discussed next.

Tensile Stress

If the resistance offered by a body is against the increase in length, the body is said to be under tensile stress.

Compressive Stress

If the resistance offered by a body is against the decrease in length, the body is said to be under compressive stress. Compressive stress is just the reverse of tensile stress.

Shear Stress

Shear stress exists when two materials tend to slide over each other in opposite directions. Note that in any typical plane of shear, the force parallel to the plane is usually known as shear stress.

Shear Stress = Shear resistance (R) / Shear area (A)

Strain

When a body is subjected to a load (force), its length changes. The ratio of the change in length to the original length of the member is called strain. If the body returns to its original shape on removing the load, the strain is called as elastic strain. If the metal remains distorted, the strain is called as plastic strain. The strain can be of three types, tensile, compressive, and shear strain.

Strain (e) = Change in Length (*dl*) / Original Length (*l*)

Elastic Limit

The maximum stress that can be applied to a material without producing permanent deformation is known as the elastic limit of the material. If the stress applied is within the elastic limit and on the removal of the stress, the material will return to its original shape and dimension.

Hooke's Law

It states that the stress is directly proportional to the strain within the elastic limit.

Stress / Strain = Constant (within the elastic limit)

Young's Modulus or Modulus of Elasticity

In case of axial loading, the ratio of intensity of tensile or compressive stress to the corresponding strain is constant. This ratio is called Young's modulus and it is denoted by E.

$$E = p/e$$

Where,

p = Stress

e = strain

Shear Modulus or Modulus of Rigidity

In case of shear loading, the ratio of shear stress to the corresponding shear strain is constant. This ratio is called shear modulus, and it is denoted by C, N, or G.

Ultimate Strength

The maximum stress that a material can withstand when load is applied is called its ultimate strength.

Factor of Safety

The ratio of the ultimate strength to the estimated maximum stress (design stress) is known as factor of safety. It is necessary that the design stress should be well below the elastic limit and to achieve this condition, the ultimate stress should be divided by a 'factor of safety'.

Lateral Strain

If a cylindrical rod is subjected to an axial tensile load, the length (l) of the rod will increase (dl) and the diameter (ϕ) of the rod will decrease ($d\phi$). In short, the longitudinal stress will not only produce a strain in its own direction, but will also produce a lateral strain. The ratio dl/l is called the longitudinal strain or linear strain, and the ratio $d\phi/\phi$ is called the lateral strain.

Poisson's Ratio

The ratio of lateral strain to the longitudinal strain is constant within the elastic limit. This ratio is called as Poisson's ratio and is denoted by 1/m. For most of the metals, the value of the 'm' lies between 3 and 4.

Poisson's ratio = Lateral Strain / Longitudinal Strain = 1/m

Bulk Modulus

If a body is subjected to equal stresses along three mutually perpendicular directions, the ratio of the direct stresses to the corresponding volumetric strain is found to be constant for a given material when the deformation is within a certain limit. This ratio is called the bulk modulus and is denoted by K.

Creep

At elevated temperatures and constant stress or load, many materials continue to deform but at a slow rate. This behavior of materials is called creep. At a constant stress and temperature, the rate of creep is approximately constant for a long period of time. After this period and after a certain amount of deformation, the rate of creep increases thereby causing fracture in the material. The rate of creep is highly dependent on both stress and temperature.

Classification of Materials

Materials are classified into three main categories: elastic, plastic, and rigid. In case of elastic materials, the deformation disappears on the removal of load. In plastic materials, the deformation is permanent. A rigid material does not undergo any deformation when subjected to an external load. However, in actual practice, no material is perfectly elastic, plastic, or rigid. The structural members are designed such that they remain in elastic conditions under the action of working loads. All engineering materials are grouped into three categories that are discussed next.

Isotropic Material

In case of Isotropic materials, the material properties do not vary with direction, which means that they have same material properties in all directions. The material properties are defined by Young's modulus and Poisson's ratio.

Orthotropic Material

In case of Orthotropic material, the material properties vary with the change in direction. They have three mutually perpendicular planes of material symmetry. The material properties are defined by three separate Young's modulus and Poisson's ratios.

Anisotropic Material

In case of Anisotropic material, the material properties vary with the change in direction, but in this case, there is no plane of material symmetry.

THEORIES OF FAILURE

The following are the theories of failure.

Von Mises Stress Failure Criterion

The von Mises stress criterion is also called Maximum Distortion Energy theory. The theory states that a ductile material starts yielding at a location when the von Mises stress becomes equal to the stress limit. In most cases, the yield strength is used as the stress limit.

Maximum Shear Stress Failure Criterion

The Maximum Shear Stress failure criterion is based on the Maximum Shear Stress theory. This theory predicts failure of a material when the absolute maximum shear stress reaches the stress limit that causes the material to yield in a simple tension test. The Maximum Shear Stress criterion is used for ductile materials.

Maximum Normal Stress Failure Criterion

This criterion is used for brittle materials. It assumes that the ultimate strength of the material in tension and compression is the same. This assumption is not valid in all the cases. For example, cracks considerably decreases the strength of the material in tension while their effect is not significant in compression because the cracks tend to close. Brittle materials do not have a specific yield point and hence it is not recommended to use the yield strength to define the stress limit for this criterion.

Self-Evaluation Test

Answer the following questions and then compare them to those given at the end of this chapter:

1. The _____ are used to eliminate the unknowns in a system.

2. An element shape is specified by _____.

3. In Autodesk Simulation Mechanical, the general process of finite element analysis is divided into three main phases: _____ , _____ , and _____.

4. In _____ analysis, the load or field conditions do not vary with respect to time.

5. The FEA is a computing technique that is used to obtain approximate solutions to the boundary value problems in engineering. (T/F)

6. The Model Analysis is used to calculate the natural frequency and mode shape of a structure. (T/F)

7. The degree of freedom is defined as the ability of a node to translate or transmit the load. (T/F)

8. A 1D element has the shape of a line or curve. (T/F)

Review Questions

Answer the following questions:

1. When the stiffness of material is the function of displacement then the behavior of material is known as _____.

2. In _____ analysis, the load or field conditions vary with respect to time.

3. The ratio of the change in length to the original length of the member is called _____.

4. The _____ states that the stress is directly proportional to the strain within the elastic limit.

5. A 3D element can have the shape of hexahedron (8 nodes), _____ , _____ , or _____ .

6. The stiffness matrix represents the resistance offered by a body to withstand the load applied. (T/F)

7. The boundary conditions are used to eliminate the unknowns in a system. (T/F)

8. In Autodesk Simulation Mechanical, the solution phase of analysis is completely automatic. (T/F)

Answers to Self-Evaluation Test

1. boundary conditions, **2.** nodes, **3.** preprocessor, solution, postprocessor, **4.** Static, **5.** T, **6.** T, **7.** T, **8.** T

Chapter 2

Introduction to Autodesk Simulation Mechanical

Learning Objectives

After completing this chapter, you will be able to:

• *Introduction to Autodesk Simulation Mechanical*
• *Understand the System Requirements to run Autodesk Simulation Mechanical*
• *Start Autodesk Simulation Mechanical*
• *Understand various screen components of Autodesk Simulation Mechanical*
• *Set Unit System*
• *Understand various Environments of Autodesk Simulation Mechanical*
• *Understand about various shortcut menus*
• *Modify the background color scheme in Autodesk Simulation Mechanical*
• *Understand the use of various hotkeys*

INTRODUCTION TO Autodesk Simulation Mechanical

Welcome to the world of Computer Aided Engineering (CAE) with Autodesk Simulation Mechanical. If you are a new user of this software package, you will be joining hands with thousands of users of this software package which provides a broad range of simulation tools to help designers and engineers to know the product performance in early stages of the design cycle. If you are familiar with the previous releases of this software, you will be able to upgrade your analysis skills with the tremendously improved version of this software.

Autodesk Simulation Mechanical is the product of Autodesk, a leading supplier of various engineering software packages that provide various tools to design, visualize, and simulate different ideas. By putting powerful Digital Prototyping technology within the reach of mainstream manufacturers, Autodesk is changing the way manufacturers think about their design processes and is helping them create more productive workflows. Autodesk approach to Digital Prototyping is unique because it is scalable, attainable, and cost-effective. These factors allow a broader group of manufacturers to realize the benefits with minimal disruption to existing workflows. It also provides the most straight forward path to create and maintain a single digital model in a multidisciplinary engineering environment.

Autodesk Simulation Mechanical is a Finite Element Analysis tool which enables critical engineering decisions to be made earlier in the design process. Using the tools in this software, designers and engineers can easily study the initial design and predict the performance of the complete digital prototype. The automatic meshing tools of this software produce high-quality elements on the first pass while working with CAD geometry. It also enables designers and engineers to directly edit a mesh for accurate placement of loads and constraints or to simplify the geometry using its modeling capabilities. Autodesk Simulation Mechanical software makes it possible to quickly validate design concepts before resources are invested in design changes or new products.

Autodesk Simulation Mechanical provides wide range of linear and nonlinear materials that allows better understanding of the real-world behavior of products and let engineers know how a product will perform in the real-world environment.

Autodesk Simulation Mechanical supports efficient workflows with today's multi-CAD environment by providing direct geometry exchange and full associativity with Autodesk Inventor, Pro/ENGINEER Creo Parametric, and other such software packages. It allows iterative design changes without redefining material, loads, constraints, or other simulation data when working with the native CAD format. You can also import 2D and 3D geometry of CAD universal file formats such as ACIS®, IGES, STEP, and STL for solid models and CDL, DXF™, and IGES for wireframe models.

As mentioned earlier, Autodesk Simulation Mechanical software provides a broad range of simulation tools which help engineers to ascertain product performance knowledge during early stages of the design cycle. It improves the collaboration, helps design better and safer products, saves time, and reduces manufacturing costs. Some of the most important types of analysis carried out using this software package are discussed next.

Linear Analysis

Linear analysis is undertaken when an object is expected to behave linearly. In other words, it follows the Hooke's Law which states that within the elastic limit stress is directly proportional to strain, refer to Figure 2-1. Also, the material that undergoes linear analysis experiences elastic deformation and will return to its original configuration or shape after the load has been removed from it. In linear analysis, the following factors are calculated:

1. Displacements and stresses due to static loads
2. Natural frequency and mode shapes of the model
3. Natural frequency and mode shapes of the model with load stiffening
4. Maximum displacement and stress due to a spectrum-type load
5. Statistical response of a system (displacements and stresses) due to random vibration, noise, or spectrum density
6. Steady state response (displacements and stresses) due to harmonic or sinusoidal load, or acceleration
7. Displacements and stresses over time due to loads that will vary in a known fashion
8. Load that causes the model to buckle due to geometric instability
9. Response of the model due to shock

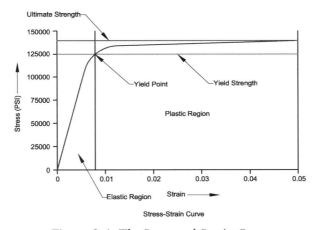

Figure 2-1 *The Stress and Strain Curve*

Non-linear Analysis

Non-linear analysis is used to analyze the behavior of an object that is loaded beyond its elastic limits and exceeds the yield strength on the stress-strain curve, refer to Figure 2-1. Also, the material that undergoes non-linear analysis experiences plastic deformation and will not return to its original configuration or shape even after the load has been removed from it. In non-linear analysis, you can calculate:

1. Displacements and stresses due to static loads
2. Displacements, velocities, accelerations, and stresses over time due to dynamic loads
3. Natural frequencies and mode shapes of the model
4. Displacements and stresses before and after the model has collapsed

Thermal Analysis

Thermal Analysis is used to determine the temperature distribution and related thermal quantities in a model. As there are three mechanisms of heat transfer, namely conduction, convection, and radiation, the thermal analysis is used to calculate the temperature distribution in a body due to all of these mechanisms. In thermal analysis, you can calculate:

1. Temperature and heat fluxes after an infinite period under steady-state conditions
2. Temperature and heat fluxes over time due to the thermal loads

SYSTEM REQUIREMENTS

The system requirements to ensure the smooth functioning of Autodesk Simulation Mechanical 2016 on 32 bit and 64 bit windows are as follows:

- Windows® 7 Home Premium, Professional, Enterprise (SP1), Ultimate; Windows® 8; Windows® 8.1; Windows Server® 2008 R2 SP1; Windows® Server 2012 operating system
- Intel® Pentium® 4, Intel Xeon®, Intel Core™, AMD Athlon™ II, or AMD Opteron™ or later (2 GHz CPU speed or higher) processor
- 2 GB RAM minimum (8 GB recommended)
- 500 GB or larger hard drive, 30 GB minimum free disk space (8 GB free disk space for installation)
- 512 MB DRAM or greater OpenGL®-capable graphics card
- 24-bit color setting at 1,280 x 1,024 or higher screen resolution
- Microsoft Office Excel, either version 2007 or higher
- Adobe Acrobat 7.0.7 or higher
- DVD drive and Mouse or any other compatible pointing device
- Internet Explorer version 7 or higher

GETTING STARTED WITH Autodesk Simulation

Install Autodesk Simulation Mechanical 2016 on your system and then double-click on the Autodesk Simulation Mechanical 2016 icon on the desktop of your computer, the system will prepare to start Autodesk Simulation Mechanical by loading all the required files. After all the required files have been loaded, the **New** or **Open** dialog box along with the initial screen of Autodesk Simulation Mechanical 2016 will be displayed, refer to Figure 2-2.

Note that if you are starting Autodesk Simulation Mechanical first time after installing it, the **Autodesk Simulation Mechanical** message window may be displayed informing you that no default modeling unit has been selected, refer to Figure 2-3. Choose the **OK** button from this window; the **Unit System** dialog box will be displayed, refer to Figure 2-4. Specify the default unit system for measurement by using the options of this dialog box. You will learn more about specifying unit system by using this dialog box later in this chapter. After specifying the unit system, choose the **OK** button; the **New** dialog box will be displayed along with the initial interface of Autodesk Simulation Mechanical, as shown in Figure 2-2.

Note

*1. The automatic display of the **New** or **Open** dialog box along with the initial screen of Autodesk Simulation Mechanical depends upon which of these dialog boxes was used in the last session of Autodesk Simulation Mechanical.*

*2. You can turn off the automatic display of the **New** or **Open** dialog box: at startup. To do so, first close the opened dialog box, if any, and then choose the **Tools** tab from the **Ribbon**. Next, choose the **Application Options** button from the **Options** panel of the **Tools** tab; the **Options** dialog box will be displayed. In this dialog box, choose the **General Information** tab and then clear the **Show file dialog on startup** check box from the **Miscellaneous options** area of the dialog box. Then choose the **Apply** button from the dialog box. Next, close the dialog box by choosing the **Close** button.*

*Figure 2-2 Initial Inteface display of **Autodesk Simulation Mechanical 2016***

*Figure 2-3 The **Autodesk Simulation Mechanical** message window*

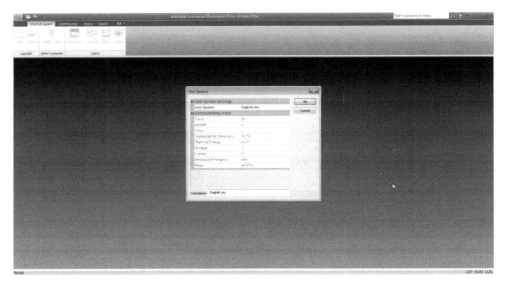

Figure 2-4 The **Unit System** *dialog box*

By using the **New** dialog box, you can create a new FEA model. In addition to creating new FEA model, you can also invoke the **Open** dialog box by choosing the **Open** button and the **Unit System** dialog box by choosing the **Override Default Units** button of the **New** dialog box. Note that if the **New** dialog box is not invoked by default, you can invoke it by choosing the **New** tool from the **Launch** panel of the **Start & Learn** tab in the **Ribbon**.

The **Open** dialog box is used to open an existing Autodesk Simulation FEA or CAD model in the Autodesk Simulation Mechanical 2016 to run the analysis. By default, the **Autodesk Simulation FEA Model (*.fem)** is selected in the **Files of type** drop-down list of the **Open** dialog box. As a result, you can open the existing Autodesk simulation FEA model by selecting the FEA model to be opened from the location where it is saved and then choosing the **Open** button from the dialog box. The *.fem* is the file extension of the Autodesk Simulation FEA model. To open a CAD model in the Autodesk Simulation Mechanical, you need to select their respective file extension from the **Files of type** drop-down list of the **Open** dialog box. You will learn more about opening the FEA and CAD models in the later chapters.

Similar to invoking the **Open** dialog box from the **New** dialog box, you can also invoke the **New** dialog box from the **Open** dialog box. Additionally, you can also invoke these dialog boxes by choosing their respective button from the **Launch** panel of the **Start & Learn** tab in the **Ribbon**.

To create a new FEA model by using the **New** dialog box, select the **FEA Model** button and then select the type of analysis to be performed on the model being created. To select the type of analysis, click on the arrow on the right in the **Choose analysis type** area of the **New** dialog box; a flyout will be displayed, refer to Figure 2-5. Now, move the cursor over the type of analysis that you want to carry out for the model; all the analyses that lie under the selected analysis type will be displayed in another flyout. Select the required type of analysis from this flyout.

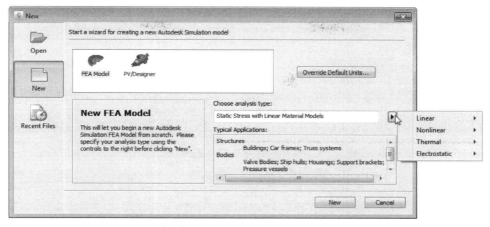

Figure 2-5 *The flyout displayed after clicking on the arrow*

After specifying the type of analysis, choose the **New** button available at the lower right corner the **New** dialog box; the **Save As** dialog box will be displayed. Browse to the location where you want to save the FEA model in your computer. Next, enter the name of the FEA model in the **File name** edit box of the dialog box and choose the **Save** button; the new FEA file is created with the specified name at the specified location. Also, the FEA Editor environment of Autodesk Simulation Mechanical is invoked for creating the FEA model, refer to Figure 2-6. Now, you can start working on creating the FEA model.

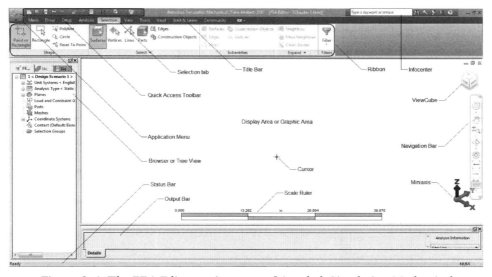

Figure 2-6 *The FEA Editor environment of Autodesk Simulation Mechanical*

It is evident from the Figure 2-6 that the interface of Autodesk Simulation Mechanical is very user-friendly. Apart from the components shown in Figure 2-6, you are also provided with various shortcut menus that are displayed on right-clicking. The type of the shortcut menu and its options will depend on where or when you are trying to access the shortcut menu. In this textbook, the shortcut menus will be discussed when they are used. Some of the important components of Autodesk Simulation Mechanical interface are discussed next.

Quick Access Toolbar

This toolbar is common to all environments of Autodesk Simulation Mechanical. Note that some of the options in this toolbar may not be available when you start Autodesk Simulation Mechanical for the first time. Figure 2-7 shows the **Quick Access Toolbar**. Some of the important tools of this toolbar are discussed next.

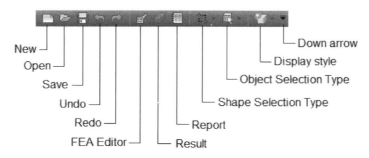

*Figure 2-7 The **Quick Access Toolbar***

Tip. *You can add or remove the tools in the **Quick Access Toolbar**. To add a tool to the **Quick Access Toolbar**, right-click on the tool to be added in the **Ribbon**; a shortcut menu will be displayed. Choose the **Add to Quick Access Toolbar** option from the shortcut menu; the chosen tool will be added. Similarly, to remove a tool from the **Quick Access Toolbar**, right-click on the tool to be removed and select the **Remove from Quick Access Toolbar** option from the shortcut menu displayed.*

Shape Selection Type Drop-down

The tools available in the **Shape Selection Type** drop-down are used to select objects by using different methods, refer to Figure 2-8. The **Point or Rectangle Select** tool is chosen by default in this drop-down. As a result, you can select an object by clicking on it. Also, you can draw a rectangular window around the objects to be selected by using this tool. The **Rectangle Select** tool is used to select multiple objects by creating a rectangle around the objects. To do so, you need to specify the two diagonally opposite corners of the rectangle in the drawing area around the objects to be selected. If you choose the **Polyline Select** tool, you can select multiple objects

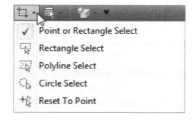

*Figure 2-8 The **Shape Selection Type** drop-down*

by creating a boundary of polyline around the objects. The **Circle Select** tool is used to select multiple objects by creating a circle around the object to be selected in the drawing area by specifying its center point and the radius. The **Reset To Point** tool is used to automatically reset the type of selection back to point selection, once you are done with selection by using the other selection method.

Object Selection Type Drop-down

The tools available in the **Object Selection Type** drop-down are used to specify the type of object to be selected, refer to Figure 2-9. The **Select Construction Objects** tool of this drop-down is used to select construction entities such as circles, arcs, lines, and splines. If you choose the **Select Surfaces** tool, you can select surfaces from the drawing area or from the **Tree View**. The **Select Vertices** tool is used to select the vertices. Similarly, you can select the entire part body by choosing the **Select Parts** tool, the edges of the part by choosing the **Select Edges** tool, and lines by choosing the **Select Lines** tool.

*Figure 2-9 The **Object Selection Type** drop-down*

In addition, you can also select the sub-entities of the selected entities or objects. For example, if a part body is selected in the drawing area using the **Select Parts** tool, you can further select its surfaces, edges, lines, construction object, or vertices. To do so, select the object whose sub-entities are to be selected and then choose the **Select Related > Surfaces/Edges/Lines/ Construction Objects/Vertices** from the **Selection Type** drop-down of the **Quick Access Toolbar**.

FEA Editor

This tool is used to invoke the FEA Editor environment of Autodesk Simulation Mechanical. In this environment, you can set or create model for analysis. By default this environment is invoked automatically on starting a new FEA file or on opening a model for analysis.

Results

This tool is used to invoke the Results environment of Autodesk Simulation Mechanical. In the Results environment, you can view the result of the analysis. Also, you can review element types, nodes, properties, loads, and so on of the model to be analyzed in this environment. You will learn more about the Results environment in later chapters.

Report

This tool is used to invoke the Report environment of Autodesk Simulation Mechanical. In the Report environment, you can review the log and summary files of the analysis report and save the analysis report in different formats such as HTML, Word, PDF, and so on. You will learn more about the Report environment in later chapters.

Ribbon and Tabs

You might have noticed that there is no command prompt in Autodesk Simulation Mechanical. The complete analysis process is carried out by invoking the tools from the tabs in the **Ribbon**. The **Ribbon** is a long bar available below the **Quick Access Toolbar**. You can customize the tabs and panels of the **Ribbon** as per your need. To do so, right-click on the **Ribbon**; a shortcut menu will be displayed. To customize the tabs, choose **Show Tabs** from this shortcut menu; a cascading menu will be displayed. Now, you can add or remove the tabs from the **Ribbon** as per your need. Note that the tick mark on the name of a tab indicates that it is available in the **Ribbon**. Similarly, to customize the panels of the tabs in the **Ribbon**, choose **Show Panels** from the shortcut menu; a cascading menu will be displayed with a list of all the panels of the active tab. You can add or remove the panels of the tabs using this list.

Autodesk Simulation Mechanical provides you with different tabs while working with different environments. This means that the tabs available from the **Ribbon** while working with the FEA Editor, Report, and Results environments will be different. You can also change the location of the **Ribbon**. By default, the location of the **Ribbon** is at the top of the screen. To change the current location of the **Ribbon**, right-click on the empty area in the **Ribbon**; a shortcut menu will be displayed. Next, choose the **Ribbon Location > Left** or **Ribbon Location > Right** from the shortcut menu displayed.

Note

*In the tabs, the tools are arranged in different panels in the **Ribbon**. Some of the panels and tools have arrows on their right, refer to Figure 2-10. These arrows are called down arrows. When you click on the down arrow of a tool, a list of additional tools will be displayed, refer to Figure 2-10.*

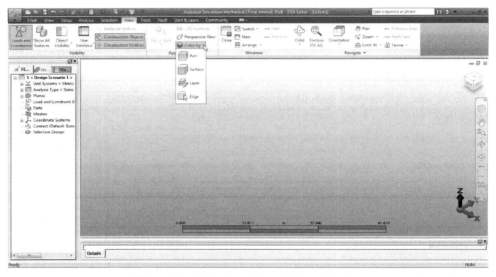

Figure 2-10   *Additional tools displayed on clicking the down arrow of a tool*

Selection Tab

This is one of the most important tabs in the **Ribbon**. All the selection tools are available in this tab. By using the tools available in this tab, you can specify the type of object you want to select and the method of selection. You will learn more about the usage of these tools in the later chapters. The **Selection** tab is shown in Figure 2-11.

Figure 2-11   *The **Selection** tab*

Setup Tab

This is another important tab of the Autodesk Simulation Mechanical. This tab will be available only in the FEA Editor environment. The tools available in this tab are used to define the boundary conditions by specifying the required constraints and the applied loads for the model.

You will learn more about the usage of these tools in the later chapters. The tools in the **Setup** tab are shown in Figure 2-12.

*Figure 2-12 The **Setup** tab*

Mesh Tab

This tab contains the tools that are required for generating mesh. Generating mesh on a component is one of the important steps carried out before proceeding with the analysis process. You will learn more about these tools in the later chapters. The tools in the **Mesh** tab are shown in Figure 2-13.

*Figure 2-13 The **Mesh** tab*

Note
*Most of the tools in the **Mesh** tab will be enabled if the model to be meshed is available or displayed in the graphic area of Autodesk Simulation Mechanical.*

Draw Tab

This tab contains the tools that are used to create and modify the geometry to be analyzed. This tab will be available only when you are in the FEA Editor environment. The tools in the **Draw** tab are shown in Figure 2-14.

*Figure 2-14 The **Draw** tab*

View Tab

The tools in this tab enable you to the control the view, orientation, appearance, and visibility of objects and view windows. This tab is available in almost all environments. The tools in the **View** tab are shown in Figure 2-15.

*Figure 2-15 The **View** tab*

Tools Tab

This tab contains tools that are mainly used for setting the preferences and for customizing the Autodesk Simulation Mechanical interface. This tab is available in almost all environments. The tools in the **Tools** tab are shown in Figure 2-16.

Figure 2-16 The *Tools* tab

Analysis Tab

This tab provides tools that are used to check the model for analysis, run the analysis, optimize the geometry, specify the type of analysis, and so on. The tools in the **Analysis** tab are shown in Figure 2-17.

Figure 2-17 The *Analysis* tab

Status Bar

The Status Bar is located at the lower most of the Autodesk Simulation Mechanical screen and is used to display status messages and various command prompts.

Output Bar

The Output Bar is located at the bottom of the graphic area and is used to view the solid meshing results, analysis log, analysis summary, and convergence plot, refer to Figure 2-18. Note that the display of convergence plot will be available on performing nonlinear analyses. You can increase the display area of the Output Bar by dragging it from its top edge.

Figure 2-18 The *Output Bar*

Navigation Bar

The Navigation Bar is located on the right of the graphics area and contains tools that are used to make the designing process easier and quicker. The navigation tools also help you to control the view and orientation of the components in the drawing window. The Navigation Bar is shown in Figure 2-19.

ViewCube

The ViewCube is available at the upper right corner of the drawing window. It is used to freely change the view of a model in 3D space. A ViewCube is a 3D navigation tool, which allows you to switch between the standard and isometric views in a single click. The ViewCube is shown in Figure 2-20.

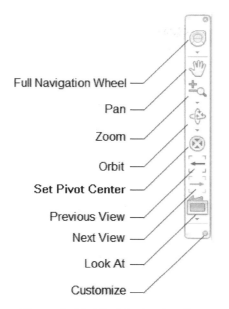

Full Navigation Wheel —

Pan —

Zoom —

Orbit —

Set Pivot Center —

Previous View —

Next View —

Look At —

Customize —

Figure 2-19 *The Navigation Bar*

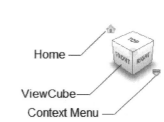

Home —

ViewCube—

Context Menu —

Figure 2-20 *The ViewCube*

Browser / Tree View

The Browser or Tree View is the most important component of the Autodesk Simulation Mechanical screen and is available below the **Ribbon** on the left in the drawing window. It keeps record of the operations performed during the analysis process in a sequence. All these operations are displayed in the form of a tree view. You can undock the Tree View by dragging it from its position. The content displayed in the Tree View is different for different environments in the Autodesk Simulation Mechanical.

Miniaxis

The Miniaxis is provided at the bottom right corner of the drawing window of the Autodesk Simulation Mechanical. It is used to display the current viewing direction of the model with respect to the three-dimensional work area. The Miniaxis is shown in Figure 2-21.

Figure 2-21 *The Miniaxis*

Scale Ruler

The Scale Ruler is provided at the bottom center of the drawing window of the Autodesk Simulation Mechanical. It is used to measure or display the length of the model relative to its current display size. The Scale Ruler is shown in Figure 2-22.

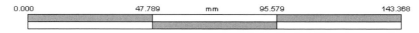

Figure 2-22 *The Scale Ruler*

SETTING THE UNIT SYSTEM

As discussed, you can set a unit system for the new FEA model by using the **New** dialog box. To do so, choose the **Override Default Units** button from the **New** dialog box; the **Unit System** dialog box will be displayed, as shown in Figure 2-23. Now, click on the cell next to the **Unit System** field in the **Unit System Settings** node of the dialog box; a down arrow will be displayed on the right of the cell. Next, click on this down arrow; a drop-down list will be displayed with a list of unit systems. You can select the required unit system from this drop-down list. If you select the **Custom** unit system, you will be allowed to customize the units. After setting the required unit system, choose the **OK** button.

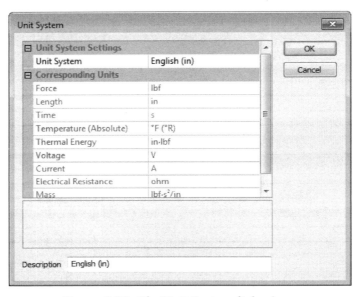

Figure 2-23 The **Unit System** dialog box

In Autodesk Simulation Mechanical, you can also change the current model unit system at any stage of the analysis process by using the Browser / Tree View. To change the model unit system, expand the **Unit System** node in the Tree View by clicking on its +sign, a list of available model and display unit systems is displayed. Select **Model Units** from the expanded **Unit System** node and right-click; a shortcut menu will be displayed. Next, select **Edit > Current Modeling Units** from the shortcut menu; the **Autodesk Simulation Mechanical** message window will be displayed. Choose the **Yes** button from this window; the **Unit System** dialog box will be displayed. Now, change the current model unit system as per your requirement and choose the **OK** button; the current model unit system will be changed.

In Autodesk Simulation Mechanical, all the inputs and results are stored in the model unit system only. However, you can display the results based on display unit system at any stage of the analysis process. By default, various display unit systems are available under the Tree View. To activate a unit system, double-click on the required display unit system available under the **Unit System** node of the Tree View. Alternatively, right-click on the required display unit system and then choose the **Activate** option from the shortcut menu. Note that in addition to the default available display unit systems, you can also add more as required. To do so, right-click on the

Unit Systems node of the Tree View and then choose the **New** option to display the **Unit System** dialog box. After specifying the required units by using this dialog box, choose the **OK** button; the new display unit system will be added and will get activated by default.

Note

When you import a CAD model, the Autodesk Simulation Mechanical identifies the unit specified for the length of the model being imported and on that basis it automatically chooses the most suitable unit system. As a result, the unit for the length dimensions remains the same as that of the CAD model, but the units for force, time, and so on may be different.

IMPORTANT ENVIRONMENTS OF AUTODESK SIMULATION MECHANICAL

Before you proceed with the analysis process in Autodesk Simulations Mechanical, it is important to understand the different environments of Autodesk Simulation Mechanical that are used in this book.

FEA Editor Environment

The FEA Editor environment is the primary environment of Autodesk Simulation Mechanical and is used to set up model for analysis. In this environment, you can create or modify the model to be analyzed. You can also change its display style, generate different mesh, define the type of analysis to be carried out, define boundary conditions, apply material, load, and so on.

Result Environment

The Result environment is used to view all the analysis results of the model. In this environment, you can also run animations to view the performance of the model. This environment is also used to check the FEA model before running an analysis by displaying its elements, nodes, properties, loads, and so on.

Note

*Autodesk Simulation Mechanical stores date and time of the last modification and analysis done on the model. If you make any change in the model geometry that can affect its analysis results then the date and time of this latest modification will be updated. In this case, on invoking the Result environment, a message window will be displayed with the message "**The existing results do not match the current model. If the changes have no affect on the results, then it is reasonable to proceed. (Changes include loads, mesh, materials, node numbering, and element numbering.) If the model has changed, the old results should not be used. Do you wish to continue?**". Choose the **No** button from this message window and run the analysis again to get the updated results based on the model modifications.*

Report Environment

The Report environment of the Autodesk Simulation Mechanical is used to create documentation for the analysis results of the FEA model. You can save the report in HTML, PDF, .doc, and so on. You can also insert the images and animations created in the Result environment into the report.

SHORTCUT MENUS

Shortcut menu is type of menu that consists some of the commonly used options in Autodesk Simulation Mechanical software. The shortcut menu will be displayed on right-clicking in the graphics area. Note that the shortcut menu contains different options for different environments of the Autodesk Simulation Mechanical. Also, if any object or entity is selected in the graphics area, the shortcut menu displays options related to the object or entity selected.

Figure 2-24 shows a shortcut menu invoked in the FEA Editor environment and Figure 2-25 shows a shortcut menu invoked in the Result environment of the Autodesk Simulation Mechanical.

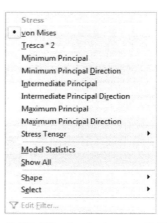

Figure 2-24 The shortcut menu displayed in the FEA Editor environment

Figure 2-25 The shortcut menu displayed in the Result environment

COLOR SCHEMES

Autodesk Simulation Mechanical allows you to use various color schemes to set the background color of the screen and for displaying the entities on the screen. Note that this book uses the white color background. To change the background color scheme to white, invoke the **Options** dialog box by choosing the **Application Options** tool from the **Options** panel of the **Tools** tab in the **Ribbon**, refer to Figure 2-26. Next, choose the **Graphics** tab from the **Options** dialog box, refer to Figure 2-26.

Select the **Background** option from the list box of the **Options** dialog box; the options used for applying the background color of the screen will be displayed in the **Background** area of the dialog box. Next, click on the **Primary background color** swatch box; the color swatches window will be displayed. Select the **White** color swatch from the color swatches window. Next, choose **Apply** to apply the new background color scheme and then choose **OK** from the dialog box. Note that all the files you open henceforth will use this color scheme. You can also apply gradient to the background color of the screen by selecting the **Gradient background color** check box and then defining the gradient background color.

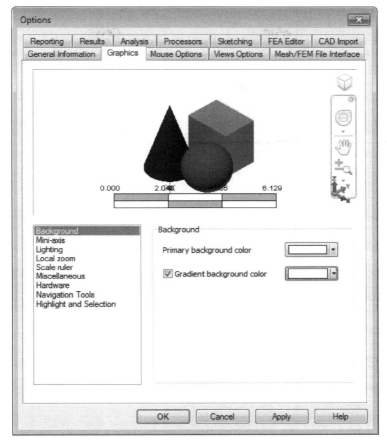

Figure 2-26 *The **Options** dialog box with **Graphics** tab chosen*

HOTKEYS

As mentioned earlier, there is no command prompt in Autodesk Simulation Mechanical. However, you can use the keys on the keyboard to invoke some tools. The keys that can be used to invoke the tools are called hotkeys. Some of the hotkeys that can be used in Autodesk Simulation Mechanical environments and their functions are mentioned in the table given next.

Table 2-1 *The hotkeys and their functions*

Hotkey	Function
CTRL+O	Invokes the **Open** dialog box
CTRL+P	Invokes the **Print** dialog box
CTRL+S	Saves the selection
CTRL+V	Pastes the selection
CTRL+X	Cuts the selection
CTRL+Y	Redo the modification

CTRL+Z	Undo the modification
CTRL+Num Pad 1	Invokes the Front view
CTRL+Num Pad 2	Invokes the Bottom view
CTRL+Num Pad 3	Invokes the Isometric view
CTRL+Num Pad 4	Invokes the Left view
CTRL+Num Pad 5	Invokes the **Enclose (Fit All)** tool
CTRL+Num Pad 6	Invokes the Right view
CTRL+Num Pad 7	Invokes the Back view
CTRL+Num Pad 8	Invokes the Top view
CTRL+Num Pad 9	Invokes the Axonometric view
Right or left arrow key	In the display area, these keys rotate the model about the vertical axis of the screen
Up or down arrow key	In the display area, these keys rotate the model about the horizontal axis of the screen
S+Arrow keys	In the display area, these keys rotate the model about the axis perpendicular to the screen
F1	Invokes the help file

Self-Evaluation Test

Answer the following questions and then compare them to those given at the end of this chapter:

1. The _____ tool is used to select construction circles, arcs, lines, splines, rectangles, and vertices that can be created by using the tools available in the **Draw** panel of the **Draw** tab in the **Ribbon**.

2. In Autodesk Simulation Mechanical, the messages and prompts are displayed at the _____.

3. In Autodesk Simulation Mechanical, you can specify the type of analysis for a model by using the **New** dialog box. (T/F)

4. In Autodesk Simulation Mechanical, you can specify a unit system for a model by using the **New** dialog box. (T/F)

5. In Autodesk Simulation Mechanical, you cannot open an existing Autodesk Simulation FEA or CAD model to run the analysis by using the **New** dialog box. (T/F)

6. The **Quick Access Toolbar** is common to all environments of Autodesk Simulation Mechanical. (T/F)

7. You can add or remove the tools in the **Quick Access Toolbar**. (T/F)

8. You cannot customize the tabs and panels of the **Ribbon**. (T/F)

Review Questions

Answer the following questions:

1. All the selection tools are available in the _____ tab.

2. The tools provided in the _____ tab are used to define the boundary conditions by specifying the constraints and loads for the model.

3. The _____ tab has the tools that are required for generating the mesh for a model.

4. The _____ environment is the primary environment of Autodesk Simulation Mechanical and is used to set up a model for analysis.

5. In Autodesk Simulation Mechanical, you can change the default or current Model unit system at any stage of the analysis process by using the Tree View. (T/F)

6. In Autodesk Simulation Mechanical, the options available in the shortcut menus depend upon the type of environment invoked. (T/F)

7. The **Draw** tab of the **Ribbon** provides the tools that are used to create and modify the geometry to be analyzed. (T/F)

8. The **Setup** tab of the **Ribbon** will be available only in the FEA Editor environment. (T/F)

Answers to Self-Evaluation Test

1. Select Construction Objects, **2.** Status Bar, **3.** T, **4.** T, **5.** F, **6.** T, **7.** T, **8.** F

Chapter 3

Importing and Exporting a Geometry

Learning Objectives

After completing this chapter, you will be able to:

- *Import CAD model in Autodesk Simulation Mechanical*
- *Understand the concept of splitting surfaces of CAD models*
- *Simplify the model geometry before importing*
- *Import FEA Model in Autodesk Simulation Mechanical*
- *Save FEA Model of Autodesk Simulation Mechanical*
- *Export FEA Model of Autodesk Simulation Mechanical into other FEA file types*
- *Understand the concept of archiving an FEA model*
- *Understand the drawing display tools*
- *Change the views of the model using the ViewCube*
- *Navigate the model using SteeringWheels*
- *Control the display of models*

INTRODUCTION

In Autodesk Simulation Mechanical, you can open the existing Autodesk simulation FEA models or CAD models to run the analysis. Autodesk Simulation Mechanical provides a geometry data translation interface to various leading CAD systems. As a result, you can directly open the CAD models saved in AutoCAD DWG (*.dwg), AutoCAD DXF (*.dxf), Autodesk Inventor (*.iam, *.ipt), Autodesk Fusion (*.dwg), SOLIDWORKS (*.sldasm, *.asm, *.sldprt, *.prt), Pro/ENGINEER (*.prt, *.asm) and many other file formats in this software. Similarly, it provides data translation interface to various other FEA software such as ABAQUS, ANSYS, Nastran, Stereolithography, Blue Ridge Numerics, and so on. In addition, you can also open the models saved in neutral file format such as *.SAT, *.STP, *.IGES, and *.STEP.

The FEA models created in Autodesk Simulation Mechanical are saved in the *.fem* file format. Also, if you open a CAD model in Autodesk Simulation Mechanical and then save it, the model will be saved in the *.fem* file format. You can export an FEA model created in Autodesk Simulation Mechanical to other FEA file formats such as ABAQUS (*.inp), ANSYS (*.cdb, *.ans), Blue Ridge Numeric (*.neu), FEMAP Neutral (*.neu), NASTRAN (*.nas, *.bdf, *.dat), PATRAN (*.pat), and SDRC Universal (*.unv).

IMPORTING A CAD MODEL

As discussed earlier, the Autodesk Simulation Mechanical provides data translation interface to CAD systems for directly opening the CAD models in it. To open a CAD model, invoke the **Open** dialog box, if not invoked already. To do so, choose the **Open** tool from the **Quick Access Toolbar** or from the **Launch** panel of the **Start & Learn** tab in the **Ribbon**; the **Open** dialog box will be invoked, as shown in Figure 3-1.

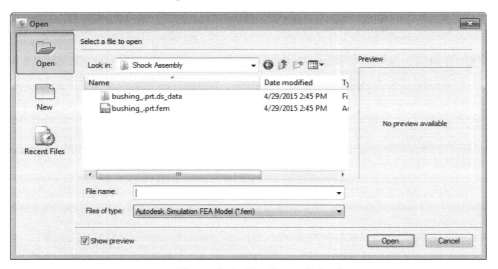

Figure 3-1 The **Open** dialog box

In this dialog box, by default, the **Autodesk Simulation FEA Model (*.fem)** file type is selected in the **File of type** drop-down list, refer to Figure 3-1. As a result, the **Open** dialog box will list only those models that have *.fem* file extension. The *.fem* is the file extension of Autodesk Simulation Mechanical. Select the required CAD file type from the **Files of type** drop-down list of the **Open** dialog box. Figure 3-2 shows the **Open** dialog box with the **Files of type** drop-down list displayed. To open the model created in Autodesk Inventor, select the **Autodesk Inventor Files (*.ipt;*.iam)** from the **Files of type** drop-down list. Next, browse to the location where the Inventor part files are saved and then select the file to be imported. After selecting the required file, choose the **Open** button from the dialog box; the **Import Inventor Work Points** message window appears, prompting you to specify whether you want to import work points along with the model. Choose the **Yes** button to import the work points while meshing as it helps to place a node at the work point of the imported model. Similar to importing work points, you can also import 3D sketches, CAD model colors, and CAD part names from the Inventor model by choosing the **Yes** button from the respective message boxes. On doing so, the **Choose Analysis Type** dialog box will be displayed, as shown in Figure 3-3. Now, select the type of analysis that you want to carry out on the model by using the options in this dialog box. After defining the analysis type, choose **OK**; the selected model will be imported and opened in Autodesk Simulation Mechanical, refer to Figure 3-4. Similarly, you can select a CAD file type of other CAD software whose model is to be imported in Autodesk Simulation Mechanical.

Note
*The message boxes displayed while importing a CAD model depend upon the radio button selected in the **Global CAD Import Options** dialog box. To invoke this dialog box, choose **Tools > Application Options** from menu bar; the **Options** dialog box will be displayed. Next, choose **CAD Import** tab > **Global CAD Import Options** from the **Options** dialog box to invoke the **Global CAD Import Options** dialog box.*

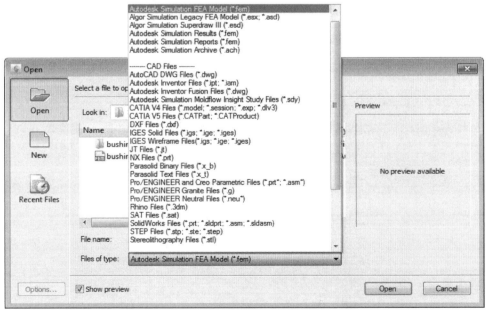

Figure 3-2 *The **Open** dialog box with the expanded view of the **Files of type** drop-down list*

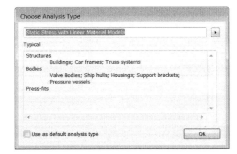

Figure 3-3 The **Choose Analysis Type** dialog box

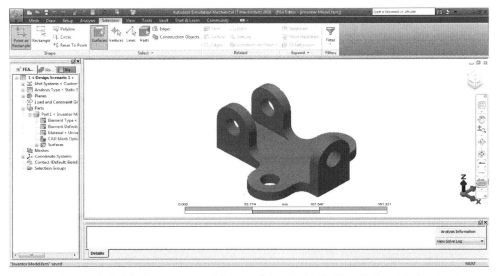

Figure 3-4 The Inventor model opened in Autodesk Simulation Mechanical

When you import any CAD model, Autodesk Simulation Mechanical identifies the unit specified for the length of the model being imported and on that basis, it automatically chooses the most suitable unit system. As a result, the unit for the length dimensions will be the same as that of the imported CAD model, but the units for force, time, and so on may not be the same. However, you can modify the default unit system by using the **Unit System** dialog box that will be displayed by right-clicking on the **Unit Systems** node of the **Tree View** and choosing the **New** option from the shortcut menu displayed, refer to Figure 3-5. The method of setting unit system is discussed in Chapter 2.

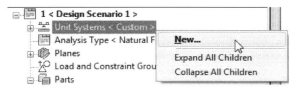

Figure 3-5 Shortcut menu displayed

Importing 3D Models from Autodesk Inventor

As discussed earlier, you can open the models created in Autodesk Inventor by selecting the **Autodesk Inventor Files (*.ipt;*.iam)** file type from the **Files of type** drop-down list in the **Open** dialog box of Autodesk Simulation Mechanical. However, if the model to be imported is created in the older version of Autodesk Inventor then it will be updated to the current version of Autodesk Simulation Mechanical and cannot be opened back in the version of Autodesk Inventor in which it was created. For example, if the model is created in Autodesk Inventor 2010 and you open it in the Autodesk Simulation Mechanical 2016 for analysis, then the version of that model will be updated from 2010 to 2016 and you cannot open it again in Autodesk Inventor 2010 version. To avoid facing such problems, you can directly import the model from Autodesk Inventor to Autodesk Simulation Mechanical.

To import the model directly from Autodesk Inventor to Autodesk Simulation Mechanical, Autodesk Inventor must be installed on your system. Next, start Autodesk Inventor and then open the model that needs to be transferred.

Once the model is opened in Autodesk Inventor, choose the **Simulation** tab in the **Ribbon**. Next, choose the **Launch Active Model** tool from the **Simulation Mechanical 2016** panel of this tab; the **Import Inventor Work Points** message window will be displayed. Choose the **Yes** button from this message window; the starting screen of Autodesk Simulation Mechanical with the **Choose Analysis Type** dialog box will be displayed, refer to Figure 3-6.

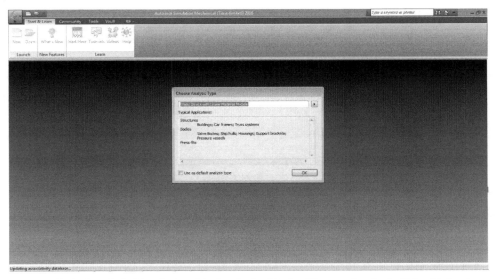

Figure 3-6 *Autodesk Simulation Mechanical with the* ***Choose Analysis Type*** *dialog box*

Tip. *You can also simplify the geometry of a model before transferring it from Autodesk Inventor to Autodesk Simulation Mechanical by suppressing some of its features such as fillets, chamfers, holes, and so on, if they are not required in the analysis process. The simplification of the Inventor model will be done by using the **Simplify Model** tool that is available in the **Simulation Mechanical 2016** panel of the **Simulation** tab in the **Ribbon**. You will learn more about simplifying the Inventor models later in this chapter.*

Select the required analysis type from the **Choose Analysis Type** dialog box. By default, the **Static Stress with Linear Material Models** analysis type is selected in this dialog box. After selecting the required analysis type, choose the **OK** button; the Inventor model will be opened in Autodesk Simulation Mechanical, as shown in Figure 3-7.

Note
If you have assigned any material property to the inventor model, then while transferring the model from Autodesk Inventor to Autodesk Simulation Mechanical, the material properties such as mass density, modulus of elasticity, poisson's ratio, thermal expansion coefficient, yield strength, ultimate tensile strength assigned to the Inventor model will also be transferred to Autodesk Simulation Mechanical.

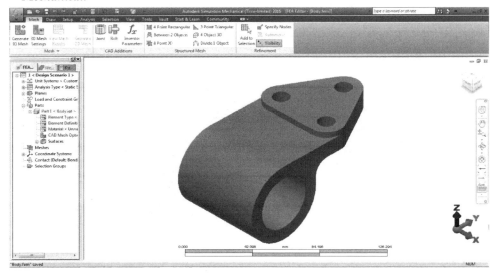

Figure 3-7 Autodesk Inventor model imported to Autodesk Simulation Mechanical

Importing 3D Models from SOLIDWORKS

Similar to importing inventor model directly from Autodesk Inventor software to Autodesk Simulation Mechanical, you can also import the SOLIDWORKS models directly from SOLIDWORKS software to Autodesk Simulation Mechanical. To import the model directly from SOLIDWORKS to Autodesk Simulation Mechanical, SOLIDWORKS software must be installed on your system. Next, start SOLIDWORKS and then open the model that needs to be transferred to Autodesk Simulation Mechanical, refer to Figure 3-8.

Once the model has been opened in SOLIDWORKS, move the cursor over the SOLIDWORKS logo that is available at the upper left corner of the screen; the SOLIDWORKS menus will be displayed, refer to Figure 3-8. Next, from the SOLIDWORKS menus, choose **Tools > Autodesk Simulation > Start Simulation**; the starting interface of Autodesk Simulation Mechanical with the **Choose Analysis Type** dialog box will be displayed, refer to Figure 3-8.

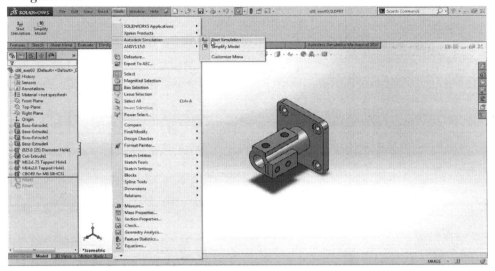

Figure 3-8 *A model opened in SOLIDWORKS*

Select the required analysis type from the **Choose Analysis Type** dialog box. By default, the **Static Stress with Linear Material Models** analysis type is selected in this dialog box. After selecting the required analysis type, choose the **OK** button from the dialog box; the SOLIDWORKS model will open in Autodesk Simulation Mechanical, as shown in Figure 3-9.

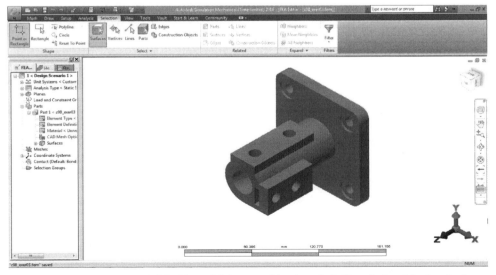

Figure 3-9 *SOLIDWORKS model imported in Autodesk Simulation Mechanical*

Note
Similar to Autodesk Inventor and SOLIDWORKS, you can also transfer a model directly to Autodesk Simulation Mechanical from other software such as Pro/Engineer/Creo Parametric, Solid Edge, and Autodesk Mechanical Desktop.

SPLITTING SURFACES OF CAD MODELS

The splitting surfaces is a process of recognizing where two parts of a model have surfaces in contact and then splitting those surfaces, if necessary, to get identical surfaces. You can specify whether you want to run the process of splitting surfaces on the CAD model being imported by choosing the **Yes** or **No** button from the **Surface Splitting** dialog box, refer to Figure 3-10. The **Surface Splitting** dialog box is displayed automatically once you import a model having two or more than two connecting components.

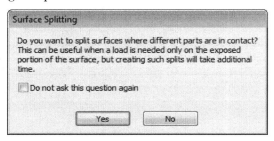

*Figure 3-10 The **Surface Splitting** dialog box*

Note that while importing a CAD model, the **Surface Splitting** dialog box may not be displayed. To display the **Surface Splitting** dialog box on importing a model, choose the **Application Options** button from the **Options** panel of the **Tools** tab in the **Ribbon**; the **Options** dialog box will be displayed, refer to Figure 3-11. Next, choose the **CAD Import** tab from the **Options** dialog box; the options related to importing a CAD geometry will be displayed, as shown in Figure 3-11.

Next, choose the **Global CAD Import Options** button from this dialog box; the **Global CAD Import Options** dialog box will be displayed, refer to Figure 3-12. If you select the **Yes** radio button next to the **Split surface on import** option in this dialog box then on importing a CAD model into Autodesk Simulation Mechanical, the contact surfaces of the model being imported will split automatically. On the other hand, if the **No** radio button is selected for the **Split surface on import** option then the process of splitting surfaces on the model being imported will not be carried out. However, if you select the **Ask** radio button then the **Surface Splitting** dialog box will be displayed each time you import a CAD model in Autodesk Simulation Mechanical.

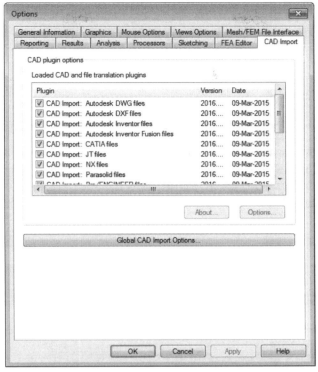

Figure 3-11 The **Options** *dialog box with the* **CAD Import** *tab chosen*

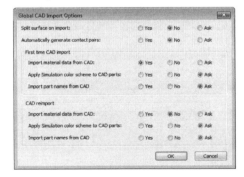

Figure 3-12 The **Global CAD Import Options** *dialog box*

SIMPLIFYING THE MODEL GEOMETRY BEFORE IMPORTING

While setting the model for analysis, most of the time you may need to first simplify the geometry to be analyzed by suppressing some of its features such as fillets, chamfers, holes, and so on. You can suppress those features that do not have major role in the analysis process or will not affect the analysis result. By suppressing some of the features, you can avoid the extra time that is involved in processes like meshing and computing results. You can simplify the geometry of the model in the software in which it is created before importing it in the Autodesk Simulation Mechanical. The process of simplifying the model geometry is discussed next.

Simplifying the Model Geometry in Autodesk Inventor

In Autodesk Inventor, you can simplify the geometry of the model before transferring it to Autodesk Simulation Mechanical. To do so, after opening the model in the Autodesk Inventor, choose the **Simplify Model** tool from the **Simulation Mechanical 2016** panel of the **Simulation** tab in the **Ribbon**; the **Model Simplification** dialog box will be displayed, as shown in Figure 3-13.

The **Model Simplification** dialog box displays a list of features of the active inventor model in the form of the tree view and is used to suppress features that are not necessary for the FEA model. To suppress a feature of a model to be imported in the Autodesk Simulation Mechanical for analysis, select it from the tree view of the **Model Simplification** dialog box. Next, drag the slider available at the top of the tree view in the dialog box toward the right upto 100% and then choose the **Apply** button from the dialog box; the selected feature will be suppressed. Figure 3-14 shows the **Model Simplification** dialog box with the fillet feature suppressed.

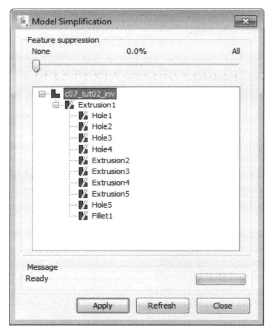

Figure 3-13 The **Model Simplification** dialog box

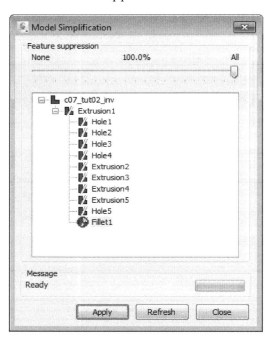

Figure 3-14 The **Model Simplification** dialog box with fillet feature suppressed

Simplifying the Model Geometry in SOLIDWORKS

In SOLIDWORKS, you can simplify the geometry of the model before transferring it to Autodesk Simulation Mechanical. To do so, open the model in SOLIDWORKS and then choose **Tools > Autodesk Simulation > Simplify Model** from the SOLIDWORKS menus; the **Model Simplification** dialog box will be displayed, refer to Figure 3-13.

To suppress a feature of a model to be imported in Autodesk Simulation Mechanical for analysis, select it from the tree view of the **Model Simplification** dialog box. Next, drag the slider available at the top of the tree view in the dialog box towards the right upto 100% and then choose the **Apply** button from the dialog box; the selected feature will be suppressed.

Note
Similar to the process of simplifying the geometry of a model in Autodesk Inventor and SOLIDWORKS before transferring them to Autodesk Simulation Mechanical, you can simplify the geometry of models in the other CAD software as well.

IMPORTING FEA MODEL

Autodesk Simulation Mechanical provides data translation interface to various leading FEA software for directly opening FEA models into it. To open an FEA model, choose the **Open** tool from the **Quick Access Toolbar** of the Autodesk Simulation Mechanical; the **Open** dialog box will be displayed. By default, in the **Files of type** drop-down list of the **Open** dialog box, the **Autodesk Simulation FEA Model (*.fem)** file type is selected. As a result, the **Open** dialog box will list only those models that have *.fem* file extension. Select the required FEA file type from the **Files of type** drop-down list of the **Open** dialog box. Next, browse to the location where the FEA model is saved. Select the required FEA model in Autodesk Simulation Mechanical and then choose the **Open** button from the dialog box; the **Unit System** dialog box will be displayed. Select the required units for the model and then choose the **OK** button; the selected FEA model will be opened in Autodesk Simulation Mechanical.

Note
*1. Similar to the process of importing the CAD and FEA models into the Autodesk Simulation Mechanical by selecting the required file type from the **Files of type** drop-down list of the **Open** dialog box, you can also import the models that are saved in neutral file format such as *.sat, *.stp, *.iges, and *.step.*

2. Autodesk Simulation Mechanical allows you to open ANSYS ANS files of version 5.4 and ANSYS CDB files of version 5, ABAQUS INP files of version 5.3, FEMAP Neutral files (NEU) of version 6.0, NASTRAN NAS, BDF and DAT files of version 2001, NASTRAN OP2 files of version 2001, PATRAN 2.5 Neutral files (PAT), SDRC I-DEAS Universal files (UNV), Stereolithography Files (STL), and Blue Ridge Numerics files (NEU).

SAVING FEA MODEL

When you open a model of any file type in Autodesk Simulation Mechanical, the system automatically saves the model in the *.fem* file extension with the same name as that of the original file at the same location. Now, after opening the model in Autodesk Simulation Mechanical, if you make any changes in it for carrying out the analysis process, you need to save those changes by choosing the **Save** button from the **Quick Access Toolbar**.

EXPORTING FEA MODEL

You can also export the FEA model of Autodesk Simulation Mechanical into the other FEA file formats such as ABAQUS (*.inp), ANSYS (*.cdb, *.ans), NASTRAN (*.nas, *.bdf, *.dat), and so on. To export Autodesk Simulation FEA model to other FEA file formats, choose the **Export > Third-Party FEA** from the **Application Menu**; the **Export** dialog box will be displayed, as shown in Figure 3-15. Now, browse to the location where you want to export the current FEA model. Next, specify a name for the FEA model in the **File name** edit box of the **Export** dialog box. After specifying the name, select the required FEA file type from the **Save as type** drop-down list of the dialog box and then choose the **Save** button; the FEA model will be exported to the selected file format. For example, if you want to export the FEA model of Autodesk Simulation

Mechanical into Nastran file format then you need to select the **NASTRAN (*.nas, *.bdf, *.dat)** file type from the **Save as type** drop-down list of the **Export** dialog box.

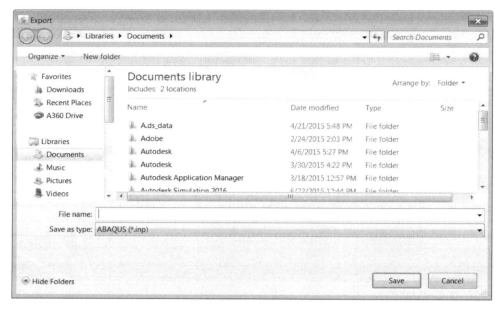

Figure 3-15 *The* **Export** *dialog box*

ARCHIVING FEA MODEL

In Autodesk Simulation Mechanical, you can save all files of an FEA model in an archive file. Archive files are compressed files which contain all the input files of the model as well as its results. As archive files are compressed files, they use less disk space. Also, compressing files of an FEA model makes it easier to store them or to transfer them from one location to another. The *.ach* is the file extension of the archive file. In Autodesk Simulation Mechanical, you can create, retrieve, repair, delete, and manage the existing archive files.

Creating an Archive File

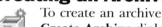

To create an archive file, choose **Archive > Create** from the **Application Menu**; the **Create Archive** dialog box will be displayed, refer to Figure 3-16. Now, browse to the location where you want to save the archive file of the currently opened FEA model. You can also specify a new name for the archive file by entering a new name in the **File name** edit box of the **Create Archive** dialog box. After specifying the location and name for the archive file, choose the **Save** button from the dialog box; the **Archive Creation Options** dialog box will be displayed, as shown in Figure 3-17. By default, the **Model only** radio button is selected in this dialog box. Note that, if you save the archive file with the **Model only** radio button selected in the **Archive Creation Options** dialog box then only the geometry of the FEA model will be saved in the archive file. However, if you select the **Model and results** radio button from the **Archive Creation Options** dialog box, all the input files of the model as well as its results will be saved in the archive file. After selecting the required radio button from the dialog box, choose the **OK** button; the archive file will be created in the specified location and the **Autodesk Simulation Mechanical** message window will be displayed on your screen informing you that the archive file has been created successfully. Choose the **OK** button from this message window.

Retrieving an Archive File

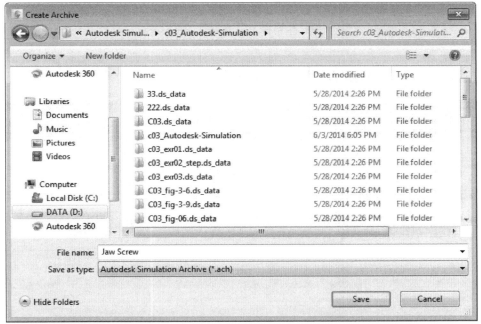

You can retrieve an archive file and open it in Autodesk Simulation Mechanical. To retrieve an existing archive file, choose **Archive > Retrieve** from the **Application Menu**; the **Extract Archive** dialog box will be displayed. Browse to the required location and then select the archive file to be retrieved. After selecting the archive file, choose the **Open button** from the dialog box; the **Browse for Folder** dialog box will be displayed. By using this dialog box, you can browse to the location where you want to extract all the data of the archived file. By default, in the **Browse for Folder** dialog box, the same folder in which the archive file is saved is selected as the folder to extract the data. Choose the **OK** button to extract the data; the selected archive file is retrieved and the model is opened in Autodesk Simulation Mechanical.

Figure 3-16 The **Create Archive** *dialog box*

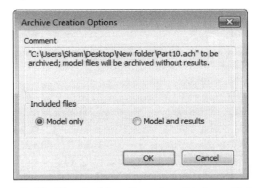

Figure 3-17 The **Archive Creation Options** *dialog box*

Note
*If the folder selected for extracting the data of an archive file already has all the data of the same archive file then on choosing the **OK** button from the **Browse for Folder** dialog box, the **Confirm File Overwrite** message window will be displayed, refer to Figure 3-18. This message window informs you that the file already exists in this folder and do you want to replace it with the existing file.*

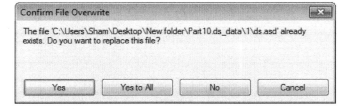

*Figure 3-18 The **Confirm File Overwrite** message window*

Repairing an Archive File

In Autodesk Simulation Mechanical, you can also repair a corrupt archive file. To do so, choose **Archive > Repair** from the **Application Menu**; the **Repair Archive** dialog box will be displayed. Browse to the required location and select the required archive file. Next, choose the **Open** button from the **Repair Archive** dialog box; the process for repairing the corrupted file will start. As soon as the repairing process is completed, an **Autodesk Simulation Mechanical** message window will be displayed with the information that the selected corrupted file has been repaired successfully. Next, choose the **OK** button from this message box to complete the process.

Managing an Existing Archive File

In Autodesk Simulation Mechanical, you can also manage an existing archive file by adding, deleting, restoring, or updating the data files of the selected archive file. To do so, choose **Archive > Manage Existing** from the **Application Menu**; the **Manage Archive** dialog box will be displayed. Browse to the location where the archive files are saved and then select the required archive file. Next, choose the **Open** button from the dialog box; the **Autodesk Simulation Archive Utility** window will be displayed which lists all the data files of the selected archive file, as shown in Figure 3-19. Now, you can add, delete, restore, or update the data files of the archive file by choosing **Edit > Add/Delete/Restore/Update** from the Menu Bar of the **Autodesk Simulation Archive Utility** window. To exit this window, choose **File > Exit** from the window.

Deleting an Archive File

To delete an existing archive file, choose **Archive > Delete** from the **Application Menu**; the **Delete Archive** dialog box will be displayed. Browse to the location where the archive files are saved. Next, select the archive file to be deleted and then choose the **Open** button from the dialog box; the **Autodesk Simulation Mechanical** message window will be displayed prompting you to delete the selected archive file. Choose the **Yes** button from this message window; the **Autodesk Simulation Mechanical** message window will be displayed again informing you that the selected archive file has been deleted. Choose the **OK** button from this message box to complete the process.

Name	Date	Time	Size	Packed	Ratio	Path
ds.asd	06/08/2015	10:52 AM	13	15	0%	tut4f.ds_data\1
ds.asd	06/08/2015	12:17 PM	13	15	0%	tut4f.ds_data\2
ds.asd	06/08/2015	02:12 PM	13	15	0%	tut4f.ds_data\3
ds.asd	06/08/2015	03:57 PM	13	15	0%	tut4f.ds_data\4
ds.asd	06/08/2015	05:58 PM	13	15	0%	tut4f.ds_data\5
ds.asd	06/12/2015	02:45 PM	13	15	0%	tut4f.ds_data\6
ds.asd	06/12/2015	02:50 PM	13	15	0%	tut4f.ds_data\7
elemcond.cdx	06/08/2015	10:58 AM	3072	151	95%	tut4f.ds_data\1\
elemcond.cdx	06/08/2015	02:54 PM	3072	151	95%	tut4f.ds_data\2\
elemcond.cdx	06/08/2015	04:31 PM	3072	151	95%	tut4f.ds_data\4\
elemcond.cdx	06/09/2015	10:25 AM	3072	151	95%	tut4f.ds_data\5\
elemcond.dbf	06/08/2015	10:58 AM	322	104	68%	tut4f.ds_data\1\
elemcond.dbf	06/08/2015	02:54 PM	322	104	68%	tut4f.ds_data\2\
elemcond.dbf	06/08/2015	04:15 PM	322	104	68%	tut4f.ds_data\4\
elemcond.dbf	06/08/2015	06:05 PM	322	104	68%	tut4f.ds_data\5\
elements.cdx	06/08/2015	10:58 AM	4096	582	86%	tut4f.ds_data\1\
elements.cdx	06/08/2015	02:54 PM	4096	582	86%	tut4f.ds_data\2\
elements.cdx	06/08/2015	04:31 PM	4096	583	86%	tut4f.ds_data\4\
elements.cdx	06/09/2015	10:25 AM	4096	583	86%	tut4f.ds_data\5\

73 files, 2681K, 468428 bytes compressed, 83%   Abort

*Figure 3-19 The **Autodesk Simulation Archive Utility** window*

UNDERSTANDING THE DRAWING DISPLAY TOOLS

The drawing display tools or navigation tools are integral part of any design or analysis software. These tools are extensively used during the design or analysis process. These tools are available in the **Navigation Bar** located on the right in the graphics window and also in the **Navigate** panel of the **View** tab in the **Ribbon**. Some of the drawing display tools in Autodesk Simulation Mechanical are discussed next and rest of these tools will be discussed in the later chapters.

Enclose (Fit All)/Zoom (Fit All)

Ribbon:	View > Navigate > Enclose (Fit All)
Navigation Bar:	Zoom flyout > Zoom (Fit All)

 This tool is used to increase the drawing display area to display all the objects in the current display view. You can invoke this tool from the **Ribbon** by choosing the **Enclose (Fit All)** button from the **Navigate** panel of the **View** tab. You can also invoke this tool from the **Navigation Bar** by choosing **Zoom (Fit All)** from the **Zoom** flyout.

Zoom

Ribbon:	View > Navigate > Zoom flyout > Zoom
Navigation Bar:	Zoom flyout > Zoom

 The **Zoom** tool is used to interactively zoom in and zoom out the drawing view. When you choose this tool, the default cursor is replaced by the zoom cursor. You can zoom out the drawing by pressing the left mouse button and dragging the cursor downward.

Similarly, you can zoom in the drawing by pressing the left mouse button and then dragging the cursor in the upward direction. You can exit this tool by choosing another tool or by pressing ESC. You can also zoom in the drawing by rolling the scroll wheel of the mouse in the downward direction. Similarly, you can zoom out the drawing by rolling the scroll wheel in the upward direction.

Window/Zoom (Window)

Ribbon:	View > Navigate > Zoom flyout > Window
Navigation Bar:	Zoom flyout > Zoom (Window)

This tool is used to define an area to be magnified and viewed in the current drawing. The area is defined using two diagonal points of a box (called window) in the graphics window. The area inscribed in the window will be magnified and displayed on the screen. You can invoke this tool from the **Ribbon** by choosing **Window** from the **Zoom** flyout in the **Navigate** panel of the **View** tab. You can also invoke this tool from the **Navigation Bar** by choosing **Zoom (Window)** from the **Zoom** flyout.

Selected/Zoom (Selected)

Ribbon:	View > Navigate > Zoom flyout > Selected
Navigation Bar:	Zoom flyout > Zoom (Selected)

This tool is used to magnify the selected entity in the graphics area upto the maximum extent and place it at the center of the graphics window. To magnify the selected entity, invoke this tool from the **Ribbon** by choosing the **Selected** option from the **Zoom** flyout in the **Navigate** panel of the **View** tab. You can also invoke this tool from the **Navigation Bar** by choosing the **Zoom (Selected)** option from the **Zoom** flyout.

Pan

Ribbon:	View > Navigate > Pan
Navigation Bar:	Pan

The **Pan** tool is used to drag the current view in the graphics window. This tool is generally used to display the contents that lie outside the display area, without actually changing the magnification scale of the current drawing. It is similar to holding a geometry and dragging it across the graphics window. You can also invoke the **Pan** tool by pressing and holding the CTRL key and the middle mouse button.

Orbit

Ribbon:	View > Navigate > Orbit flyout > Orbit
Navigation Bar:	Orbit flyout > Orbit

The **Orbit** tool is used to rotate a model freely about any axis. It is useful when you want to rotate a model to any position. It is a transparent tool as it can be invoked inside any other command. You can invoke this tool by choosing the **Orbit** tool from the **Navigate** panel in the **View** tab or from the **Orbit** flyout in the **Navigation Bar**. On invoking this tool, the cursor will change into an orbit cursor. You can orbit the geometry into the graphics area by pressing the left mouse button and dragging the orbit cursor.

Constrained Orbit/Orbit (Constrained)

Ribbon:	View > Navigate > Orbit flyout > Constrained Orbit
Navigation Bar:	Orbit flyout > Orbit (Constrained)

 This is one of the most important tools used for rotating the model around the vertical axis about a pivot point. When the **Constrained Orbit** tool is invoked, the cursor changes to the orbit cursor. You can press the left mouse button and drag the cursor to rotate the model towards left or right around the vertical axis about a previously defined pivot point. You can also define a new pivot point about which you want to rotate the model vertically. To define a pivot point, choose the **Set Pivot Center** tool from the **Navigation Bar** and click anywhere on the model in the graphics area. You can invoke this tool from the **Ribbon** by choosing the **Constrained Orbit** option from the **Orbit** flyout in the **Navigate** panel of the **View** tab. You can also invoke this tool from the **Navigation Bar** by choosing **Orbit (Constrained)** from the **Orbit** flyout.

> **Note**
> *While using the **Orbit (Constrained)** tool, the vertical axis about which the model will rotate will be normal to the Top face of the ViewCube.*

Previous View

Ribbon:	View > Navigate > Previous View
Navigation Bar:	Previous View

The **Previous View** tool in the **Navigate** panel of the **View** tab in the **Ribbon** is used to view the previous orientations of the model. You can also invoke this tool from the **Navigation Bar**.

Next View

Ribbon:	View > Navigate > Next View
Navigation Bar:	Next View

The **Next View** tool in the **Navigate** panel of the **View** tab in the **Ribbon** is used to activate the view that was current before you chose the **Previous View** tool. You can also invoke this tool from the **Navigation Bar**.

CHANGING THE VIEW USING THE VIEWCUBE

 Autodesk Simulation Mechanical provides you with an option to change the view of a 3D model freely in 3D space using the ViewCube. A ViewCube is a 3D navigation tool which allows you to switch between the standard and isometric views in a single click. By default, it is in the inactive state, as shown in Figure 3-20. When you move the cursor closer to the ViewCube, it gets activated, refer to Figure 3-21.

Figure 3-20 *Inactive ViewCube*

Figure 3-21 *Active ViewCube*

Note
*You can control the visibility of the ViewCube by using the **Ribbon**. To do so, click on the down arrow displayed on the right of the **User Interface** tool in the **Visibility** panel of the **View** tab in the **Ribbon**; a drop-down list will be displayed, as shown in Figure 3-22. Select or clear the **ViewCube** check box from this drop-down list to display or hide the ViewCube, respectively in the graphics window.*

The faces, vertices, and edges of the ViewCube are called clickable areas. If you place the cursor on any of the clickable areas of the ViewCube, the corresponding area will be highlighted. Click on the required area to orient the model such that the clicked area and the model become parallel to the screen. If you press and drag the left mouse button over the ViewCube, it will provide a visual feedback of the current viewpoint of the model. When you right-click on the ViewCube, a shortcut menu will be displayed, as shown in Figure 3-23. The options in this shortcut menu are discussed next.

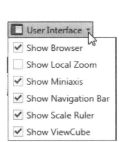

Figure 3-22 *The **User Interface** drop-down list*

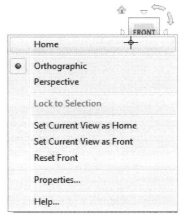

Figure 3-23 *The shortcut menu displayed after right-clicking on the ViewCube*

Home
It is used to display the default view of the model. On choosing this option, you can switch over to the home or isometric view of the model.

Orthographic
Choose this option to display the model in the orthographic view.

Perspective

Choose this option to display the model in the perspective view.

Lock to Selection

If you choose this option in a particular view, then that view gets locked and further manipulation of the view of the object will occur with respect to the locked position.

Set Current View as Home

This option is used to set the current view as the default or home view.

Set Current View as Front

This option is used to set the current view of the model as front view. When you select this option, the current view of the model will become the front view.

Reset Front

This option is used to reset the front view to the default setting.

Properties

Choose this option to invoke the **ViewCube Properties** dialog box, refer to Figure 3-24. The options in this dialog box are discussed next.

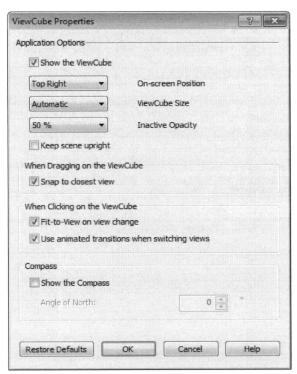

*Figure 3-24 The **ViewCube Properties** dialog box*

Show the ViewCube

By default, this check box is selected. As a result, the ViewCube will be displayed in the graphics window.

On-Screen Position

This drop-down list is used to display the on-screen position of the ViewCube. You can select the required option such as **Top Right**, **Bottom Right**, **Top Left**, or **Bottom Left** from the **On-Screen Position** drop-down list to set the position of ViewCube at any corner of the window. By default, the **Top Right** option is selected in this drop-down list. As a result, the ViewCube is displayed at the top right corner of the graphic window by default.

ViewCube Size

This drop-down list is used to set the size of the ViewCube. You can select the required option such as **Automatic**, **Tiny**, **Small**, **Medium**, or **Large** from the **ViewCube Size** drop-down list to set the size of the ViewCube. By default, the Automatic option is selected in this drop-down list. As a result, the size of the ViewCube is automatically adjusted according to the display area of the window.

Inactive Opacity

This drop-down list is used to set the opacity of the ViewCube in the inactive state. You must have noticed that when you move the cursor over the ViewCube in the graphics window, all additional controls of the ViewCube are displayed with full opacity. However, when the cursor is at some distance from the ViewCube, the ViewCube will be displayed with low opacity. This is because by default, the **50%** option is selected in the **Inactive Opacity** drop-down list. As a result, the opacity of the ViewCube in the inactive stage is set to fifty percent by default.

Keep scene upright

If this check box is selected then on clicking the edges, corners, or faces of the ViewCube, the ViewCube attempts to turn such that the upside-down orientations of the scene are avoided.

Snap to closest view

If the **Snap to closest view** check box is selected in the **When Dragging on the ViewCube** area of the dialog box, the view point will snap to one of the fixed views when the model is rotated using the ViewCube.

When Clicking on the ViewCube Area

The check boxes in this area are used for setting the preferences while clicking on the ViewCube.

Fit-to-View on view change

By default, the **Fit-to-View on view change** check box is selected. As a result, on clicking on the edges, corners, or faces of the ViewCube, the orientation of the model will be changed accordingly. Also, it zooms out or zooms in the current display to fit the model into the graphics window. If this check box is clear, the model will not fit in the current graphics window when the orientation of the model is changed using ViewCube.

Use animated transitions when switching views

By default, the **Use animated transitions when switching views** check box is selected. As a result, on changing the view of the model by using the ViewCube, an animated transition is displayed to help you visualize the spatial relationship between views.

Compass Area

This area is used to set the preference for the default display of the compass. If you select the **Show the Compass** check box, the compass will be displayed around the ViewCube in the graphics window, as shown in Figure 3-25. The **Angle of North** spinner of this area is used to set the angle between the **FRONT** face of the ViewCube and the **N** (North direction) of the compass.

Figure 3-25 *The ViewCube with Compass*

NAVIGATING THE MODEL USING STEERINGWHEELS

Ribbon:	View > Navigate > Steering Wheels flyout > Full Navigation Wheel
Navigation Bar:	Steering Wheels flyout > Full Navigation Wheel

 SteeringWheels is one of the most useful and convenient way to navigate the model. To display SteeringWheels, choose the **Full Navigation Wheel** tool from the **Navigation Bar**; the SteeringWheels will be displayed attached to the cursor, as shown in Figure 3-26. The SteeringWheels are the tracking tools that are divided into different wedges. Each wedge represents a single navigation tool such as **PAN**, **ORBIT**, **ZOOM**, **REWIND**, **LOOK**, **CENTER**, **WALK**, and **UP/DOWN**. You can activate any wedge of the SteeringWheels by pressing and holding the cursor over it. The SteeringWheels travels along with the cursor to provide a quick access to common navigation controls. Right-click on the SteeringWheels; a shortcut menu will be displayed. Different types of SteeringWheels are available in this shortcut menu. You can choose any type of SteeringWheels by clicking on it. You can also invoke different types of SteeringWheels by selecting the corresponding option from the **SteeringWheels** flyout in the **Navigation Bar**, refer to Figure 3-27. To display the **SteeringWheels** flyout, click on the down arrow available below the active **SteeringWheels** tool in the **Navigation Bar**, refer to Figure 3-27. You can also invoke the SteeringWheels from the **Navigate** panel of the **View** tab in the **Ribbon**.

Figure 3-26 *The **Full Navigation Wheel** SteeringWheels*

Figure 3-27 *The **SteeringWheels** flyout*

CONTROLLING THE DISPLAY OF MODELS

Autodesk Simulation Mechanical allows you to control the display of the models by setting various display modes and setting the camera type. The options for controlling the display of the models are discussed next.

Setting the Visual Styles

Ribbon: View > Appearance > Visual Style flyout

Visual style of a model determines the display of edges and face of a model in the graphics window. You can set the visual style for the solid models by using the **Visual Style** flyout provided in the **Appearance** panel of the **View** tab, refer to Figure 3-28. Various visual styles available in this flyout are discussed next.

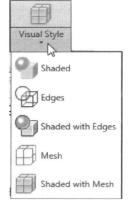

Shaded

 This style is used to shade the faces of a model with standard materials and colors. To apply this visual style, choose the **Shaded** tool from the **Visual Style** flyout of the **Appearance** panel in the **View** tab of the **Ribbon**. In this style, the visibility of the visible and hidden edges is turned off.

Edges

Figure 3-28 Tools in the Visual Style flyout

 This style is used to display only the model edges with the shading turned off. To apply this visual style, choose the **Edges** tool from the **Visual Style** flyout of the **Appearance** panel in the **View** tab of the **Ribbon**. In this style, the visibility of both the visible and hidden edges is turned on and these edges are displayed as solid lines.

Shaded with Edges

 In this style, the model appears shaded with all external edges clearly visible. To apply this visual style, choose the **Shaded with Edges** tool from the **Visual Style** flyout of the **Appearance** panel in the **View** tab of the **Ribbon**. In this style, the standard materials and colors are assigned to the model.

Mesh

This style is used to display the model with only the meshing lines visible. To apply this visual style, choose the **Mesh** tool from the **Visual Style** flyout of the **Appearance** panel in the **View** tab of the **Ribbon**.

Shaded with Mesh

In this style, the model appears shaded with meshing line visible. To apply this visual style, choose the **Shaded with Mesh** tool from the **Visual Style** flyout of the **Appearance** panel in the **View** tab of the **Ribbon**.

Setting the Camera Type

By default, the models are displayed in the orthographic camera type. You can change the camera type from the default orthographic to the perspective camera. To change the camera view to perspective, choose the **Perspective** tool from the expanded **Appearance** panel; the model will be displayed in the perspective camera.

TUTORIALS

To perform the tutorials of this chapter, download the input files of tutorials used in this chapter from *www.cadcim.com*. The complete path for downloading the file is as follows:

Textbooks > CAE Simulation > Autodesk Simulation Mechanical > Autodesk Simulation Mechanical 2016 for Designers > Input Files > c03_simulation_2016_input.zip

Extract the downloaded *c03_simulation_2016_input* zipped file and save it at the location *C:\ Autodesk Simulation Mechanical\c03*.

Tutorial 1

In this tutorial, you will import the Inventor assembly model shown in Figure 3-29 to Autodesk Simulation Mechanical. Select the default analysis type that is **Static Stress with Linear Material Models** to carry out the analysis process. After opening the assembly, you will change its visual style to the **Shaded with Edges** style. You will also use the SteeringWheels to navigate the model.   **(Expected time: 20 min)**

Figure 3-29 *The Knuckle joint*

The following steps are required to complete this tutorial:

a.   Download the zipped file containing input files of Chapter 3.
b.   Start Autodesk Simulation Mechanical.
c.   Set the **Split surface on import** setting to **Ask**.
d.   Open the Inventor model.
e.   Change the visual style of the model to Shaded with Edges style.
f.   Navigate the model using the SheeringWheels.
g.   Save and close the model.

Importing the Inventor Assembly File

1. Double-click on the shortcut icon of **Autodesk Simulation Mechanical 2016** on the desktop of your computer; the initial screen of Autodesk Simulation is displayed with either **Open** or **New** dialog box.

 Note
*As discussed earlier, when you start Autodesk Simulation Mechanical, by default the initial screen of Autodesk Simulation Mechanical is displayed with either **Open** or **New** dialog box depending upon the dialog box invoked in the last session of Autodesk Simulation Mechanical.*

*You can turn off the automatic display of dialog box at the start up of Autodesk Simulation Mechanical. To do so, first close the opened dialog box, if any, and then choose the **Application Options** button from the **Options** panel of the **Tools** tab; the **Options** dialog box will be displayed. In this dialog box, choose the **General Information** tab and then clear the **Show file dialog on startup** check box from the **Miscellaneous options** area. Next, choose the **Apply** button and then the **Close** button from the dialog box.*

In this tutorial, before opening the model in the Autodesk Simulation, you need to set the splitting surfaces setting to **Ask** so that whenever you import a model, the system asks you whether you want to split the surfaces of the model.

2. Close the **Open** or **New** dialog box, if displayed, by choosing the **Cancel** button from it.

3. Choose the **Tools** tab of the **Ribbon** and then choose the **Application Options** button from the **Options** panel of the **Tools** tab; the **Options** dialog box is displayed.

4. Choose the **CAD Import** tab from the **Options** dialog box; the options related to importing the CAD models are displayed in the dialog box.

5. Choose the **Global CAD Import Options** button from the dialog box; the **Global CAD Import Options** dialog is displayed, refer to Figure 3-30.

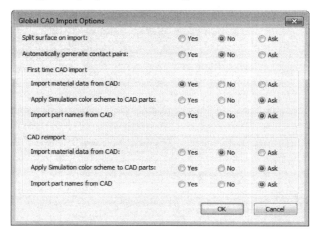

*Figure 3-30 The **Global CAD Import Options** dialog box*

6. Select the **Ask** radio button from the **Split surface on import** area and then choose the **OK** button from the dialog box. Next, exit the **Options** dialog box by choosing **OK**.

   Now, you need to open the inventor model of the Knuckle joint assembly in Autodesk Simulation Mechanical.

7. Choose the **Open** button from the **Launch** panel of the **Start & Learn** tab in the **Ribbon**; the **Open** dialog box is displayed.

   As mentioned in the tutorial description that the assembly model to be used for this tutorial is created in Autodesk Inventor. Therefore, you need to select the file extension of the assembly file created in Autodesk Inventor from the **Files of type** drop-down list of the **Open** dialog box to open the Inventor assembly in the Autodesk Simulation Mechanical.

8. Select the **Autodesk Inventor Files(*.ipt; *.iam)** file type from the **Files of type** drop-down list of the **Open** dialog box.

9. Browse to the location *C:\Autodesk Simulation Mechanical\c03\Tut01*; the inventor files of this tutorial are displayed in the **Open** dialog box.

10. Select the **Knuckle Joint.iam** file from the **Open** dialog box and then choose the **Open** button; the **Updating model** dialog box is displayed. Note that this dialog box is displayed only when you open the selected file for the first time.

11. Choose the **Yes** button from this dialog box; the **Surface Splitting** dialog box is displayed, as shown in Figure 3-31.

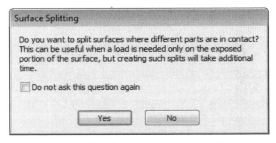

Figure 3-31 The **Surface Splitting** *dialog box*

12. Choose the **Yes** button from the **Surface Splitting** dialog box; either some message windows or the **Choose Analysis Type** dialog box may be displayed. If message windows are displayed, choose the **Yes** button from all the message windows; the **Choose Analysis Type** dialog box is displayed with the **Stress with Linear Material Models** analysis type selected by default, as shown in Figure 3-32.

13. Accept the default selection and choose the **OK** button from the dialog box; the Inventor assembly file is opened into Autodesk Simulation Mechanical, as shown in Figure 3-33. As soon as the model is opened in Autodesk Simulation Mechanical, a copy of the model is automatically saved as an FEA model with the file extension *.fem* in the same folder where the Inventor model is saved.

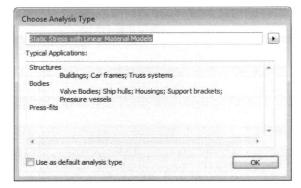

Figure 3-32 *The **Choose Analysis Type** dialog box*

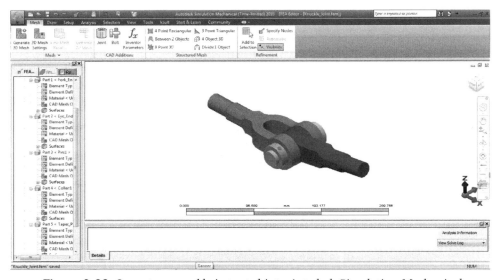

Figure 3-33 *Inventor assembly imported into Autodesk Simulation Mechanical*

Changing the Visual Style

In this section, you need to change the default visual style of the assembly to the Shaded with Edges visual style.

1. Choose the **View** tab of the **Ribbon**; all tools that are grouped together into the **View** tab of the **Ribbon** are displayed.

2. Click on the down arrow available at the bottom of the **Visual Style** button in the **Appearance** panel; the **Visual Style** flyout is displayed, as shown in Figure 3-34.

3. Choose the **Shaded with Edges** tool from this flyout; the current visual style of the assembly is changed to Shaded with Edges style, as shown in Figure 3-35.

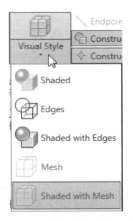

Figure 3-34 *The **Visual Style** flyout*

Figure 3-35 *The assembly displayed in Shaded with Edges visual style*

Navigating the Assembly by Using the SteeringWheel

Now, you need to navigate the assembly by using the SteeringWheels.

1. Choose the **Full Navigation Wheel** tool from the **Navigation Bar**; the Steering Wheels attached with the cursor is displayed, as shown in Figure 3-36.

2. Move the SteeringWheels over the assembly by moving the cursor.

3. Move the cursor over the **ORBIT** wedge of the SteeringWheels; the **ORBIT** wedge is highlighted. Next, press and hold the left mouse button; the **Orbit** tool is activated. Now, drag the cursor; the assembly rotates about the **PIVOT** point, refer to Figure 3-37.

Figure 3-36 *The SteeringWheels*

Figure 3-37 *The **Orbit** tool activated by using the SteeringWheels*

4. Move the cursor over the **PAN** wedge of the SteeringWheels; the **PAN** wedge is highlighted. Next, press and hold the left mouse button; the **Pan** tool is activated. Now, drag the cursor; the assembly moves along the cursor.

5. Move the cursor over the **REWIND** wedge of the SteeringWheels and press and hold the left mouse button; the previous views are displayed in small boxes in a row. Drag the cursor along the row and select a view.

6. Move the SteeringWheels over the assembly by moving the cursor. Next, move the cursor over the **CENTER** wedge; a pivot is attached to the cursor. Next, release the left mouse button over the assembly at the point where you want to create a pivot point. Note that as soon as you release the left mouse button, a pivot point is created in the assembly and moved to the center of the graphics area.

7. Move the cursor over the **ZOOM** wedge of the SteeringWheels and press and hold the left mouse button; the **Zoom** tool is activated. Now, drag the cursor in the upward/downward direction to zoom in/out the graphics area.

8. Click on the cross mark at the top right corner of the SteeringWheels to exit this tool. Alternatively, right-click and choose the **Close Wheel** option from the shortcut menu displayed to exit the **SteeringWheels** tool.

9. Move the cursor over the ViewCube to activate it and then choose the **Home** button;   the current view of the assembly is changed to the isometric view.

Saving the Model
Now, you need to save changes made in the FEA model.

1. Choose the **Save** button from the **Quick Access Toolbar**; the changes made in the   assembly are saved.

2. Choose **Close** from the **Application Menu** to close the assembly.

Tutorial 2

In this tutorial, you will import a SOLIDWORKS assembly file shown in Figure 3-38 into Autodesk Simulation Mechanical. Select the default analysis type which is **Static Stress with Linear Material Models**. After opening the assembly, you need to change the views of the assembly by using the ViewCube.   **(Expected time: 15 min)**

The following steps are required to complete this tutorial:

a. Download the zipped file containing files of Chapter 3.
b. Start Autodesk Simulation Mechanical.
c. Set the **Split surfaces on import** setting **Yes** or **No** to **Ask.**
d. Navigate the model using the ViewCube.
e. Save and close the model.

Figure 3-38 SLODWORKS assembly

Importing the SOLIDWORKS Assembly File

1. Start Autodesk Simulation Mechanical by double-clicking on the shortcut icon of **Autodesk Simulation Mechanical 2016** on the desktop of your computer.

   Now, you need to open the SOLIDWORKS assembly into the Autodesk Simulation Mechanical.

2. Choose the **Open** button from the **Launch** panel of the **Start & Learn** tab in the **Ribbon**; the **Open** dialog box is displayed.

 Note
*In this textbook, the automatic display of the **Open** or **New** dialog box on the start up of Autodesk Simulation Mechanical has been turned off.*

   As mentioned in the tutorial description that the assembly model being used for this tutorial is created in SOLIDWORKS. Therefore, you need to select the file extension of the assembly file created in SOLIDWORKS from the **Files of type** drop-down list of the **Open** dialog box to open the SOLIDWORKS assembly in Autodesk Simulation Mechanical.

3. Select the **SOLIDWORKS Files (*.prt; *.sldprt; *.asm; *.sldasm)** file type from the **Files of type** drop-down list of the **Open** dialog box.

4. Browse to the location *C:\Autodesk Simulation Mechanical\c03\Tut02*; the SOLIDWORKS assembly file of this tutorial is displayed in the **Open** dialog box.

5. Select the **Table.SLDASM** file and then choose the **Open** button from the **Open** dialog box; the **Updating model** dialog box is displayed. Note that this dialog box is displayed only when you open the older version of the model for the first time.

6. Choose the **Yes** button from this dialog box; the **Surface Splitting** dialog box is displayed.

7. Choose the **No** button from the **Surface Splitting** dialog box; either some message windows or the **Choose Analysis Type** dialog box may be displayed. If message windows are displayed, choose the **No** button from the **Simulation Mechanical Color Palette (reimport)** window and **Yes** button from the other message window(s); the **Choose Analysis Type** dialog box is displayed with the **Stress with Linear Material Models** analysis type selected by default in it, as shown in Figure 3-39.

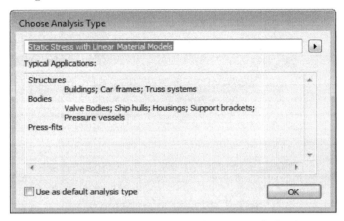

*Figure 3-39 The **Choose Analysis Type** dialog box*

Note

*The **Surface Splitting** dialog box will be displayed only if the **Ask** radio button in the **Split surface on import** area of the **Global CAD Import Options** dialog box is selected. If the **Yes** or **No** radio button form **Split surface on import** area is selected in this dialog box then after choosing the **Open** button from the **Open** dialog box, the **Choose Analysis Type** dialog box will be displayed directly.*

6. Accept the default selection and choose the **OK** button from the dialog box; the SOLIDWORKS assembly file is opened in the Autodesk Simulation Mechanical, as shown in Figure 3-40.

Changing Views of the Assembly by Using the ViewCube

Now, you need to change the views of the assembly by using the ViewCube.

1. In case, the ViewCube is hidden or not visible in the graphics area, you need to first turn on its visibility. To do so, click on the down arrow displayed on the right of the **User Interface** tool of the **Visibility** panel in the **View** tab of the **Ribbon**; the **User Interface** flyout is displayed, refer to Figure 3-41.

2. Select the **Show ViewCube** check box from this flyout; the ViewCube is displayed at the upper right corner of the graphics area.

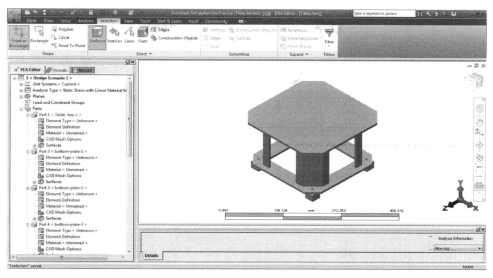

Figure 3-40 *SLODWORKS assembly imported into Autodesk Simulation Mechanical*

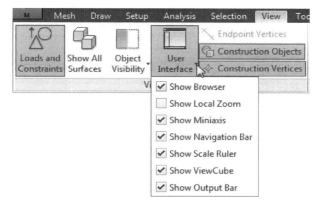

Figure 3-41 *The **User Interface** flyout*

Note

If the ViewCube is already displayed at the upper right corner of the graphics area then you can skip step 1 and 2.

3. Move the cursor over the **FRONT** clickable area of the ViewCube; the **FRONT** clickable area is highlighted, refer to Figure 3-42.

4. Click on the **FRONT** clickable area to display the front view of the model; the model and clicked area become parallel to the screen. Figure 3-43 shows the Front view of the model displayed after clicking on the **FRONT** clickable area of the ViewCube.

Note

*If you do not get the orientation view as shown in the Figure 3-43 then you can set it so by using the ViewCube. To do so, click on the down arrow available next to the ViewCube and then choose the **Set Current View as Front** option to make the current view as the front view.*

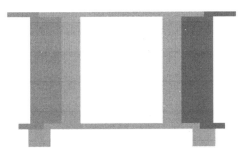

Figure 3-42 The ViewCube with
FRONT *clickable area highlighted*

Figure 3-43 The Front view of the model

5. Choose the **Home** button of the ViewCube; the current view of the model is changed to the isometric view.

6. Move the cursor over the top right corner of the ViewCube, refer to Figure 3-44 and when it is highlighted, press and hold the left mouse button and drag the cursor. As the cursor moves, the model re-orients to give you a better view of the model.

7. Choose the **Home** button of the ViewCube again to display the model in the isometric view.

Figure 3-44 The ViewCube

Saving the Model

Now, you need to save the changes made in the FEA model.

1. Choose the **Save** button from the **Quick Access Toolbar**; the changes made in the assembly are saved.

2. Choose **Close** from the **Application Menu** to close the assembly.

Tutorial 3

In this tutorial, you will import a STEP file of the model, as shown in Figure 3-45. You need to select the analysis type which is **Static Stress with Linear Material Models**. After opening the assembly, you need to change the views of the assembly by using the ViewCube.

(Expected time: 15 min)

The following steps are required to complete this tutorial:

a. Download the zipped file containing files of Chapter 3.
b. Start Autodesk Simulation Mechanical.

c. Change the **Split surfaces on import** setting **Yes** or **No** to **Ask**.
d. Open the SOLIDWORKS model.
e. Navigate the model using the ViewCube.
f. Save and close the model.

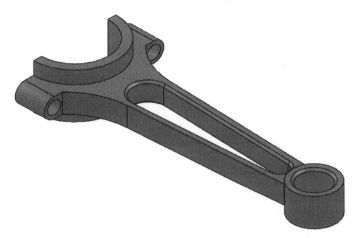

Figure 3-45 STEP file model

Importing the SOLIDWORKS Assembly File

1. Start Autodesk Simulation Mechanical by double-clicking on the shortcut icon of **Autodesk Simulation Mechanical 2016** on the desktop of your computer.

   Now, you need to open the SOLIDWORKS assembly into Autodesk Simulation Mechanical.

2. Choose the **Open** button from the **Launch** panel of the **Start & Learn** tab in the **Ribbon**; the **Open** dialog box is displayed.

   Note
   *In this textbook, the automatic display of the **Open** or **New** dialog box on the start up of Autodesk Simulation Mechanical has been turned off.*

   As mentioned in the tutorial description, the model being used for this tutorial is a STEP file. Therefore, you need to select the file extension of the STEP file format from the **Files of type** drop-down list of the **Open** dialog box.

3. Select the **STEP Files (*.stp; *.ste; *.step)** file type from the **Files of type** drop-down list of the **Open** dialog box.

4. Browse to the location *C:\Autodesk Simulation Mechanical\c03\Tut03*; the STEP file of this tutorial is displayed in the **Open** dialog box.

5. Select the **C03_Tut03** file and then choose the **Open** button from the **Open** dialog box; the **Choose Analysis Type** dialog box is displayed with the **Stress with Linear Material Models** analysis type selected by default in it, as shown in Figure 3-46.

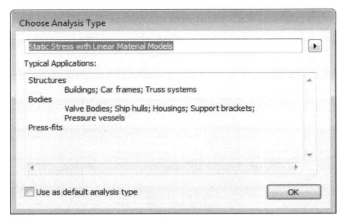

Figure 3-46 *The **Choose Analysis Type** dialog box*

6. Accept the default selection and choose the **OK** button from the dialog box; the STEP file is opened into the Autodesk Simulation Mechanical, as shown in Figure 3-47.

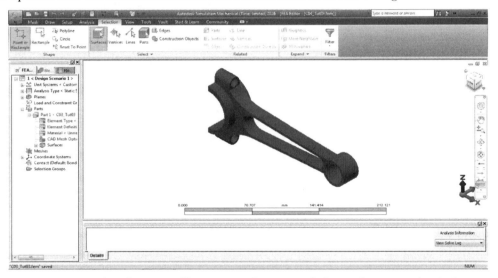

Figure 3-47 *The STEP file is imported into Autodesk Simulation Mechanical*

Changing Views of the Assembly by Using the ViewCube

Now, you need to change the views of the assembly by using the ViewCube.

1. Press and hold the middle mouse button and then drag the cursor; the model changes its orientation as you drag the cursor. Orient the model such that it is displayed similar to one shown in Figure 3-48.

2. Move the cursor over the top right corner of the ViewCube, refer to Figure 3-49 and then click the left mouse button when it is highlighted; the model gets reoriented and gives you a better view.

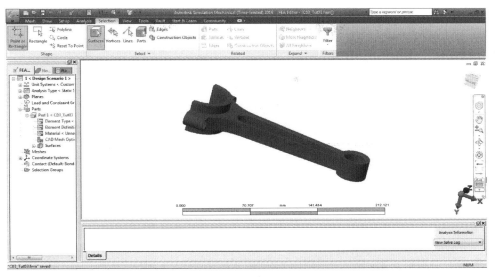

Figure 3-48 *The model after changing its default orientation*

Figure 3-49 *The ViewCube*

Saving the Model

Now, you need to save the changes made in the FEA model.

1. Choose the **Save** button from the **Quick Access Toolbar**; the changes made in the assembly are saved.

2. Choose **Close** from the **Application Menu** to close the assembly.

Self-Evaluation Test

Answer the following questions and then compare them to those given at the end of this chapter:

1. The _____ is the file extension of Autodesk Simulation Mechanical.

2. The _____ is a process of recognizing contact surfaces of a model and then splitting them to get identical surfaces, if necessary.

3. The _____ dialog box displays the list of all the features of the active model in the form a tree view, so that you can suppress some of the features of the model that are not necessary for the FEA model.

4. The _____ files are compressed files which contain numerous input files of the model as well as the results of the model.

5. In Autodesk Simulation Mechanical, you can directly import CAD model from the software in which it is created. (T/F)

6. To import a model directly from Autodesk Inventor to Autodesk Simulation Mechanical, Autodesk Inventor must be installed on your system. (T/F)

7. You cannot simplify the geometry of an inventor model before transferring it from Autodesk Inventor to Autodesk Simulation Mechanical. (T/F)

8. By default, the **Static Stress with Linear Material Models** analysis type is selected in the **Choose Analysis Type** dialog box. (T/F)

9. You cannot export the FEA model of Autodesk Simulation Mechanical into any other FEA file type. (T/F)

10. In Autodesk Simulation Mechanical, you can save the files of an FEA model to an archive file. (T/F)

Review Questions

Answer the following questions:

1. Which of the following dialog boxes is used to select an analysis type?

    (a) **Choose Analysis Type**   (b) **Choose Analysis**
    (c) **Select Analysis Type**   (d) None of these

2. Which of the following dialog boxes is displayed when you choose **Archive > Create** from the Application Menu of Autodesk Simulation Mechanical?

    (a) **Archive**   (b) **Create Archive**
    (c) **Create**   (d) **Create Arch**

3. Which of the following tools is used to view the previous orientation of the model?

    (a) **Previous**   (b) **Previous Orientation**
    (c) **Previous View**   (d) None of these

4. Which of the following tools is used to increase the drawing display area to display all the objects in the current display view?

    (a) **Enclose (Fit All)**   (b) **Fit All**
    (c) **Zoom (Fit All)**   (d) Both a and c

5. Which of the following dialog boxes is displayed when you choose **Archive > Retrieve** from the Application Menu?

    (a) **Extract Archive**   (b) **Archive Retrieve**

    (c) **Retrieve Archive**   (d) None of these

6. In Autodesk Simulation Mechanical, you can turn on or off the visibility of the ViewCube. (T/F)

7. The SteeringWheels are the tracking tools that are divided into different wedges. (T/F)

8. In Autodesk Simulation Mechanical, you cannot export a FEA model to the file extension of ANSYS. (T/F)

9. When you open a model of any file type in Autodesk Simulation Mechanical, the system automatically saves the model with its original name and at original place with the *.fem* file extension. (T/F)

10. Autodesk Simulation Mechanical provides with data translation interface to various leading FEA software for directly opening FEA models created in other FEA software. (T/F)

EXERCISES

The input files used in the exercises are available in their respective chapter folder, where the input files of the tutorials are available. You can download the input files of the exercises used in this chapter from www.cadcim.com, if not already downloaded. The complete path for downloading the files is given below:

Textbooks > CAE Simulation > Autodesk Simulation Mechanical > Autodesk Simulation Mechanical 2016 for Designers > Input Files > c03_simulation_2016_input.zip

Exercise 1

In this exercise, you will import the *c03_exr01.igs* file from the **03_simulation_2016_input** extracted folder to Autodesk Simulation Mechanical. The model for this exercise is shown in Figure 3-50. Select the default analysis type which is **Static Stress with Linear Material Models**. After opening the assembly, you need to change the view of the model to Front view by using the ViewCube and then save the FEA model at the location *C:/Autodesk Simulation Mechanical/ c03\Exr01*.   **(Expected time: 10 min)**

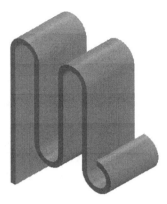

Figure 3-50 IGS model for Exercise 1

Exercise 2

In this exercise, you will import the *c03_exr02.STEP* file from the **03_simulation_2016_input** extracted folder to Autodesk Simulation Mechanical. The model is shown in Figure 3-51. Select the default analysis type which is **Static Stress with Linear Material Models** to carry out the analysis process. After opening the model, navigate it using the SteeringWheels. Save the FEA model at the location *C:\Autodesk Simulation Mechanical\c03\Exr02*.

(Expected time: 10 min)

Figure 3-51 STEP model for Exercise 2

Exercise 3

In this exercise, you will import the *c03_exr03.IGS* file from the **03_simulation_2016_input** extracted folder to Autodesk Simulation Mechanical. The model is shown in Figure 3-52. Select the default analysis type which is **Static Stress with Linear Material Models** to carry out the analysis process. After opening the model, change its visual style to the Shaded with Edges style. Also, save the FEA model at the location *C:\Autodesk Simulation Mechanical\c03\Exr03*.

(Expected time: 10 min)

Figure 3-52 *IGS model for Exercise 3*

Answers to Self-Evaluation Test

1. .fem, **2.** Surface splitting, **3. Model Simplification**, **4.** archive **5.** T, **6.** T, **7.** F, **8.** T, **9.** F, **10.** T

Chapter 4

Creating and Modifying a Geometry

Learning Objectives

After completing this chapter, you will be able to:
- *Understand the concept of creating 2D geometry of an object*
- *Understand different selection methods*
- *Modify spacing between the grid lines*
- *Create entities such as lines, circles, rectangles, arcs, and splines*
- *Edit and modify the existing 2D sketched geometries*
- *Trim the unwanted entities of the sketch*
- *Extend the unwanted entities of the sketch*
- *Create intersection in the sketched entities*
- *Divide the sketched entities into multiple segments*
- *Modify the attributes of the sketch drawn*

INTRODUCTION

In Autodesk Simulation Mechanical, in addition to importing 3D solid geometry and performing meshing operations, you can also create 2D and 3D sketched geometries that represent 3D solid objects. The tools to create 2D and 3D sketch geometries are available in the **Draw** panel of the **Draw** tab in the **Ribbon**. After creating the sketched geometry to be meshed, you can further modify it by using the tools available in the **Modify** panel of the **Draw** tab in the **Ribbon**. You can create a 2D sketched geometry in any of the default planes (XY, YZ, or XZ) and generate mesh. In addition to the default planes, you can also create additional planes for creating 2D sketched geometry. A 2D sketched geometry created in any plane can be meshed in the FEA Editor environment of Autodesk Simulation Mechanical by using the **Generate 2D Mesh** tool. Note that a geometry can represent one or more than one part and for generating mesh, each part of the geometry within a plane must be enclosed in a single continuous area. The geometry where two parts meet must be identical for meshes to match up. In Autodesk Simulation, a sketched geometry can be represented by Truss, Beam, Membrane, 2D, Brick, Plate, Tetrahedron, and other elements. As discussed earlier, a sketched geometry created in any plane can be meshed in the FEA Editor environment. However, a geometry with 2D elements can only be created in the YZ plane for generating meshing. Figure 4-1 shows a geometry consisting of two parts and Figure 4-2 shows the same geometry after meshing.

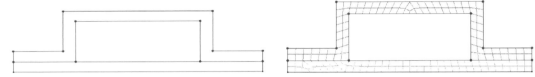

Figure 4-1 *A geometry consisting of two parts*   *Figure 4-2* *The geometry after meshing*

Before you learn about various sketching tools used for creating sketched geometries, it is important to understand the different selection methods.

SELECTION METHODS

In Autodesk Simulation Mechanical, there are several selection methods for selecting objects: Point or Rectangle selection method, Rectangle selection method, Polyline selection method, and Circle selection method. By using these selection methods, you can select surfaces, vertices, lines, parts, edges, and construction objects. All these selection methods are discussed next.

Point or Rectangle Selection Method

By using this method, you can select objects or entities by clicking the left mouse button or by drawing a temporary rectangular window around the objects to be selected. You can select surfaces, vertices, lines, parts, edges, and construction objects by using this method.

To select surfaces by using the Point or Rectangle selection method, choose the **Selection** tab in the **Ribbon** and then choose the **Point or Rectangle** tool from the **Shape** panel of the **Selection** tab. Next, choose the **Surfaces** tool from the **Select** panel of the **Selection** tab to select the surfaces. Move the cursor over the surface that you want to select; the surface will be highlighted, refer to Figure 4-3. Click the left mouse button to select the highlighted surface. You can also

select multiple surfaces of a model by using the
CTRL key. Similarly, using this method, you can
select vertices, lines, parts, edges, and
construction objects by choosing the respective
tool from the **Select** panel of the **Selection** tab.
For example, to select the edges of a model by
using this method, choose the **Edges** tool from
the **Shape** panel or to select the construction
objects, choose the **Construction Objects** tool.

In this selection method, to select geometries,
you can also draw a rectangular window around

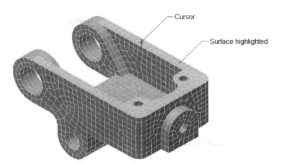

Figure 4-3 *The surface highlighted*

the geometries to be selected. To do so, activate the **Point or Rectangle** tool and then choose
the required tool such as **Surfaces**, **Vertices**, **Lines**, **Parts**, **Edges**, or **Construction Objects** from
the **Select** panel of the **Selection** tab. Next, press and hold the left mouse button to specify the
first corner of the rectangle window and then move the cursor; the preview of the rectangular
window will be displayed. Next, specify the diagonally opposite corner by clicking the left mouse
button; all the objects, depending upon the tool selected in the **Select** panel that are completely
inside the rectangular window will be selected.

Rectangle Selection Method

In this selection method, you can select objects or entities by drawing a rectangular window
around the objects to be selected. The objects that are completely enclosed within the rectangular
window drawn will be selected. However, the objects that lie partially inside the boundaries of
the window will not be selected. This method helps in selecting multiple objects at a time. You
can select surfaces, vertices, lines, parts, edges, and construction objects by using this method.

To select surfaces by using this method, choose the **Rectangle** tool from the **Shape** panel of the
Selection tab. Next, choose the **Surfaces** tool from the **Select** panel of the **Selection** tab to select
the surfaces. Specify the first corner of the rectangular window by clicking the left mouse button
in the drawing area and then move the cursor to specify the opposite corner, refer to Figure 4-4.
Note that as you move the cursor, a rubber band rectangular window of continuous line will be
displayed. The size of the window changes as you move the cursor. Specify the opposite corner
of the rectangular window; the surfaces that are enclosed in this window will be selected, refer
to Figure 4-5.

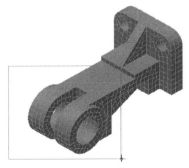

Figure 4-4 *Rectangular window being drawn*

Figure 4-5 *The enclosed surfaces selected*

Similarly, you can also select vertices, lines, parts, edges, or construction objects by choosing the corresponding tool from the **Select** panel of the **Selection** tab in the **Ribbon**.

Polyline Selection Method

By using the Polyline selection method, you can select objects or entities by drawing a closed irregular polygon profile around the objects to be selected. This method of selection is similar to the Rectangle selection method with the only difference that by using this method, you can define a closed profile of any shape created by the polyline. You can specify the selection area by specifying points around the objects that you want to select. Similar to the Rectangle selection method, the objects to be selected by using this method should be completely enclosed within the closed polygon profile. You can select surfaces, vertices, lines, parts, edges, and construction objects by using this method.

To select surfaces by using the Polyline selection method, choose the **Polyline** tool from the **Shape** panel of the **Selection** tab. Next, choose the **Surfaces** tool from the **Select** panel of the **Selection** tab to select the surfaces. Specify the start point of the first line entity of the closed polygon by clicking the left mouse button in the drawing area and then move the cursor to specify the end point. As soon as you move the cursor, a rubber band line will be attached to the cursor. The length of the rubber band line changes as you move the cursor. Specify the end point of the first line entity; a line entity will be drawn between the two specified points. Move the cursor away from the endpoint of the drawn line entity and you will notice that another line is attached to the cursor. The start point of this line is the endpoint of the last line and the length of this line can be increased or decreased by moving the cursor. Specify a point in the drawing area as the endpoint of the second line entity; the second line entity will be drawn. Now, a new rubber-band line is displayed starting from the endpoint of the last line. This is an ongoing process and you can draw as many continuous line entities as required to create a polygon by specifying points on the screen using the left mouse button. To complete the process of creating a polygon profile, you need to specify the end point of the last line entity to the start point of the first line entity drawn. All the surfaces enclosed inside the polygon drawn will be selected in the drawing area. Figure 4-6 shows a polygon profile being created and Figure 4-7 shows the same model with all the enclosed surfaces selected.

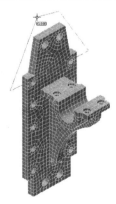

Figure 4-6 A Polygon being drawn   *Figure 4-7* The enclosed surfaces selected

Similarly, you can also select vertices, lines, parts, edges, and construction objects by choosing the respective tool from the **Select** panel of the **Selection** tab in the **Ribbon**.

Circle Selection Method

In the Circle selection method, you can select objects or entities by drawing a circle around the objects to be selected. In this method, you need to specify a selection area by drawing a circle around the objects that you want to select. On doing so, the objects that are completely enclosed within the circle drawn will be selected. You can select surfaces, vertices, lines, parts, edges, and construction objects by using this method.

To select surfaces by using the Circle selection method, choose the **Circle** tool from the **Shape** panel of the **Selection** tab; the Circle selection method will be invoked. Next, choose the **Surfaces** tool from the **Select** panel of the **Selection** tab to select the surfaces. Specify the center point of the circle and then move the cursor to define its radius. The radius of the circle changes as you move the cursor. Click to define the radius of the circle; all the surfaces enclosed inside the circle drawn will be selected in the drawing area. Figure 4-8 shows a circle being drawn and Figure 4-9 shows the same model with all the enclosed surfaces selected.

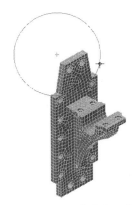
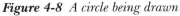
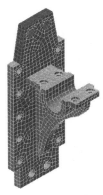

Figure 4-8 A circle being drawn   *Figure 4-9 Surfaces enclosed inside circle are selected*

Similarly, you can also select vertices, lines, parts, edges, and construction objects by choosing the respective tool from the **Select** panel of the **Selection** tab in the **Ribbon**.

CREATING 2D SKETCHED GEOMETRY

You can create 2D sketched geometries by using the tools available in the **Draw** panel, refer to Figure 4-10. For creating 2D sketched geometries, invoke a new FEA model session of Autodesk Simulation Mechanical by selecting the **FEA Model** option from the **New** dialog box. After selecting the **FEA Model** option, select the **New** button from this dialog box; the **Save As** dialog box will be displayed. Save the FEA Model to the desired location by using the **Save As** dialog box; FEA Editor environment will be displayed. Next, you need to invoke the sketching environment by selecting a sketching plane. You can select any of the default planes (XY, YZ, and XZ) as the sketching plane from the **Tree View**. In the **Tree View** of the FEA Editor environment, all the default planes are available under the **Planes** node. To invoke the sketching environment, select the required plane from the **Tree View** and right-click; a shortcut menu will be displayed, refer to Figure 4-11. Choose the **Sketch** option from the shortcut menu; the selected plane will become parallel to the screen and the sketching environment will be invoked with the grid lines, as shown in Figure 4-12. You can also invoke the sketching environment by double-clicking on the required plane in the **Tree View**.

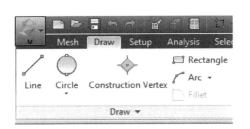

Figure 4-10 *The **Draw** panel*

Figure 4-11 *Shortcut menu displayed*

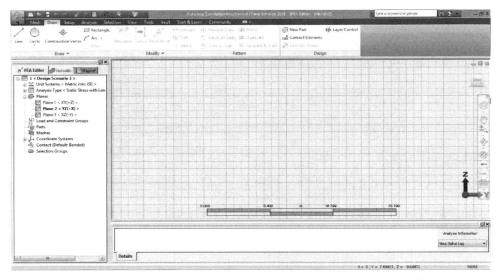

Figure 4-12 *The sketching environment invoked after selecting the YZ plane as the sketching plane*

By default, the spacing between the grid lines is set to 1 unit (unit of the current document). You can modify the spacing between the grid lines as per your requirement.

The procedure for modifying the grid spacing and creating the sketched entities by using the tools of the **Draw** panel of the **Draw** tab are discussed next.

Modifying the Grid Spacing

Grid lines are the reference lines on the screen at a predefined spacing, refer to Figure 4-12. In Autodesk Simulation Mechanical, by default, the spacing between the grid lines is set to 1 unit (unit of the current document). These lines can be used as reference lines for creating a drawing. It also gives you an idea about the size of the geometry drawn in the drawing area. You can change the distance between the grid lines as per your requirement by using the **Options** dialog box.

To invoke the **Options** dialog box, choose the **Tools** tab in the **Ribbon** and then choose the **Application Options** tool; the **Options** dialog box will be displayed. In this dialog box, choose the **Sketching** tab; the options available in the **Sketching** tab will be displayed in the dialog box, refer to Figure 4-13. The **Major X Spacing** and the **Major Y Spacing** edit boxes of the **Grid** area are used to define the desired grid spacing along the X and Y axes. By default, the value 1 unit is set in these edit boxes. After specifying the required values in these edit boxes, choose the **Apply** button; the spacing between the grid lines will be modified as per the value entered. Next, exit from the dialog box by choosing the **Cancel** button.

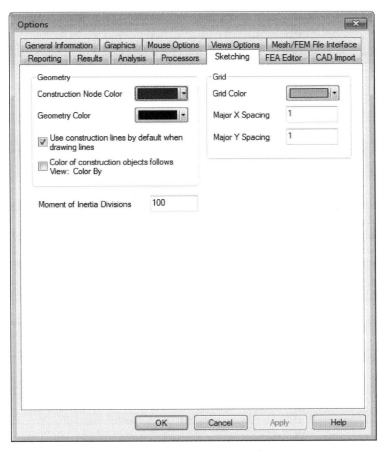

Figure 4-13 The **Options** *dialog box*

Drawing Lines

Ribbon: Draw > Draw > Line

The **Line** tool is used to create lines. Lines are one of the basic sketching entities available in Autodesk Simulation. In general terms, a line is defined as the distance between two points. In Autodesk Simulation, you can draw a line of any length and at any angle. To draw a line, choose the **Line** tool from the **Draw** panel of the **Draw** tab in the **Ribbon**; the **Define Geometry** dialog box will be displayed, as shown in Figure 4-14. Also, you will be prompted to specify the coordinate points for the start point of the line.

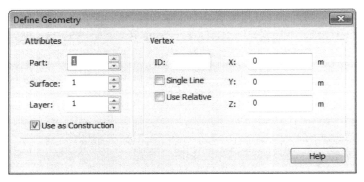

*Figure 4-14 The **Define Geometry** dialog box*

Specify the coordinates of the start point of the line in the **X**, **Y**, and **Z** edit boxes of the **Vertex** area in the **Define Geometry** dialog box and then press **ENTER**; the start point of the line will be specified. Also, a rubber band line will be attached to the cursor and you will be prompted to specify the endpoint of the line. Specify the coordinates of the endpoint of the line in the **X**, **Y**, and **Z** edit boxes of the dialog box and then press ENTER; the line will be drawn between two specified points, refer to Figure 4-15. Now, move the cursor away from the endpoint of the line drawn. You will notice that another rubber band line is attached to the cursor, refer to Figure 4-15. The start point of this line is the endpoint of the last line and the length of this line can be increased or decreased by moving the cursor. In this way, you can create continuous chain of lines by using the **Line** tool. To exit from the **Line** tool, press ESC twice.

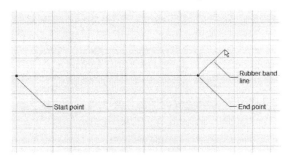

Figure 4-15 A line created by specifying two points

Note
You can also specify points for creating entities of a geometry by clicking on the vertices in the drawing area. You will learn more about creating vertices later in this chapter. Alternatively, you can also specify points for creating entities directly by clicking in the drawing area.

The remaining options of the **Define Geometry** dialog box are discussed next.

Attributes Area

The **Attributes** area of the **Define Geometry** dialog box is used to specify attributes such as part number, surface number, and layer number by using their respective spinners available in this area.

By default, the **Use as Construction** check box in the **Attributes** area of this dialog box is selected. As a result, the entities drawn will act as construction entities and will not be a part of mesh. The construction entities are used to create wire frames for generating meshes. They also act as an aid to build a mesh. On clearing the **Use as Construction** check box, the entities drawn will not be the construction entities and act as regular lines.

Vertex Area

The **X**, **Y**, and **Z** edit boxes of the **Vertex** area are used to specify the X, Y, and Z coordinates of a point. The **ID** edit box of this area is used to specify the ID of the vertex. As soon as you specify the ID of the vertex, the coordinates value of that ID will automatically be filled in the **X**, **Y**, and **Z** edit boxes. To find the ID of a vertex, hover the cursor over the vertex in the graphic area; a pop-up window will be displayed with its ID and coordinate points, refer to Figure 4-16.

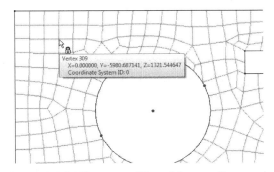

Figure 4-16 *The vertex ID and its coordinate points*

If the **Single Line** check box is not selected in the **Vertex** area of the dialog box, you can create a continuous chain of lines. On selecting this check box, you can create only individual lines. By default, the **Use Relative** check box is cleared in the dialog box. As a result, the coordinates entered in the **X**, **Y**, and **Z** edit boxes will measure from the origin (X=0, Y=0, and Z=0). However, on selecting the **Use Relative** check box, the displacements along the X, Y, and Z axes (DX, DY, and DZ) are measured with reference to the previous point, not the origin.

Figure 4-17 shows a sketch created on YZ plane using the **Line** tool.

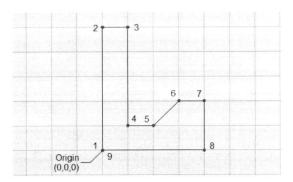

Figure 4-17 *Sketch created using the **Line** tool*

The coordinates of each point of the sketch shown in Figure 4-17 are given in the table below.

Table 4-1 *Coordinate points for drawing sketch shown in Figure 4-17*

Points	X coordinate	Y coordinate	Z coordinate
First point	0	0	0
Second point	0	0	5
Third point	0	1	5
Fourth point	0	1	1
Fifth point	0	2	1
Sixth point	0	3	2
Seventh point	0	4	2
Eighth point	0	4	0
Ninth point	0	0	0

 Note
In Table 4-1, all the coordinate values are measured with respect to the origin.

Drawing Circles

In Autodesk Simulation Mechanical, the tools for drawing clicles by different methods are grouped together in the **Circle** drop-down. To invoke this drop-down, click on the down arrow available at the bottom of the active circle tool in the **Draw** panel of the **Draw** tab in the **Ribbon**, refer to Figure 4-18. The methods of drawing circles are discussed next.

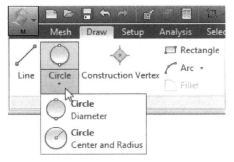

Figure 4-18 *The **Circle** drop-down*

Drawing Circles by Specifying the Diameter Points

Ribbon: Draw > Draw > Circle drop-down > Circle - Diameter

The **Circle - Diameter** tool is used to draw a circle by specifying two points on the circumference of the circle that determine the diameter of the circle. To draw a circle by specifying two points on the circumference, choose the **Circle - Diameter** tool from the **Circle** drop-down in the **Draw** panel; the **Define a Circle By Diameter Points** dialog box will be displayed, as shown in Figure 4-19. Also, you will be prompted to specify the first diameter point of the circle.

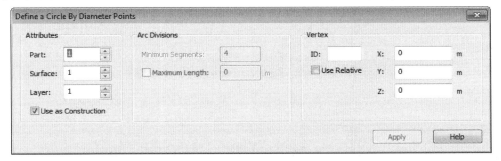

*Figure 4-19 The **Define a Circle By Diameter Points** dialog box*

Specify the coordinates of the first diameter point of the circle in the **X, Y,** and **Z** edit boxes and then choose the **Apply** button from the dialog box; the first diameter point of the circle will be specified. Also, a rubber band circle will be attached to the cursor, refer to Figure 4-20 and you will be prompted to specify the second diameter point of the circle. Specify the coordinates for the second diameter point of the circle and choose **Apply**; a circle will be drawn, refer to Figure 4-21. Press ESC to exit from the tool.

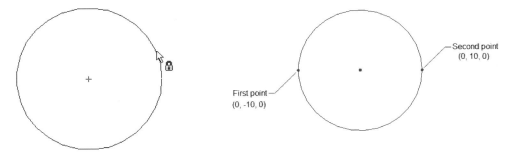

Figure 4-20 A rubber band circle attached with the cursor after specifying the first point

*Figure 4-21 The circle drawn by using the **Circle - Diameter** tool*

The options available in the **Attributes** and **Vertex** areas of this dialog box are same as discussed earlier. The options in the **Arc Divisions** area of the dialog box are discussed next.

Arc Divisions Area

The **Minimum Segments** edit box of the **Arc Divisions** area of the **Define a Circle By Diameter Points** dialog box is used to specify the minimum number of line segments into which the circle will be divided. Note that each line segment will be of equal length. You can also control the lengths of the line segments. To do so, select the **Maximum Length** check box and then specify the maximum length of line segments in the **Maximum Length** edit box. The value entered in the **Maximum Length** edit box will define the limit for the length of line segments. Note that these options will be enabled only when the **Use as Construction** check box of the dialog box is cleared.

Drawing Circle by Specifying the Center Point and Radius

Ribbon: Draw > Draw > Circle drop-down > Circle - Center and Radius

The **Circle - Center and Radius** tool is used to draw a circle by specifying a center point of a circle and a point on its circumference. This is most widely used method for drawing circles. Note that the point on the circumference of the circle will define the radius of the circle. To draw a circle using this tool, choose the **Circle - Center and Radius** tool from the **Circle** drop-down; the **Define a Circle By Center and Radial Point** dialog box will be displayed. Also, you will be prompted to specify the center point of the circle.

Specify the coordinates of the center point of the circle in the **X**, **Y**, and **Z** edit boxes and then press ENTER; the center point of the circle will be specified and you will be prompted to specify a radial point for the circle. Also, a rubber band circle will be attached to the cursor, refer to Figure 4-22. Now, specify the coordinates for the radial point of the circle and press ENTER; a circle will be drawn, refer to Figure 4-23. The options available in this dialog box are same as discussed earlier. Press ESC to exit from the dialog box.

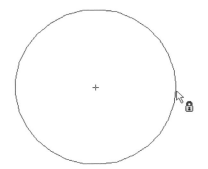

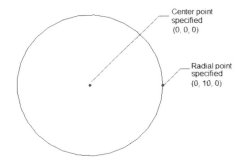

Figure 4-22 *A rubber band circle attached to the cursor after specifying the first point*

Figure 4-23 *The circle drawn by using the* **Circle - Center and Radius** *tool*

Drawing Rectangles

Ribbon: Draw > Draw > Rectangle

You can draw a rectangle by specifying its two diagonally opposite corners by using the **Rectangle** tool. To draw a rectangle, choose the **Rectangle** tool from the **Draw** panel of the **Draw** tab in the **Ribbon**; the **Add Rectangle** dialog box will be displayed, as shown in Figure 4-24. Also, you will be prompted to specify the first corner of the rectangle.

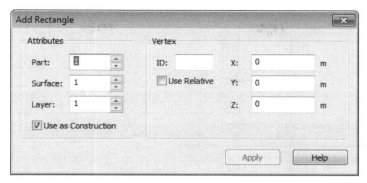

*Figure 4-24 The **Add Rectangle** dialog box*

Specify coordinates for the first corner of the rectangle in the **X**, **Y**, and **Z** edit boxes of the dialog box and then press ENTER; the first corner of the rectangle will be specified and you will be prompted to specify the second diagonally opposite corner of the rectangle. Specify the coordinates for the diagonally opposite corners of the rectangle in the edit boxes of the dialog box and then press ENTER; a rectangle will be created, refer to Figure 4-25. The options in the **Add Rectangle** dialog box are same as discussed earlier.

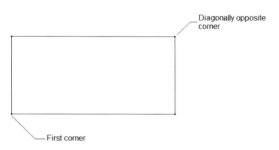

Figure 4-25 A rectangle created

Drawing Arcs

In Autodesk Simulation, you can draw arcs by using four tools: **Arc - Three Points**, **Arc - Center and Endpoints**, **Arc - Angle and Endpoints**, and **Arc - Radius and Endpoints**. All these tools are grouped together in the **Arc** drop-down in the **Draw** panel. To invoke this drop-down, choose the down arrow on the right of the active arc tool, refer to Figure 4-26. The methods used to create arcs using these tools are discussed next.

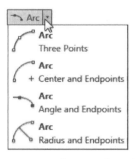

*Figure 4-26 The **Arc** drop-down*

Drawing Three Point Arcs

Ribbon:	Draw > Draw > Arc drop-down > Arc - Three Points

The three point arcs are the ones that are drawn by defining the start point, endpoint, and a point on the circumference or the periphery of the arc. To create three point arcs, choose the **Arc - Three Points** tool from the **Arc** drop-down of the **Draw** panel; the **Define Arc using Three Points** dialog box will be displayed and you will be prompted to specify the start point of the arc. The options of this dialog box area are same as discussed earlier.

Enter the X, Y, and Z coordinates of the start point of the arc in their respective edit boxes of the dialog box and then press ENTER; the start point of the arc will be specified in the graphic area. Also, you will be prompted to specify the end point of the arc. Enter the X, Y, and Z coordinates of the endpoint of the arc and press ENTER; the endpoint of the arc will be specified and you will be prompted to specify the midpoint of the arc. Now, when you move the cursor, you will notice that a rubber band arc will be attached to it. Enter the coordinates of the midpoint of the arc and then press ENTER; an arc will be created, refer to Figure 4-27.

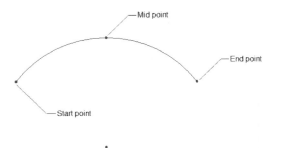

Figure 4-27 Arc created by specifying its three points

Drawing Center Point Arcs

Ribbon: Draw > Draw > Arc drop-down > Arc - Center and Endpoints

The center point arcs are the arcs that are drawn by defining the center point, start point, and endpoint of the arc. To create a center point arc, choose the **Arc - Center and Endpoints** tool from the **Arc** drop-down of the **Draw** panel; the **Define Arc using Center and Endpoints** dialog box will be displayed, as shown in Figure 4-28. Also, you will be prompted to specify the center point of arc.

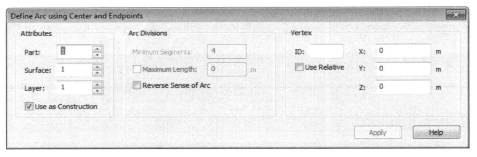

Figure 4-28 The **Define Arc using Center and Endpoints** dialog box

Enter the X, Y, and Z coordinates of the center point of the arc to be drawn in their respective edit boxes and then press ENTER; the center point of the arc will be specified in the graphic

area. Also, you will be prompted to specify the start point of the arc. Enter the X, Y, and Z coordinates of the start point of the arc and press ENTER; the start point of the arc will be specified and you will be prompted to specify the endpoint of the arc. Note that after specifying the start point of the arc, if you move the cursor in the clockwise direction, the resulting arc will be drawn in the clockwise direction. To draw an arc in the counterclockwise direction, select the **Reverse Sense of Arc** check box available in the **Arc Divisions** area of the dialog box and move the cursor in the counterclockwise direction. Next, specify the end point of the arc by entering its coordinates in the **X**, **Y**, and **Z** edit boxes of the dialog box and then press ENTER; the arc will be created, refer to Figure 4-29.

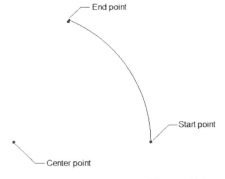

Figure 4-29 An arc created by specifying its center, start, and end points

Drawing Angle Arcs

Ribbon: Draw > Draw > Arc drop-down > Arc - Angle and Endpoints

The angle arcs are the arcs that are drawn by defining the angle, start point, and end point of the arc. To create an angle arcs, choose the **Arc - Angle and Endpoints** tool from the **Arc** drop-down of the **Draw** panel; the **Define Arc Using Angle and Endpoints** dialog box will be displayed, as shown in Figure 4-30. The options in this dialog box are the same as those discussed earlier except the **Angle** edit box. The **Angle** edit box of the **Arc Divisions** area of the dialog box is used to specify the angle of the arc. Note that the angle of arc entered in the **Angle** edit box will be measured with respect to the horizontal axis, refer to Figure 4-31.

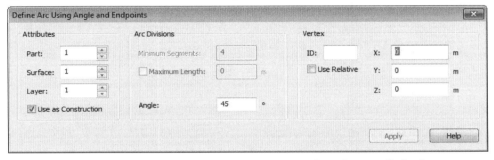

*Figure 4-30 The **Define Arc Using Angle and Endpoints** dialog box*

Enter angle of the arc in the **Angle** edit box and then specify coordinates of the start point of the arc in the **X**, **Y**, and **Z** edit boxes, respectively. After specifying the coordinates, press ENTER; the start point of the arc will be specified and you will be prompted to specify the end point of the arc. Enter the coordinates of the end point and then press ENTER; the end point of the arc will be specified and the preview of the arc with respect to the angle specified in the **Angle** edit box will be displayed. Note that after specifying the end point of the arc, the direction of the arc being created can flip either side on moving the cursor. Next, click to specify the direction of arc creation. Figure 4-32 shows an arc created by specifying its angle, start point, and end point.

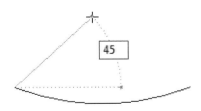

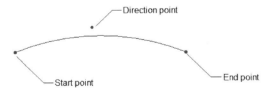

Figure 4-31 The angle of arc measured from the horizontal axis .

Figure 4-32 An arc created by specifying angle, start point, and end point

Drawing Radius Arcs

Ribbon: Draw > Draw > Arc drop-down > Arc - Radius and Endpoints

The radius arcs are the ones that are drawn by defining the radius, start point, and end point of the arc. To create radius arcs, choose the **Arc - Radius and Endpoints** tool from the **Arc** drop-down; the **Define Arc Using Radius and Endpoints** dialog box will be

displayed. The **Radius** edit box of the **Arc Divisions** area of the dialog box is used to specify the radius of the arc. The other options of this dialog box are same as discussed earlier.

Enter radius for the arc in the **Radius** edit box and then specify the coordinates of the start point and the end point of the arc. On specifying the end point of the arc, the preview of the arc will be displayed in the graphics area and the direction of the arc will change on moving the cursor. Click to specify the direction of arc. To display the other half of the arc, you can select the **Reverse Sense of Arc** check box from the **Arc Divisions** area of the dialog box. Figure 4-33 shows an arc created with the **Reverse Sense of Arc** check box cleared and Figure 4-34 shows an arc with the **Reverse Sense of Arc** check box selected.

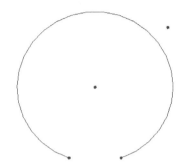

Figure 4-33 *Arc created with the **Reverse Sense of Arc** check box cleared*

Figure 4-34 *Arc created with the **Reverse Sense of Arc** check box selected*

Drawing Construction Vertices

Ribbon: Draw > Draw > Construction Vertex

The construction vertices are used as reference points for creating lines, arcs, meshes, and so on. To create a construction vertex, choose the **Construction Vertex** tool from the **Draw** panel of the **Draw** tab in the **Ribbon**; the **Define Construction Vertex** dialog box will be displayed, refer to Figure 4-35. Also, you will be prompted to specify the coordinates of the point.

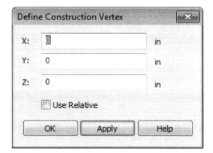

Figure 4-35 *The **Define Construction Vertex** dialog box*

Specify the coordinates of the point in their respective edit boxes of the dialog box and then press ENTER; a respective construction vertex will be created and you will be prompted again to specify the coordinates of the point. It indicates that you can continue creating vertices by specifying their coordinate points in the dialog box. Once all the vertices are created, choose the **OK** button from the dialog box.

Note
You can also create construction vertices on the surfaces of a solid CAD model. A construction vertex created on the solid CAD model will force the surface mesher to create a node at that location.

Creating Fillets

Ribbon: Draw > Draw > Fillet

A fillet creates a tangent arc at the intersection of two sketched entities. It trims or extends the entities to be filleted, depending on the geometry of the sketched entity. You can also apply a fillet to two nonparallel lines. To create a fillet, choose the **Fillet** tool from the **Draw** panel of the **Draw** tab in the **Ribbon**; the **Add Fillet** dialog box will be displayed, as shown in Figure 4-36. Also, you will be prompted to select the first object to be used in generating the fillet.

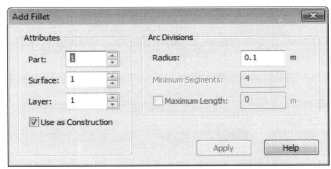

*Figure 4-36 The **Add Fillet** dialog box*

Ensure that the **Construction Objects** tool is chosen in the **Selection** panel. Select the first object from the drawing area; the selected object will be highlighted in a different color and you will be prompted to select the second object to be used in generating the fillet. Select the second object from the drawing area; the preview of the fillet with the default radius value will be displayed in the drawing area. You can change the default fillet radius by entering a new value in the **Radius** edit box of the **Arc Divisions** area of the dialog box. Enter the required radius value in the **Radius** edit box and then choose the **Apply** button from the dialog box; the fillet will be created.

Figure 4-37 shows the intersecting entities before and after applying the fillet. You can also select the non-intersecting entities for creating a fillet. In this case, the selected entities will be extended to form a fillet, as shown in Figure 4-38.

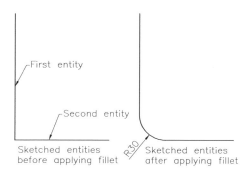

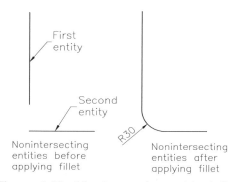

Figure 4-37 *Intersecting entities before and after applying a fillet*

Figure 4-38 *Non-intersecting entities before and after applying a fillet*

Creating Tangent Lines

Ribbon: Draw > Draw > Tangent Line

The **Tangent Line** tool is used to create tangent lines between the selected objects. To do so, expand the **Draw** panel of the **Draw** tab by clicking on the down arrow available in its title bar and then choose the **Tangent Line** tool, refer to Figure 4-39. Next, choose the **Tangent Line** tool; you will be prompted to select an object or point for creating tangents.

Select the first object to be used in creating tangent lines, refer to Figure 4-40; you will be prompted to select the second object or point that will be used for creating the tangent. Select the second object for creating tangent lines; a preview of all the possible tangent lines between the selected objects is displayed in the drawing area, refer to Figure 4-41. Also, you will be prompted to select the tangent lines that you want to retain. Select the tangent lines to be retained from the preview of the tangent lines, refer to Figure 4-41. To select more than one tangent line, press the CTRL key and select the tangent lines one by one. After selecting the tangent lines, release the CTRL key and press ENTER; the selected tangent lines will be created between the selected objects, refer to Figure 4-42.

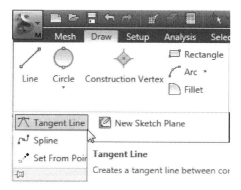

Figure 4-39 *The* **Tangent Line** *tool in the expanded* **Draw** *panel*

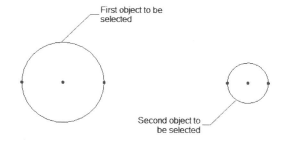

Figure 4-40 *Objects to be selected for creating tangent lines*

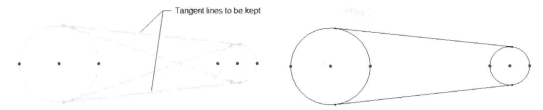

Figure 4-41 *Preview of the tangent lines displayed between the selected objects*

Figure 4-42 *Tangent lines created between the selected objects*

Creating Splines

Ribbon: Draw > Draw > Spline

The **Spline** tool is used to create splines. To create a spline, choose the **Spline** tool from the expanded **Draw** panel of the **Draw** tab in the **Ribbon**; the **Add Spline** dialog box will be displayed, as shown in Figure 4-43. Also, you will be prompted to select a control point that will help define the spline.

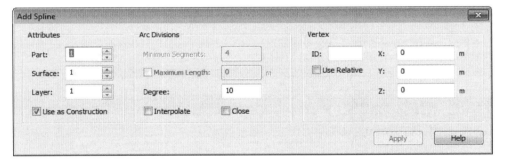

Figure 4-43 *The **Add Spline** dialog box*

Specify the degree of curvature for creating the spline in the **Degree** edit box of the **Add Spline** dialog box. Note that the degree of a spline cannot be more than the number of control points used to draw the spline. After specifying the degree of spline, you need to specify its control points. Specify the coordinates for the first control point of the spline in their respective edit boxes in the **Add Spline** dialog box and then press ENTER. Similarly, specify the other control points of the splines. After specifying all the control points of the spline, choose the **Apply** button; the spline will be created. By default, the **Interpolate** check box in the dialog box is cleared. As a result, points defined for creating the spline will act as poles of the spline. Figure 4-44 shows a spline with the **Interpolate** check box cleared. On selecting the **Interpolate** check box, the spline will pass through all the defined control points. Figure 4-45 shows a spline created with the **Interpolate** check box selected.

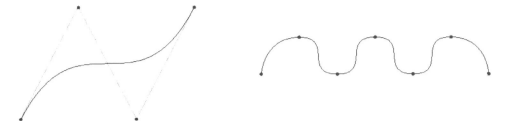

Figure 4-44 *Spline created with the **Interpolate** check box cleared*

Figure 4-45 *Spline created with the **Interpolate** check box selected*

On selecting the **Close** check box in the dialog box, you can create a closed spline, refer to Figure 4-46.

Figure 4-46 *The closed spline created by selecting the **Close** check box*

EDITING AND MODIFYING GEOMETRY

Editing and modifying geometry is a very important part in any design or manufacturing program. You need to edit and modify the sketches during various stages of a design. Autodesk Simulation Mechanical provides you with a number of tools that can be used to edit and modify the sketched entities. These tools are discussed next.

Trimming Sketched Entities

Ribbon:	Draw > Modify > Trim

The **Trim** tool is used to trim the unwanted entities in a sketch. You can use this tool to trim a line, arc, circle, spline, and so on. To trim an entity, choose the **Trim** tool from the **Modify** panel of the **Draw** tab in the **Ribbon**; you will be prompted to select the boundary item. Select the boundary object which refers to the cutting object for trimming an entity. As soon as you select the boundary or the cutting object, you will be prompted to select an item to modify. Select the object to be trimmed from the drawing area; the cursor will be changed to the trim cursor and you will be prompted to select the segment of the object to be trimmed. Select the segment of the object to be trimmed from the drawing area; the selected segment will be trimmed. Also, you will again be prompted to select the boundary item. It indicates that you can continue trimming the other unwanted entities of the sketch. To exit the **Trim** tool, press the ESC key. Figure 4-47 shows a sketch with an unwanted segment of the entity to be trimmed and Figure 4-48 shows the same sketch after trimming the entity.

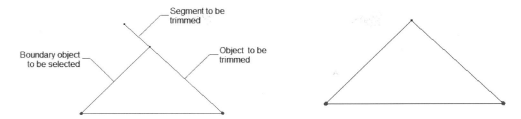

Figure 4-47 *Sketch before trimming*   **Figure 4-48** *Sketch after trimming*

Extending Sketched Entities

Ribbon:	Draw > Modify > Extend

The **Extend** tool is used to extend the sketched entity upto the intersection of the selected entity. To extend an entity upto its intersection with another selected sketched entity, choose the **Extend** tool from the **Modify** panel of the **Draw** tab in the **Ribbon**; you will be prompted to select the boundary object. Select the boundary object from the drawing area upto which you want to extend an entity. As soon as you select the boundary, you will be prompted to select an item to modify. Select the object to be extended from the drawing area; the selected entity will be extended upto the selected boundary entity. Also, you will be prompted again to select the boundary item. It indicates that you can continue extending the other entities of the sketch. To exit from the **Extend** tool, press the ESC key. Figure 4-49 shows a sketch with an entity to be extended and Figure 4-50 shows the same sketch after extending the entity.

Figure 4-49 *Sketch before extending the entity*   **Figure 4-50** *Sketch after extending the entity*

Intersecting Sketched Entities

Ribbon:	Draw > Modify > Intersect

The **Intersect** tool is used to intersect one entity with another entity. To intersect entities, choose the **Intersect** tool from the **Modify** panel of the **Draw** tab in the **Ribbon**; the **Intersect Lines** window will be displayed at the lower left corner of the drawing area, as shown in Figure 4-51. Also, you will be prompted to select an intersecting line.

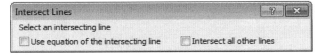

Figure 4-51 *The **Intersect Lines** window*

Select an intersecting line from the drawing area; you will be prompted to select another intersecting line or press ENTER to select lines to intersect. You can select multiple intersecting lines from the drawing area. After selecting the intersecting lines, press the ENTER key; you will be prompted to select lines to intersect. Select lines to be intersected from the drawing area. Once you select all the lines to be intersected, press ENTER; the selected lines will intersect with respect to the selected intersecting lines. Figure 4-52 shows the line to be intersected and the intersecting lines. Figure 4-53 shows the resultant sketch. Note that the line to be intersected in Figure 4-52 will be divided into three lines after intersecting, refer to Figure 4-53.

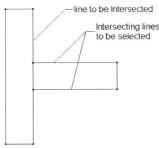

Figure 4-52 Sketch before intersecting

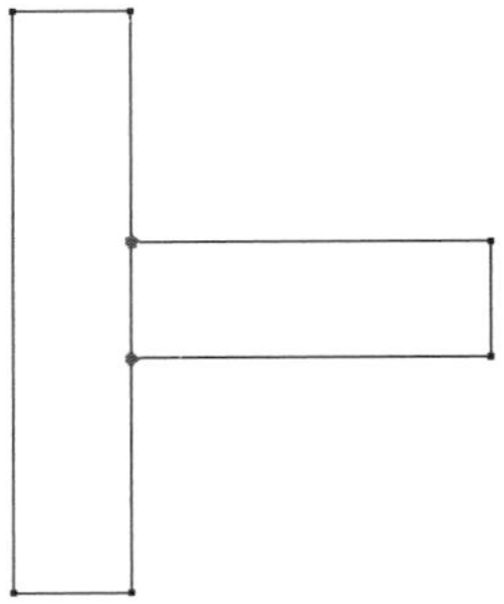

Figure 4-53 Sketch after intersecting

Dividing Sketched Entities

Ribbon: Draw > Modify > Divide

The **Divide** tool is used to divide sketched entity or entities into specified divisions. You can select single or multiple sketched entities to be divided. To divide the selected entity/ entities, select the sketched entity or entities to be divided from the drawing area; the selected sketched entities will be highlighted and the **Divide** tool will be enabled in the **Modify** panel of the **Draw** tab in the **Ribbon**. Next, choose the **Divide** tool; the **Divide Lines** dialog box will be displayed, as shown in Figure 4-54.

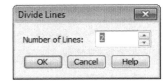

*Figure 4-54 The **Divide Lines** dialog box*

Specify the number of divisions for an entity in the **Number of Lines** edit box of the **Divide Lines** dialog box. Next, choose the **OK** button; all the selected entities of the sketch will be divided into specified number of divisions. Figure 4-55 shows a sketched entity to be divided and Figure 4-56 shows the resultant entity after its division.

Figure 4-55 Sketched entity to be divided

Figure 4-56 Sketched entity after the division

Changing Attributes

The **Attributes** tool is used to change the attributes of an entity. As discussed earlier, while drawing a sketched entity using the tools available in the **Draw** panel, its attributes such as the part, surface, and layer number will be assigned. You can further change or modify the already assigned attributes of an entity by using the **Attributes** tool. To change the attributes of a line, select the line from the drawing area and then choose the **Attributes** tool from the **Modify** panel of the **Draw** tab in the **Ribbon**; the **Line Attributes** dialog box will be displayed, as shown in Figure 4-57. You can also invoke the **Line Attributes** dialog box by double-clicking on the line. This dialog box can also be invoked by right-clicking on the line and choosing the **Edit Attributes** option from the shortcut menu displayed. By using the options available in this dialog box, you can change the part, surface, and layer number assigned to the selected line. You can also clear or select the **Construction object** check box of the dialog box to convert the selected line to the regular line or the construction line.

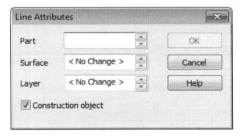

*Figure 4-57 The **Line Attributes** dialog box*

To change the attributes of arcs, circles, and splines, select the required entity from the drawing area and right-click; a shortcut menu will be displayed. Choose the **Edit** option from the shortcut menu; the respective dialog box of the selected entity will be displayed. You can change the attributes of the selected entity by using the options of the invoked dialog box. You can also double-click on an entity to display the respective dialog box for editing the attributes.

In addition to the editing attributes of arcs, circles, and splines, you can also modify their geometry by redefining the points that were defined while creating them. To do so, right-click on an entity (arc, circle, or spline) and choose the **Edit** option from the shortcut menu displayed; a dialog box will be displayed. The name of the dialog box depends upon the type of entity you have selected to invoke it. For example, if you right-click on a spline and choose the **Edit** option from the shortcut menu displayed, then the **Add Spline** dialog box will be displayed. Once the dialog box is invoked, on pressing the ESC key, you can undo through the existing defined points and specify new points. Similarly, if a three point arc is defined, select the three point arc from the drawing area and right-click to display a shortcut menu. Next, choose the **Edit** option from the shortcut menu displayed; the **Define Arc using Three Points** dialog box will be displayed. Now, on pressing the ESC key once, you can undo to the third defined point of the arc. As a result, you can redefine the third new point by specifying its new coordinates in their respective edit boxes in the dialog box.

Mirroring Sketched Entities

Ribbon: Draw > Pattern > Mirror

*Figure 4-58 The **Mirror Objects** dialog box*

The Mirror tool is used to create the mirror image of the sketched entities about a mirroring line. To create the mirror image of an entity, select the entity to be mirrored from the drawing area and then choose the **Mirror** tool from the **Pattern** panel of the **Draw** tab in the **Ribbon**; the **Mirror Objects** dialog box will be displayed, as shown in Figure 4-58. Also, you will be prompted to select a mirroring line.

Choose the **Pick** button from the **Mirror Objects** dialog box; the **Mirror Objects** dialog box will be closed. Now, select a mirroring line from the drawing area. As soon as you select mirroring object/line, the **Mirror Objects** dialog box will be displayed again. Choose the **OK** button from the dialog box; the mirror image of the selected entity will be created. You can also select multiple entities to be mirrored by pressing the SHIFT or CTRL key. If you do not want to keep the original entities in the drawing area after creating their mirrored image, you can select the **Delete original objects** check box from the **Mirror Objects** dialog box. On selecting this check box, the original entity will be deleted after its mirrored image is created. Figure 4-59 shows the sketched entities to be mirrored using the mirroring line and Figure 4-60 shows the resulting mirrored image of the selected entities.

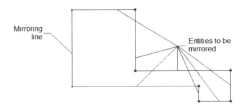

Figure 4-59 Entities to be mirrored and the mirroring line

Figure 4-60 Sketch after mirroring the selected entities

Moving, Rotating, Scaling, and Copying Sketched Entities

Ribbon: Draw > Pattern > Move or Copy
                     Draw > Pattern > Rotate or Copy
                     Draw > Pattern > Scale or Copy

In Autodesk Simulation Mechanical, you can move, rotate, scale, and copy the selected entities by using the **Move or Copy**, **Rotate or Copy**, or **Scale or Copy** tool. To move, rotate, scale, and copy the selected entities, choose the **Move or Copy**, or **Rotate or Copy**, or **Scale or Copy** tool from the **Pattern** panel of the **Draw** tab in the **Ribbon**; the **Move, Rotate, Scale or Copy** dialog box will be displayed, refer to Figure 4-61.

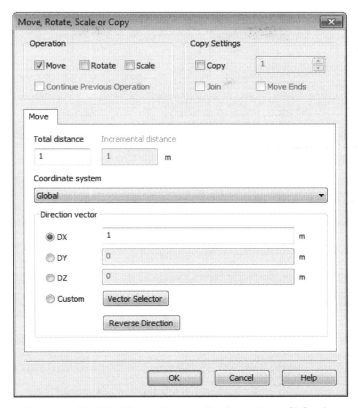

Figure 4-61 The *Move, Rotate, Scale or Copy* dialog box

Note
*The **Move or Copy**, **Rotate or Copy**, and **Scale or Copy** tools will be enabled only in the **Pattern** panel of the **Draw** tab in the **Ribbon**, if the entities to be modified are selected in the drawing area.*

*Depending upon the tool chosen to invoke the **Move**, **Rotate**, **Scale** or **Copy** dialog box, the respective check box in the **Operation** area of the dialog box will be selected. For example, if you choose the **Move or Copy** tool to invoke the dialog box, the **Move** check box will be selected in the **Operation** area of the dialog box.*

The options in the **Move, Rotate, Scale or Copy** dialog box are discussed next.

Operation Area

The **Operation** area of the dialog box is used to specify the type of operation you want to perform on the selected entities. On selecting the **Move** check box in this area, you can move the selected entities up to a specified distance. On selecting the **Rotate** check box in this area, you can rotate the selected entities up to the specified angle. To increase or decrease the scale factor of the selected entities, you need to select the **Scale** check box in this area. Note that depending upon the check box selected in the **Operation** area of the dialog box, the respective options of the selected check box will be displayed in the tab available at the lower portion of the dialog box. For example, if the **Move** check box is selected, the **Move** tab will be displayed in the dialog box which contains options for moving the selected entities, refer to Figure 4-61.

You can also perform more than one operation at a time on the selected entities by selecting the check boxes respective to those operations from the **Operation** area of the dialog box. For example, to move and rotate the selected entities, select the **Move** and **Rotate** check boxes from the **Operation** area of the dialog box. On selecting both the check boxes, the **Move** and **Rotate** tabs will be displayed in the **Move, Rotate, Scale or Copy** dialog box, as shown in Figure 4-62.

Figure 4-62 The **Move, Rotate, Scale or Copy** dialog box with the **Move** and **Rotate** tabs displayed

Copy Settings Area

The **Copy Settings** area of the dialog box is used to create multiple copies of the selected entities while moving, rotating, or scaling. To create copies of the selected entities during an operation, select the **Copy** check box from the **Copy Settings** area of the **Move, Rotate, Scale or Copy** dialog box; the **Copy** edit box on the right of the **Copy** check box will be enabled. Specify the number of copies of the selected entities to be created in the **Copy** edit box. The options available in the **Move**, **Rotate**, and **Scale** tabs of the dialog box are discussed next.

Move Tab

The **Move** tab will be available only if the **Move** check box is selected in the **Operation** area of the dialog box. The options available in the **Move** tab are used to specify the direction vector, total distance, and increment distance between two instances for the selected entities to be moved.

The **Total distance** edit box of the **Move** tab is used to specify the total distance by which you want to move the selected entities. The **Incremental distance** edit box of the **Move** tab is used to specify the incremental distance between two instances of the selected entities. Note that the **Incremental distance** edit box of the **Move** tab is enabled only when the **Copy** check box in the **Copy Settings** area is selected and you are creating multiple copies of the selected entities.

You can select the required coordinate system from the **Coordinate system** drop-down list of the **Move** tab with respect to which you want to move the selected entities. This drop-down list displays the list of all the coordinate systems available in the current drawing area. By default, the **Global** coordinate system is selected in this drop-down list. If you have created a user-defined coordinate system as per your requirement, the name of that coordinate system will also be displayed in this drop-down list.

Tip. *To create a new coordinate system in the FEA Editor environment of Autodesk Simulation Mechanical, select the* **Coordinate Systems** *node from the* **Tree View** *and then right-click; a shortcut menu will be displayed. Choose the* **New** *option from the shortcut menu; the* **Creating Coordinate System Definition** *dialog box will be displayed. Specify the type of local coordinate system from the* **Coordinate System Type** *drop-down list of the dialog box. Next, specify the coordinates for X, Y, and Z directions in their corresponding fields to define the coordinate system and then choose the* **OK** *button.*

The **Direction Vector** area of the **Move** tab is used to specify the direction vector and the distance by which you want to move the selected entities. Select the required radio button, DX, DY, or DZ from the **Direction Vector** area of the **Move** tab to specify the direction vector. On selecting the direction vector radio button, the corresponding edit box of the selected direction vector will be enabled. In this edit box, you can enter the required distance value by which you want to move the selected entities in the specified direction. In addition to the default direction vector, you can also specify a new user-defined direction vector as per your requirement by selecting the **Custom** radio button from the **Direction Vector** area of the **Move** tab.

To specify the user-defined direction vector, select the **Custom** radio button from the **Direction Vector** area of the dialog box and then choose the **Vector Selector** button available in front of the **Custom** radio button; the dialog box will be disappeared. Also, you will be prompted to select the vertex to specify the origin or start point for the direction vector. Specify the origin for the direction vector using the left mouse button; you will be prompted to select the vertex to specify the end point for the direction vector. Specify the end point for the direction vector; the **Move, Rotate, Scale or Copy** dialog box will be displayed again. You can also reverse the direction of creation by choosing the **Reverse Direction** button from the dialog box. After specifying all the parameters, choose the **OK** button from the dialog box.

Rotate Tab

The options available in the **Rotate** tab are used to specify the total angle of rotation, incremental angle between two instances, axis of rotation, and the center point of rotation. The **Total angle** edit box of the **Rotate** tab is used to specify the total angle of rotation. By using the **Increment angle** edit box of this tab, you can specify the incremental angle between two instances. Note that the **Incremental angle** edit box of the **Rotate** tab is enabled when the **Copy** check box of the **Copy Settings** area of the dialog box is selected for creating multiple copies of the selected entities.

The options available in the **Rotation axis** area of the **Rotate** tab are used to specify the axis of rotation and the options in the **Rotation center point** area are used to specify the center point of rotation for the entities to be rotated. Figure 4-63 shows a sketch before rotation and Figure 4-64 shows a sketch after rotating it. The angle of rotation in this case is 45 degrees.

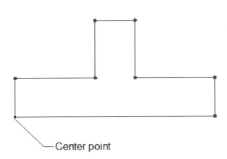

Figure 4-63 *Sketch before rotating*   ***Figure 4-64*** *Sketch after rotating*

Scale Tab

The options available in the **Scale** tab are used for scaling the selected entity. The **Scale type and factor** area of the **Scale** tab is used to specify the type of scaling to be performed and the scale factor for the entity to be scaled. The options available in the **Scale direction vector** area of the dialog box are used to specify the direction vector for scaling the entity. The **Fixed point** area of the dialog box is used to specify the coordinates of the fixed or base point about which you want to scale the selected entity. Figure 4-65 shows a scaled copy of the original sketch. The scale factor in this case is 2.

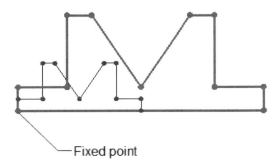

Figure 4-65 *The scaled copy of the original sketch*

TUTORIALS

Tutorial 1

In this tutorial, you will create the geometry of the model shown in Figure 4-66. The geometry is shown in Figure 4-67. Note that the geometry consists of two parts. The dimensions shown in Figure 4-67 are given only for reference. You need to use the **YZ(+X)** plane as the sketching plane for creating the geometry. All the dimensions are in inches.

(Expected time: 30 min)

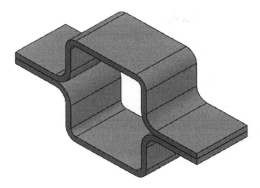

Figure 4-66 *The assembly for Tutorial 1*

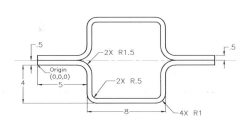

Figure 4-67 *The geometry of the model*

The following steps are required to complete this tutorial:

a. Start Autodesk Simulation Mechanical and a new FEA Model environment.
b. Set the units of the documents.
c. Create the bottom part of the geometry by using the **Line** tool.
d. Fillet the sketched entities using the **Fillet** tool.
e. Mirror the sketched entities to create the top part of the geometry.
f. Save and close the model.

Starting Autodesk Simulation Mechanical

In this section, you need to open a new session of FEA in Autodesk Simulation Mechanical.

1. Create a new folder with the name **c04** at the location *C:\Autodesk Simulation Mechanical*. Next, create another folder with the name **Tut01** folder inside the *c04* folder.

2. Start Autodesk Simulation Mechanical 2016 by double-clicking on the shortcut icon of **Autodesk Simulation Mechanical 2016** available on the desktop of your computer.

3. Choose the **New** button from the **Launch** panel of the **Getting Started** tab of the **Ribbon**; the **New** dialog box is displayed.

Note
*In this tutorial, the automatic display of the **New** or **Open** dialog box on the start up of Autodesk Simulation Mechanical has been turned off. In case the **New** dialog box is displayed on the start up of Autodesk Simulation Mechanical, you can skip the step 2. At the same time, if the **Open** dialog box is displayed, you can choose the **New** button from the **Open** dialog box to invoke the **New** dialog box.*

4. Select the **FEA Model** option from the **New** dialog box and then choose the **New** button; the **Save As** dialog box is displayed.

5. Browse to the location *C:\Autodesk Simulation Mechanical\c04\Tut01* and then enter **c04_tut01** in the **File name** edit box of the dialog box. Next, choose the **Save** button; a new FEA file

with the name *c04_tut01* is saved at the specified location and a new FEA model environment of Autodesk Simulation Mechanical is invoked.

Setting the Units of the Documents

Now, you need to change the default Model unit system to English (in).

1. Expand the **Unit Systems** node available in the **Tree View** by clicking on its **+** sign, refer to Figure 4-68.

2. Double-click on the **Display Units < English (in) >** unit system from the expanded **Unit Systems** node of the **Tree View**; the unit system for the current session is changed to English (in) and the **Display Units < English (in) >** unit system is highlighted in the **Tree View**.

Creating the Bottom Part of the Geometry

As it is evident from Figures 4-66 and 4-67 that the geometry of the model consists of two parts. You will first create the bottom part of the geometry and then its mirror image to create the top part of the geometry. Before you start drawing the geometry, you need to specify the sketching plane on which you want to create the geometry. As mentioned in the tutorial description, you need to select the **YZ(+X)** plane as the sketching plane for drawing the geometry.

1. Expand the **Planes** node available in the **Tree View** by clicking on its **+** sign; refer to Figure 4-69.

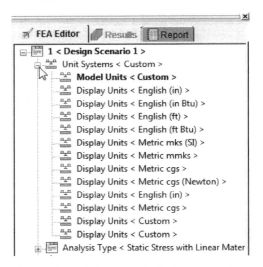

Figure 4-68 *The expanded view of the **Unit Systems** node in the **Tree View***

Figure 4-69 *The expanded view of the **Planes** node in the **Tree View***

2. Double-click on the **YZ(+X)** plane available in the **Planes** node of the **Tree View**; the sketching environment is invoked and the selected **YZ(+X)** plane becomes parallel to the screen.

3. Choose the **Draw** tab of the **Ribbon**; all the tools in this tab are displayed.

4. Choose the **Line** tool from the **Draw** panel of the **Draw** tab; the **Define Geometry** dialog box is displayed, as shown in Figure 4-70.

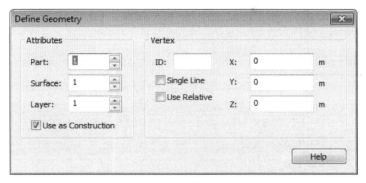

*Figure 4-70 The **Define Geometry** dialog box*

5. Make sure that value **1** is entered in the **Part**, **Surface**, and **Layer** edit boxes of the dialog box. Also, ensure that **0**, **0**, **0** is entered in the **X**, **Y**, and **Z** edit boxes of the dialog box, respectively.

6. Press the ENTER key; the start point of the line is specified at the origin (0,0,0) and you are prompted to specify the end point of the line.

7. Enter **0, 5, 0** in the **X, Y**, and **Z** edit boxes, respectively in the **Define Geometry** dialog box and then press the ENTER key; the second point of the line is specified in the drawing area and you are prompted to specify the end point of the line again. Also, a rubber band line is attached to the cursor, refer to Figure 4-71. Note that on moving the cursor, the length of the rubber band line will increase or decrease.

 Note
*The display of the sketching plane with grid lines has been turned off for clarity. To turn off the display of the sketching plane, right-click on sketching plane in the **Tree View**; a shortcut menu will be displayed. Choose the **Visibility** option from the shortcut menu.*

8. Enter **0, 5, -4** in the **X, Y**, and **Z** edit boxes, respectively in the **Define Geometry** dialog box and then press the ENTER key; the second line is drawn, refer to Figure 4-72.

Figure 4-71 Sketch after creating the first entity   *Figure 4-72 Sketch after creating the second entity*

9. Enter **0, 13, -4** in the **X**, **Y**, and **Z** edit boxes, respectively in the **Define Geometry** dialog box and then press ENTER; the next line is drawn.

10. Similarly, specify the coordinate values (**0, 13, 0**), (**0, 18, 0**), (**0, 18, 0.5**), (**0, 12.5, 0.5**), (**0, 12.5, -3.5**), (**0, 5.5, -3.5**), (**0, 5.5, 0.5**), (**0, 0, 0.5**), and (**0, 0, 0**). Figure 4-73 shows the geometry after creating all the line entities of the bottom part.

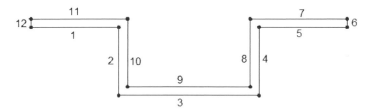

Figure 4-73   *The geometry after creating all the line entities of the bottom part*

Note

In Figure 4-73, line entities have been numbered for better understanding while creating fillets.

Filleting the Sketched Entities

Next, you need to remove sharp corners of the sketch by applying fillets.

1. Choose the **Fillet** tool from the **Draw** panel of the **Draw** tab; the **Add Fillet** dialog box is displayed, as shown in Figure 4-74.

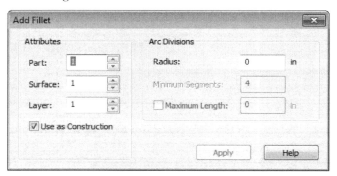

Figure 4-74   *The **Add Fillet** dialog box*

2. Make sure the value **1** is entered in the **Part**, **Surface**, and **Layer** edit boxes of the dialog box. Next, enter **1** in the **Radius** edit box of the dialog box.

3. Move the cursor in the drawing area and select lines 1 and 2, refer to Figure 4-73; a preview of the fillet with radius of 1 inch is displayed in the drawing area. Next, choose the **Apply** button from the dialog box; the fillet of radius 1 inch is created between the selected lines, refer to Figure 4-75.

Figure 4-75 *Fillet of radius 1 inch applied between lines 1 and 2*

4. Select lines 2 and 3, refer to Figure 4-73, and then choose the **Apply** button; the fillet of radius 1 inch is applied between lines 2 and 3. Similarly, apply fillet of radius 1 inch between lines 3 and 4, and 4 and 5. Next, apply fillet of radius 1.5 inch between the lines 7 and 8, and 10 and 11. Also, apply fillet of radius 0.5 inch between the lines 8 and 9, and 9 and 10. Next, exit the dialog box. Figure 4-76 shows the geometry after creating part 1.

Figure 4-76 *The geometry after completing the bottom part*

Mirroring the Sketched Entities

Next, you need to create the mirror image of the bottom part to create the top part of the geometry, by using the **Mirror** tool.

1. Choose the **Selection** tab of the **Ribbon** to display tools related to the selection.

2. Choose the **Construction Objects** tool from the **Select** panel and the **Rectangle** tool from the **Shape** panel of the **Selection** tab in the **Ribbon**.

3. Draw a rectangular window across all the sketched entities by specifying its two diagonally opposite corners, refer to Figure 4-77. Note that as you specify the second corner of the window, all the entities inside the window are selected.

4. Choose the **Draw** tab and then choose the **Mirror** tool from the **Pattern** panel of the **Draw** tab in the **Ribbon**; the **Mirror Objects** dialog box is displayed, as shown in Figure 4-78.

5. Choose the **Pick** button from the **Mirror Objects** dialog box; the dialog box disappears and you are prompted to select the mirroring object.

6. Select the line 11 of the sketch, refer to Figure 4-79; the **Mirror Objects** dialog box is displayed again and the name of the selected line is displayed in the **Pick** selection area of the dialog box, refer to Figure 4-80.

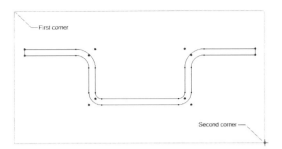

Figure 4-77 *A window drawn for selection*

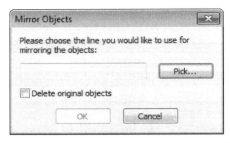

Figure 4-78 *The **Mirror Objects** dialog box*

Figure 4-79 *The line to be selected as the mirroring object*

Figure 4-80 *The name of the selected entity is displayed in the **Pick** selection area*

 Note
*If you have not assigned a name to the line entities of the sketch, then the name of the selected line will be displayed in the **Pick** selection area of the dialog box as **Unnamed**.*

7. Choose the **OK** button from the dialog box; the mirror image of the selected entities is created and all the entities of the mirror images are selected, refer to Figure 4-81. Do not click anywhere in the graphics area.

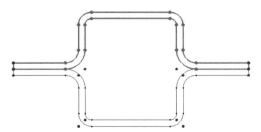

Figure 4-81 *The entities of the mirror image are selected*

8. Right-click in the drawing area and choose the **Edit Attributes** option from the shortcut menu displayed; the **Line Attributes** dialog box is displayed, as shown in Figure 4-82.

9. Set the value **2** in the **Part** edit box of the **Line Attributes** dialog box and then choose the **OK** button; the second part is created. Also, the **Part 2** is added to the **Tree View**. Figure 4-83 shows the final geometry after creating both the parts.

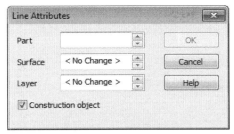

Figure 4-82 *The **Line Attributes** dialog box*

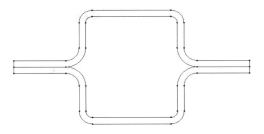

Figure 4-83 *The final geometry*

Saving and Closing the File

1. Choose the **Save** button from the **Quick Access Toolbar** to save the file and then choose **Close** from the **Application Menu** to close the file.

Tutorial 2

In this tutorial, you will create the geometry of the assembly shown in Figure 4-84. The geometry of the assembly is shown in Figure 4-85. Note that the geometry consists of two parts. You need to select the **YZ(+X)** plane as the sketching plane for creating the geometry. All the dimensions are in inches.   **(Expected time: 20 min)**

Figure 4-84 *The assembly for Tutorial 2*

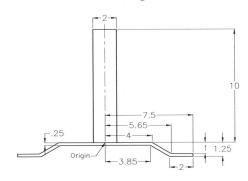

Figure 4-85 *The geometry of the assembly*

The following steps are required to complete this tutorial:

a. Start Autodesk Simulation Mechanical and a new FEA Model environment.
b. Set the units of the documents.
c. Create the top part of the geometry by using the **Rectangle** tool.
d. Create the bottom part of the geometry by using the **Line** and **Mirror** tools.
e. Save and close the model.

Starting Autodesk Simulation Mechanical

You need to start a new session of FEA model in Autodesk Simulation Mechanical.

1. Create a new folder with the name **Tut02** at the location *C:\Autodesk Simulation Mechanical/ c04*.

2. Start Autodesk Simulation Mechanical if not started already.

3. Choose the **New** button from the **Launch panel** of the **Getting Started** tab of the **Ribbon**; the **New** dialog box is displayed.

4. Select the **FEA Model** option from the **New** dialog box and then choose the **New** button; the **Save As** dialog box is displayed.

5. Browse to the location *C:\Autodesk Simulation Mechanical\c04\Tut02* and then enter **c04_tut02** in the **File name** edit box of the dialog box. Next, choose the **Save** button; a new FEA file with the name *c04_tut02* is saved at the specified location and a new FEA model environment of Autodesk Simulation Mechanical is invoked.

Setting the Units of the Documents

Now, you need to change the default Model unit system to English (in).

1. Expand the **Unit Systems** node available in the **Tree View** by clicking on its **+** sign.

2. Double-click on the **Display Units < English (in) >** unit system from the expanded **Unit Systems** node of the **Tree View**; the unit system for the current session is changed to English (in) and the **Display Units < English (in) >** unit system is highlighted in the **Tree View**.

Creating the Top Part of the Geometry

It is evident from Figures 4-84 and 4-85 that the geometry of the assembly consists of two parts. Before you start creating the geometry, you need to specify the sketching plane on which you want to create the geometry. As mentioned in the tutorial description, you will select the **YZ (+X)** plane as the sketching plane for drawing the geometry.

1. Expand the **Planes** node available in the **Tree View** by clicking on its **+** sign. Next, double-click on the **YZ(+X)** plane available in the expanded **Planes** node of the **Tree View**; the sketching environment is invoked and the selected **YZ(+X)** plane becomes parallel to the screen.

2. Choose the **Draw** tab of the **Ribbon**; all the tools of the **Draw** tab are displayed.

3. Choose the **Rectangle** tool from the **Draw** panel of the **Draw** tab; the **Add** ▭ Rectangle **Rectangle** dialog box is displayed, as shown in Figure 4-86.

4. Make sure that value **1** is entered in the **Part**, **Surface**, and **Layer** edit boxes of the dialog box.

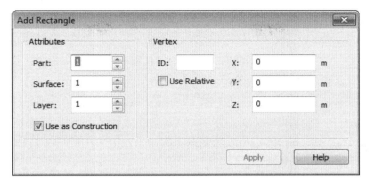

*Figure 4-86 The **Add Rectangle** dialog box*

5. Enter **0**, **-1**, **0** in the **X**, **Y**, and **Z** edit boxes of the dialog box, respectively and then press ENTER; the first corner of the rectangle is specified and a rubber band rectangle is attached to the cursor. Also, you are prompted to specify the diagonally opposite corners of the rectangle.

6. Enter **0**, **1**, **10** in the **X**, **Y**, and **Z** edit boxes, respectively in the **Add Rectangle** dialog box and then press ENTER; a rectangle is created.

7. Choose the **Apply** button and then close the **Add Rectangle** dialog box. Figure 4-87 shows the geometry after creating its top part.

Figure 4-87 Geometry after creating the top part

Creating the Bottom Part of the Geometry

After creating the top part of the geometry, you need to create the bottom part of the geometry by using the **Line** and **Mirror** tools.

1. Choose the **Line** tool from the **Draw** panel of the **Draw** tab; the **Define Geometry** dialog box is displayed.

2. Enter **2** in the **Part**, **Surface**, and **Layer** edit boxes, respectively of the **Define Geometry** dialog box. Also, make sure that **0**, **0**, **0** is entered in the **X**, **Y**, and **Z** edit boxes, respectively of the dialog box.

3. Press the ENTER key; the start point of the line is specified at the origin (0,0,0) and you are prompted to specify the endpoint of the line.

4. Enter **0**, **4**, **0** in the **X**, **Y**, and **Z** edit boxes of the **Define Geometry** dialog box, respectively and then press the ENTER key; the second point of the line is specified in the drawing area and you are prompted to specify the endpoint of the line again. Also, a rubber band line is attached to the cursor.

   On moving the cursor, the length of the rubber band line increases or decreases.

5. Enter **0, 5.65, -1** in the **X, Y**, and **Z** edit boxes of the **Define Geometry** dialog box, respectively and then press the ENTER key; the line is drawn, refer to Figure 4-88.

6. Enter **0, 7.5, -1** in the **X, Y**, and **Z** edit boxes of the dialog box, respectively and then press ENTER; the line is drawn.

7. Similarly, specify the coordinates (**0, 7.5, -1.25**), (**0, 5.5, -1.25**), (**0, 3.85, -0.25**), (**0, 0, -0.25**), and (**0, 0, 0**) to complete the right side of the second part. Next, close the dialog box. Figure 4-89 shows the geometry after creating right side of the second part.

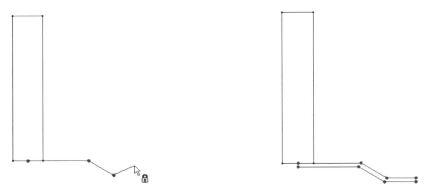

Figure 4-88 *Sketch of the second part being drawn*

Figure 4-89 *Geometry after creating the right side of the second part*

The remaining half of the second part will be created by using the **Mirror** tool.

8. Choose the **Selection** tab of the **Ribbon** to display selection tools.

9. Choose the **Construction Objects** tool from the **Select** panel and the **Point or Rectangle** tool from the **Shape** panel of the **Selection** tab in the **Ribbon**.

10. Select all the entities of the second part except the mirroring line by pressing and holding the CTRL key, refer to Figure 4-90.

11. Choose the **Draw** tab and then choose the **Mirror** tool from the **Pattern** panel of the **Draw** tab in the **Ribbon**; the **Mirror Objects** dialog box is displayed, as shown in Figure 4-91.

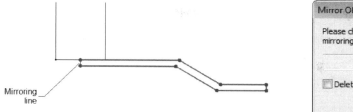

Figure 4-90 *Entities selected for mirroring*

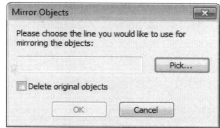

Figure 4-91 *The* *Mirror Objects* *dialog box*

12. Choose the **Pick** button from the **Mirror Objects** dialog box; the dialog box disappears and you are prompted to select the mirroring object.

13. Move the cursor over the mirroring line, refer to Figure 4-90 and then select it using the left mouse button; the **Mirror Objects** dialog box is displayed again and the name of the selected line is displayed in the **Pick** selection box of the dialog box.

14. Choose the **OK** button from the dialog box; the mirroring image of the selected entities is created, as shown in Figure 4-92.

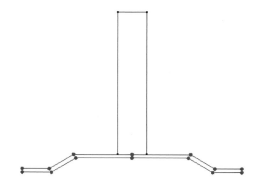

Figure 4-92 *Geometry after mirroring the entities*

After mirroring the sketched entities, you need to delete the mirroring line.

15. Click anywhere in the drawing area to exit the existing selection set and then select the mirror line.

16. Press the **Delete** key to delete the selected mirroring line. Figure 4-93 shows the final geometry.

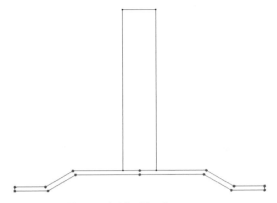

Figure 4-93   *Final geometry*

Saving and Closing the File

1. Choose the **Save** button from the **Quick Access Toolbar** to save the file and then choose **Close** from the **Application Menu** to close the file.

Tutorial 3

In this tutorial, you will create the structure shown in Figure 4-94. The structure contains members of two different cross-section angles: 8 X 8 X 1 and 6 X 6 X 1, refer to Figure 4-94. Therefore, you will create the structure in two different layers. The dimensions are given in inches.

(Expected time: 35 min)

The coordinates of each joint of the structure shown in Figure 4-94 are given in the following table.

Table 4-2   *Coordinate points for drawing structure shown in Figure 4-94*

Points	X coordinate	Y coordinate	Z coordinate
1	0	0	0
2	400	0	0
3	300	250	0
4	250	500	0
5	250	600	0
6	150	600	0
7	150	500	0
8	100	250	0
9	500	500	0
10	375	500	0
11	-100	500	0
12	25	500	0

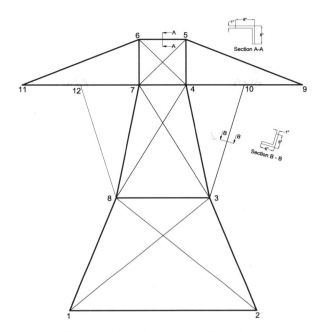

Figure 4-94 *Structure for Tutorial 3*

The following steps are required to complete this tutorial:

a. Start Autodesk Simulation Mechanical and a new FEA Model environment.
b. Set the units of the documents.
c. Create Structure Members of Cross-Section 8 X 8 X 1 by using the **Line** tool.
d. Create Structure Members of Cross-Section 6 X 6 X 1 by using the **Line** tool.
e. Save and close the model.

Starting Autodesk Simulation Mechanical

Start a new session of FEA model in Autodesk Simulation Mechanical.

1. Create a new folder with the name **Tut03** at the location *C:\Autodesk Simulation Mechanical/ c04*.

2. Start Autodesk Simulation Mechanical if not started already.

3. Invoke the **New** dialog box if not invoked by default.

4. Select the **FEA Model** option from the **New** dialog box and then choose the **New** button; the **Save As** dialog box is displayed.

5. Browse to the location *C:\Autodesk Simulation Mechanical\c04\Tut03* and then enter **c04_tut03** in the **File name** edit box of the dialog box. Next, choose the **Save** button; a new FEA file with the name *c04_tut03* is saved at the specified location and a new FEA model environment of Autodesk Simulation Mechanical is invoked.

Setting the Units of the Documents

Now, you need to change the default Model unit system to English (in).

1. Expand the **Unit Systems** node available in the **Tree View** and then set the units of the current session to English (in).

   It is evident from Figure 4-94 that the structure contains members of two different cross-section angles. Therefore, you need to create members of different cross sections with different surface and layout attributes. It will help you to identify members of different cross-sections easily while defining the cross-sections of members in later chapters.

Creating Structure Members of Cross-Section 8 X 8 X 1

You need to invoke the sketching environment by selecting the **XY(+Z)** plane as the sketching plane for creating the structure members.

1. Expand the **Planes** node available in the **Tree View** by clicking on its **+** sign. Next, double-click on the **XY(+Z)** plane available in the expanded **Planes** node of the **Tree View**; the sketching environment is invoked and the selected XY plane becomes parallel to the screen.

2. Choose the **Draw** tab of the **Ribbon**; all the tools of the **Draw** tab are displayed.

3. Choose the **Line** tool from the **Draw** panel of the **Draw** tab; the **Define Geometry** dialog box is displayed, as shown in Figure 4-95.

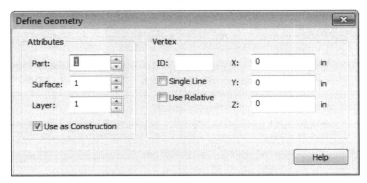

*Figure 4-95 The **Define Geometry** dialog box*

4. Make sure that value **1** is entered in the **Part**, **Surface**, and **Layer** edit boxes of the dialog box. Next, clear the **Use as Construction** check box in the **Attributes** area of the dialog box.

5. Make sure that **0, 0, 0** is entered in the **X, Y,** and **Z** edit boxes of the dialog box. Next, press ENTER; the first point of the line is specified and a rubber band line is attached to the cursor. Also, you are prompted to specify the second point of the line.

6. Enter **400, 0, 0** in the **X, Y,** and **Z** edit boxes of the dialog box, respectively and then press ENTER; a line is created between the two specified points.

7. Enter **300, 250, 0** in the **X, Y,** and **Z** edit boxes of the dialog box, respectively and then press ENTER; the third point is specified in the drawing area.

8. Enter **250, 500, 0** in the **X, Y,** and **Z** edit boxes of the dialog box, respectively and then press ENTER; the fourth point is specified in the drawing area.

9. Enter **250, 600, 0** in the **X, Y,** and **Z** edit boxes, respectively and then press ENTER; the fifth point is specified in the drawing area.

10. Enter **150, 600, 0** in the **X, Y,** and **Z** edit boxes, respectively and then press ENTER; the sixth point is specified in the drawing area.

11. Enter **150, 500, 0** in the **X, Y,** and **Z** edit boxes, respectively and then press ENTER; the seventh point is specified in the drawing area.

12. Enter **100, 250, 0** in the **X, Y,** and **Z** edit boxes, respectively and then press ENTER; the eighth point is specified in the drawing area.

13. Make sure that **0, 0, 0** is entered in the **X, Y,** and **Z** edit boxes and then press ENTER; the structure is created, as shown in Figure 4-96.

14. Choose the **Enclose (Fit All)** tool from the **Navigate** panel of the **View** tab in the **Ribbon**; the entire structure now fits completely inside the screen, refer to Figure 4-96. Next, press the ESC key to terminate the creation of the continuous line.

15. Move the cursor toward the upper right corner of the structure and specify the start point of the new structure member.

16. Enter **500, 500, 0** in the **X, Y,** and **Z** edit boxes, respectively and then press ENTER; the ninth point is specified in the drawing area.

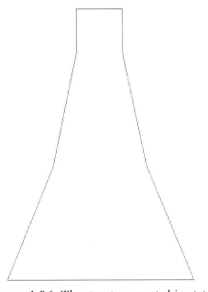

17. Enter **375, 500, 0** in the **X, Y,** and **Z** edit boxes, respectively and then press ENTER; the tenth point is specified in the drawing area.

Figure 4-96 The structure created in step 13

18. Enter **250, 500, 0** in the **X, Y,** and **Z** edit boxes, respectively and then press ENTER; the structure is created, as shown in Figure 4-97. Next, press the ESC key to terminate the creation of continuous members.

19. Move the cursor toward the upper left corner of the structure and specify the start point of the new structure member.

20. Enter **-100**, **500**, **0** in the **X**, **Y**, and **Z** edit boxes, respectively and then press ENTER; the eleventh point is specified in the drawing area.

21. Enter **25**, **500**, **0** in the **X**, **Y**, and **Z** edit boxes, respectively and then press ENTER; the twelfth point is specified in the drawing area.

22. Enter **150**, **500**, **0** in the **X**, **Y**, and **Z** edit boxes, respectively and then press ENTER; the structure is created. Next, press the ESC key to terminate the creation of continuous structure members.

23. Enter **100**, **250**, **0** in the **X**, **Y**, and **Z** edit boxes, respectively and then press ENTER; the start point of the new structure member is specified.

24. Enter **300**, **250**, **0** in the **X**, **Y**, and **Z** edit boxes, respectively and then press ENTER; the member is created between the specified points. Next, press the ESC key to terminate the creation of structure members.

25. Enter **150**, **500**, **0** in the **X**, **Y**, and **Z** edit boxes, respectively and then press ENTER; the start point of the new structure member is specified.

26. Enter **250**, **500**, **0** in the **X**, **Y**, and **Z** edit boxes, respectively and then press ENTER; the member is created between the specified points, refer to Figure 4-98. Next, press the ESC key to terminate the creation of structure members.

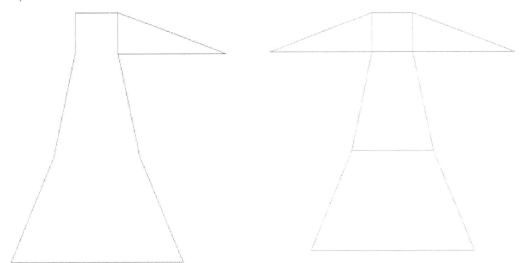

Figure 4-97 *The structure created in step 18*   ***Figure 4-98*** *The structure created in step 26*

Creating Structure Members of Cross-Section 6 X 6 X 1

Now, you will create structure members of cross section 6 X 6 X 1. You need to define different surface and layout attributes for this structure members.

1. Set the value in the **Surface** and **Layer** spinners to **2** in the **Attributes** area of the **Define Geometry** dialog box.

2. Enter **0, 0, 0** in the **X, Y,** and **Z** edit boxes, respectively and then press ENTER; the start point of the new structure member is specified at the origin.

3. Enter **300, 250, 0** in the **X, Y,** and **Z** edit boxes, respectively and then press ENTER; a new structure member is created between the specified points, refer to Figure 4-99.

4. Similarly, create the remaining structure members by specifying their coordinates points. Figure 4-99 shows the final structure with all its members. Next, exit the **Line** tool by pressing the ESC key twice.

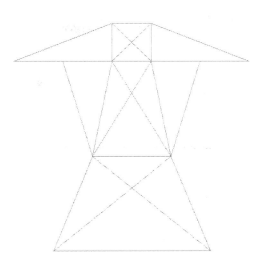

Figure 4-99 The final structure of Tutorial 3

Saving and Closing the File

1. Choose the **Save** button from the **Quick Access Toolbar** to save the file and then choose **Close** from the **Application Menu** to close the file.

Self-Evaluation Test

Answer the following questions and then compare them to those given at the end of this chapter:

1. In Autodesk Simulation Mechanical, you can draw arcs by using four tools _____, _____, _____, and _____.

2. To create continuous chain of lines by using the **Line** tool, the _____ check box should be cleared in the **Vertex** area of the **Define Geometry** dialog box displayed on invoking the **Line** tool.

3. The **Attributes** area of the **Define Geometry** dialog box is used to specify attributes such as _____, _____, and _____ .

4. To draw a circle by specifying two points on the circumference of the circle, choose the _____ tool from the **Circle** drop-down list of the **Draw** panel of the **Draw** tab in the **Ribbon**.

5. When you choose the **Circle - Diameter** tool, the _____ dialog box is displayed.

6. To draw a circle by specifying the center point and a point on its circumference, choose the _____ tool from the **Circle** drop-down list of the **Draw** panel in the **Draw** tab of the **Ribbon**.

7. The _____ tool is used to create a rectangle by specifying its two diagonally opposite corners.

8. To invoke the sketching environment for creating sketch of an object, you need to select the required sketching plane from the **Tree View**. (T/F)

9. In Autodesk Simulation Mechanical, you can draw a line of any length and at any angle by using the **Line** tool. (T/F)

10. In Autodesk Simulation Mechanical, you cannot edit spline created by using the **Spline** tool. (T/F)

Review Questions

Answer the following questions:

1. Which of the following tools is used to extend the sketched entity upto the intersection of the selected entity?

    (a) **Trim**   (b) **Extend**
    (c) **Extend Entity**   (d) None of these

2. Which of the following edit boxes of the **Divide Lines** dialog box is used to specify the number of divisions for the selected entity?

    (a) **Number of Divide**   (b) **Divide Lines**
    (c) **Number of Lines**   (d) None of these

3. Which of the following tools is used to create tangent lines between the selected objects?

    (a) **Tangent Line**   (b) **Tan Lines**
    (c) **Tangent**   (d) **Line**

4. The _____ tool is used to create a mirrored copy of the selected sketched entities.

5. On selecting the _____ check box from the **Add Spline** dialog box, you can create a closed spline.

6. The _____ tool is used to create a tangent arc at the intersection of two selected sketched entities.

7. On choosing the **Arc - Angle and Endpoints** tool from the **Arc** drop-down of the **Draw** panel in the **Draw** tab of the **Ribbon**, the _____ dialog box is displayed.

8. In Autodesk Simulation Mechanical, you cannot trim or delete the unwanted entities of a sketch. (T/F)

9. In Autodesk Simulation Mechanical, you cannot move, rotate, scale, and copy the selected entities by using the **Move or Copy** tool. (T/F)

EXERCISE

Exercise 1

In this exercise, you will create the geometry of the assembly shown in Figure 4-100. The geometry of the assembly is shown in Figure 4-101. Note that the geometry consists of two parts. You need to select the **XY(+Z)** plane as the sketching plane for creating the geometry. All the dimensions are given in inches.   (**Expected time: 30 min**)

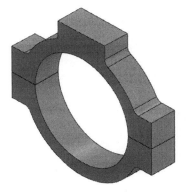

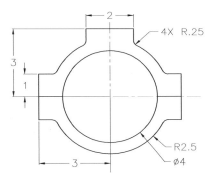

Figure 4-100 *The assembly for Exercise 1*   *Figure 4-101* *The geometry of the assembly*

Answers to Self-Evaluation Test

1. Arc - Three Points, Arc - Center and Endpoints, Arc - Angle and Endpoints, and Arc - Radius and Endpoints, **2.** Single Line, **3.** Part number, Surface number, Layer number, **4.** Circle - Diameter, **5.** Define a Circle By Diameter Points, **6.** Circle - Center and Radius, **7.** Rectangle, **8.** T, **9.** T, **10.** F

Chapter 5

Meshing-I

Learning Objectives

After completing this chapter, you will be able to:
- *Understand the concept of meshing*
- *Specify the settings required for generating 3D mesh*
- *Generate 3D solid mesh*
- *Generate Midplane type mesh*
- *Generate Plate/Shell type mesh*
- *View the meshing result*
- *Understand the concept of unmatched and multi-matched feature lines*
- *Eliminate unmatched and multi-matched feature lines from the model*
- *Generate 2D mesh*
- *Specify the settings required for generating 2D mesh*

INTRODUCTION TO MESHING

As stated earlier that Autodesk Simulation Mechanical needs a finite element model to carry-out the analysis process. A finite element model is the one that has small and discrete elements of a defined size created by using meshing process. Like all simulation software, in Autodesk Simulation Mechanical, meshing is a process of dividing a solid model into various small number of elements. All the elements of a meshed model or finite element model are connected to each other through nodes and each node may have two or more elements connected. Figures 5-1 and 5-2 show a model before and after meshing.

Figure 5-1 *A model before meshing*   *Figure 5-2* *The model after meshing*

Meshing a model is an important part of the pre-processing in any FEA software. Autodesk Simulation Mechanical has a user-friendly interface that allows you to control the geometric properties of the mesh and generate an effective mesh. An effective mesh provides you maximum accuracy in less computational time and without any error. Autodesk Simulation Mechanical is also provided with the default settings for generating the mesh as per the geometry of the model. However, most of the time you need to modify the default settings for generating an effective mesh. Figure 5-3 shows a finite element model generated with default mesh settings and Figure 5-4 shows the same model with user specified mesh settings.

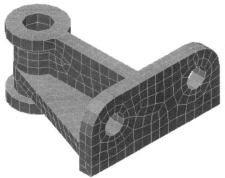

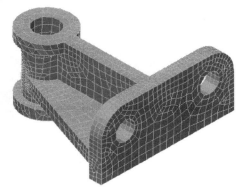

Figure 5-3 *Finite element model with default mesh settings*   *Figure 5-4* *Finite element model with user specified mesh settings*

GENERATING MESH

In Autodesk Simulation Mechanical, you can generate an effective mesh in the FEA Editor environment. In the FEA Editor environment, you can generate solid, mid-plane, or surface mesh on a 3D model. You can also generate a structured mesh and generate mesh on 2D sketches in the FEA Editor environment of the Autodesk Simulation Mechanical. The tools that are used for generating different types of mesh are grouped together into different panels in the **Mesh** tab of the **Ribbon** in the FEA Editor environment.

GENERATING 3D MESH

Ribbon:	Mesh > Mesh > Generate 3D Mesh

The **Generate 3D Mesh** tool is used to generate 3D mesh with default parameters. This tool is available in the **Mesh** panel of the **Mesh** tab in the **Ribbon**. On choosing the **Generate 3D Mesh** tool, the **Meshing Progress** window will be displayed on the screen, as shown in Figure 5-5. This window displays the progress of the meshing process. Once the meshing is complete, the **View Mesh Results** window will be displayed with the message **Would you like to view the meshing results at this time?**, refer to Figure 5-6. If you choose the **No** button, the **View Mesh Results** window will be closed and the meshed model will be displayed in the screen with the default mesh settings. However, on choosing the **Yes** button from the **View Mesh Results** window, the **Meshing Results** window will be displayed, as shown in Figure 5-7. You can view the meshing results in this window. You will learn more about viewing meshing results by using this window later in this chapter. You can specify the mesh settings for a model, as required, by using the **3D Mesh Settings** tool of the **Mesh** panel in the **Mesh** tab. The procedure of specifying 3D mesh settings using this tool is discussed next.

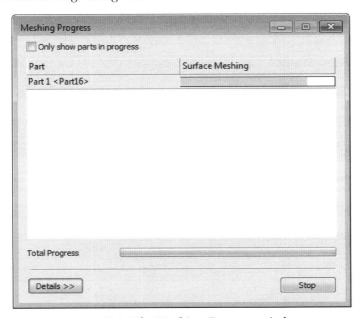

Figure 5-5 *The **Meshing Progress** window*

Note
*The **Generate 3D Mesh** tool is used to generate mesh as per the settings specified in the **Model
Mesh Setting** dialog box that will be displayed on choosing the **3D Mesh Settings** tool from the
Mesh panel of the **Mesh** tab.*

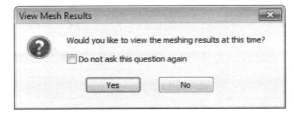

*Figure 5-6 The **View Mesh Results** window*

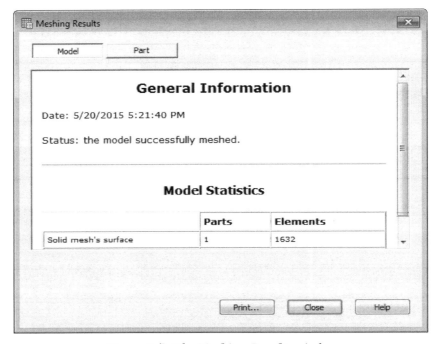

*Figure 5-7 The **Meshing Results** window*

Specifying 3D Mesh Settings

Ribbon: Mesh > Mesh > 3D Mesh Settings

In Autodesk Simulation Mechanical, you can specify the settings of the mesh to be
generated on a 3D model by using the **3D Mesh Settings** tool. On choosing the **3D
Mesh Settings** tool from the **Mesh** panel of the **Mesh** tab in the **Ribbon**, the **Model
Mesh Settings** dialog box will be displayed, as shown in Figure 5-8. The options in
this dialog box are discussed next.

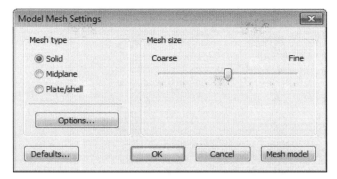

*Figure 5-8 The **Model Mesh Settings** dialog box*

Solid

The **Solid** radio button is selected by default in the **Mesh type** area of the dialog box. This radio button is used to mesh a model with solid elements such as brick or tetrahedral. You can specify various parameters of a solid element such as element type, aspect ratio, warp angle, volume-to-length ratio, and so on by using the **Options** button of the **Model Mesh Settings** dialog box. To do so, first select the **Solid** radio button in the **Mesh type** area of the dialog box and then choose the **Options** button; another **Model Mesh Settings** dialog box will be displayed, as shown in Figure 5-9. The options available in this modified dialog box are discussed next.

 Note
*The options available in the **Model Mesh Settings** dialog box displayed after choosing the **Options** button depend upon the radio button selected in the **Mesh type** area of the **Model Mesh Settings** dialog box.*

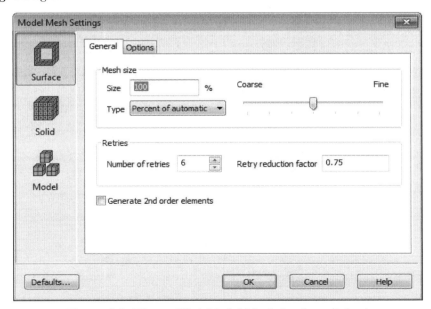

*Figure 5-9 The modified **Model Mesh Settings** dialog box*

Tip. *While generating a solid mesh by selecting the **Solid** radio button from the **Mesh type** area of the dialog box, first a surface mesh is generated by meshing all surfaces of a model and forms a watertight solid. Once the surface mesh is generated, solid mesh will be generated by filling its volume with solid elements. However, generating solid mesh after generating the surface mesh depends upon whether the **Perform solid meshing at time of analysis** check box is selected or not. This check box is located in the **General** tab corresponding to the **Model** button in the **Model Mesh Settings** dialog box. If this check box is selected then the solid mesh will be generated at the time of analysis only. If this check box is cleared then the solid mesh will be generated immediately after generating the surface mesh.*

Surface

The **Surface** button located on the left side of the **Model Mesh Settings** dialog box is chosen by default. As a result, on the right side of the dialog box, two tabs **General** and **Options** will be displayed, refer to Figure 5-9. The options available in these tabs are used to specify the parameters for the surface mesh and are discussed next.

General Tab: The options available in the **General** tab are used to specify the size of the elements to be generated. If the **Percent of automatic** option is selected from the **Type** drop-down list in the **Mesh size** area of the **General** tab then you can specify the size of the elements in terms of percentage of the default mesh size in the **Size** edit box. Note that the default mesh size will be calculated based on the dimensions of the model to be meshed. You can also drag the slider available on the right side of the dialog box to specify the mesh size. However, if the **Absolute mesh size** option is selected from the **Type** drop-down list, you can directly specify the approximate size of the elements in the **Size** edit box.

In Autodesk Simulation Mechanical, if a valid surface mesh cannot be created with the specified mesh size, the mesh size will be reduced automatically and mesh will be generated on the model again. This process will be repeated until a valid surface mesh is created or it will retry depending on the value specified in the **Number of retries** spinner in the **Retries** area of the **General** tab. In other words, the **Number of retries** spinner is used to specify the number of retries performed automatically in order to achieve valid surface mesh. By default, the value **6** is set in the **Number of retries** spinner. You can change this default value as per the requirement of the design. You can also specify a factor by which you want to reduce the size of the mesh every time while performing retry in order to achieve the valid mesh in the **Retry reduction factor** edit box of the **Retries** area.

Options Tab: The options available in the **Options** tab are used to specify the number of elements along curved features and curved edges of the model. You can also control the quadrilateral elements while splitting them into triangular elements by using the options available in the **Splitting quadrangles into triangles** area of the **Options** tab. Note that all these options of the **Options** tab will be available only if the **Use automatic geometry-based mesh size function** check box is cleared in the **Default meshing options** area of the **General** tab corresponding to the **Model** button of the modified **Model Mesh Settings** dialog box.

Solid

On choosing the **Solid** button from the modified **Model Mesh Settings** dialog box, five tabs, namely **General**, **Quality**, **Options**, **Tetrahedra**, and **Advanced** will be displayed in the dialog box, refer to Figure 5-10. The options available in these tabs are used for specifying the mesh element type, aspect ratio, warp angle, volume-to-length ratio, microholes, properties of tetrahedral meshes, boundary layer meshes, and so on. The options available in these tabs are discussed next.

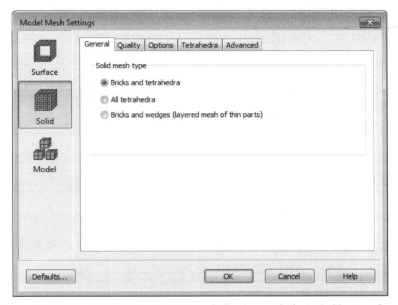

*Figure 5-10 The **Model Mesh Settings** dialog box with the **Solid** button chosen*

General Tab: This tab allows you to select the required type of solid mesh for the model to be meshed. By default, the **Bricks and tetrahedra** radio button is selected in the **Solid mesh type** area of this tab. As a result, with the default settings, the solid mesh will have as many 8-node brick elements as possible with respect to the geometry of the model and the mesh element size. Also, if necessary, it may also consist of 6-node wedge, 5-node pyramid, or 4-node tetrahedral elements at the center of the model. The **Bricks and tetrahedra** mesh type is used to create high quality mesh with less number of elements. Figure 5-11 shows the four possible geometrical configurations that can be used to create a brick element.

If you select the **All tetrahedra** radio button from the **Solid mesh type** area of this tab, the solid mesh generated will have only 4-node tetrahedra elements. If you select the **Bricks and wedges (layered mesh of thin parts)** radio button, the mesh will be created as per the user-defined number of solid elements through the thickness. This option is suitable for thin parts with relatively constant thickness. You can specify the user-defined number of solid elements through the entire thickness in the **User specified maximum thickness** edit box of the **Layered Mesh** tab. This tab will be displayed on selecting the **Bricks and wedges (layered mesh of thin parts)** radio button from the dialog box.

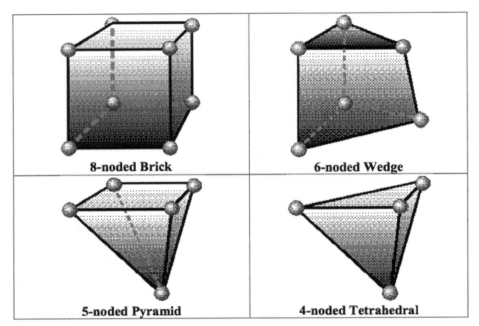

Figure 5-11 *Brick element geometry configurations*

Quality Tab: This tab allows you to control the aspect ratio and warp angle of the solid elements. Note that this tab will not be available when the **Bricks and wedges (layered mesh of thin parts)** option is selected in the **General** tab.

By default, the **Automatic enforcement** radio button is selected in the **Maximum aspect ratio** area of the **Quality** tab, as shown in Figure 5-12. As a result, the aspect ratio will be calculated based on the surface mesh. You can increase or decrease the value of the aspect ratio by using the slider available below the **Automatic enforcement** radio button in the **Quality** tab. Note that the smaller aspect ratio will result in better accuracy.

On selecting the **Upper limit** radio button from the **Maximum aspect ratio** area of the **Quality** tab, the edit box next to it will be enabled. This edit box allows you to specify the maximum aspect ratio. Note that all solid elements will be created within the limit of maximum aspect ratio specified in the edit box. In other words, the **Upper limit** radio button of the **Maximum aspect ratio** area is used to specify the upper limit for the aspect ratio. No solid elements will cross the limit of the specified aspect ratio. If you select the **None** radio button from the **Maximum aspect ratio** area of the **Quality** tab, no restrictions will be applied to the aspect ratios of the elements.

The **Include maximum warp angle constraint** check box in the **Quality** tab is used to specify the maximum limit for the warp angle to be constrained so that the warp angle of internal faces of the solid elements is constrained within the maximum limit specified. When you select this check box, an edit box will be enabled on its right. You can specify the maximum limit of the warp angle for the internal faces of the solid elements in this edit box.

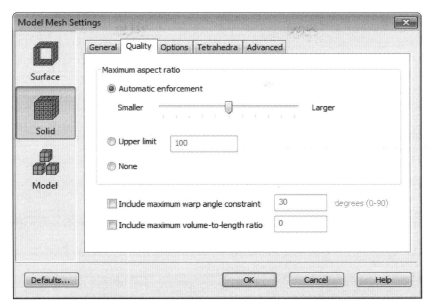

Figure 5-12 *The Automatic enforcement radio button selected in the Model Mesh Settings dialog box*

The **Include maximum volume-to-length ratio** check box in the **Quality** tab is used to specify the maximum limit for volume to length ratio. When you select this check box, the edit box located on its right is enabled. In this edit box, you can specify the maximum limit for the ratio of the cube root of the volume of an element to the length of the longest edge.

Options Tab: The **Allow microholes** check box of the **Options** tab is used to eliminate the errors or warnings that occur due to the microholes which are very small in size compared to the volume of the part while creating a solid mesh with high quality elements. Figure 5-13 shows the **Model Mesh Settings** dialog box with the **Options** tab chosen.

Tetrahedra Tab: The options available in this tab are used to specify the size of the tetrahedral elements. If you select the **Fraction of mesh size** option from the **Target edge length based on** drop-down list in the **Tetrahedral meshing options** area, the value entered in the **Target edge length** edit box will be multiplied by the surface mesh size to calculate the size of the tetrahedral elements. If you select the **Absolute mesh dimension** option from the **Target edge length based on** drop-down list, then the value entered in the **Target edge length** edit box will be used as the size of the tetrahedral elements. Figure 5-14 shows the **Model Mesh Settings** dialog box with the **Tetrahedra** tab chosen.

You can also define the transitions of mesh size from smaller to larger area of the model using the **Transition rate** edit box of the **Tetrahedral meshing options** area. The value entered in the **Transition rate** edit box will be the ratio of the average edge length of adjacent elements.

Note
*The value entered in the **Transition rate** edit box must be greater than one. In this case, the greater the value in this edit box, the lower will be the quality of elements.*

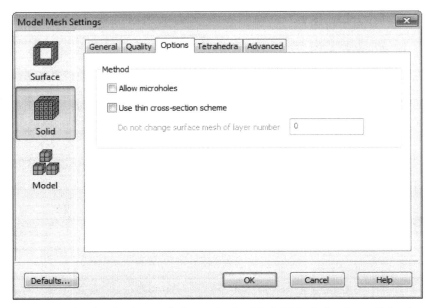

Figure 5-13   *The **Options** tab chosen in the **Model Mesh Settings** dialog box*

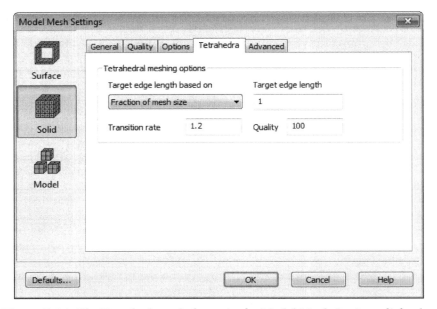

Figure 5-14   *The **Tetrahedra** tab chosen in the **Model Mesh Settings** dialog box*

The **Quality** edit box of this tab is used to define the upper limit for the aspect ratio of the elements.

Advanced Tab: If the **Provide detailed status information** check box is selected in the **Advanced** tab, the output with detailed meshing information will be created as a log file. This log file with detailed meshing information will be used in order to determine the reason of failure occurred in the mesh, if any. Figure 5-15 shows the **Model Mesh Settings** dialog box with the **Advanced** tab chosen.

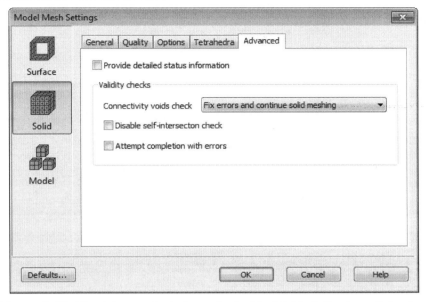

Figure 5-15 *The* ***Advanced*** *tab chosen in the* ***Model Mesh Settings*** *dialog box*

In the process of generating the solid mesh, errors may be found in the surface mesh. If the **Fix errors and continue solid meshing** option is selected in the **Connectivity voids check** drop-down list in the **Validity checks** area of the **Advanced** tab then the meshing process will automatically correct the surface mesh. However, if the **Do not fix errors** option is selected in the **Connectivity voids check** drop-down list and errors are found during the solid meshing process then the further process of meshing will stop.

 Note
The surface mesh must be corrected or made error free before generating the solid mesh.

If the **Disable self-intersection** check box is selected in the **Advanced** tab, then while generating the solid mesh, the intersection checks for the surface mesh will be disabled. On selecting the **Attempt completion with errors** check box in the **Advanced** tab, a solid mesh will be generated regardless of errors in the surface mesh. This may result in voids throughout the model.

Model

 On choosing the **Model** button, the **General** tab will be displayed in the **Model Mesh Settings** dialog box. The options available in this tab depend on whether the **Use VCAD** check box is selected or not. This check box is available in the expanded **Mesh** panel of the **Mesh** tab in the **Ribbon**, refer to Figure 5-16. Figure 5-17 shows the options displayed in the **General** tab of the **Model Mesh Settings** dialog box which is displayed on choosing the **Model** button, when the **Use VCAD** check box is selected in the **Mesh** panel. Figure 5-18 shows the options displayed in the **General** tab of the same dialog box when the **Use VCAD** check box is cleared.

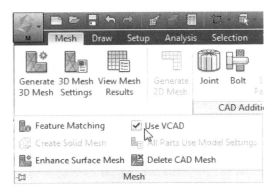

Figure 5-16 The **Use VCAD** check box selected in the **Mesh** penal

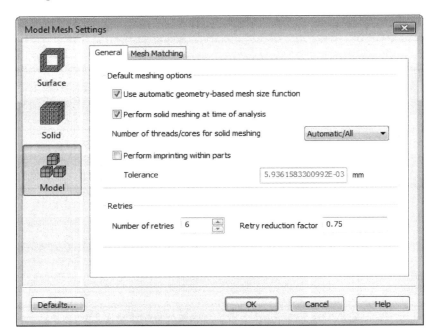

Figure 5-17 The options displayed in the **Model Mesh Settings** dialog box when
the **USE VCAD** tool is active

Tip. *VCAD refers to Virtual CAD and represents CAD model in the system memory. When
you choose the* **Use VCAD** *tool, the virtual CAD mesh engine will be used for meshing.
However, if this tool is not chosen, the legacy mesh engine will be used. The legacy mesh
engine will be recommended only if there is a problem with the virtual CAD mesh engine
for a particular model.*

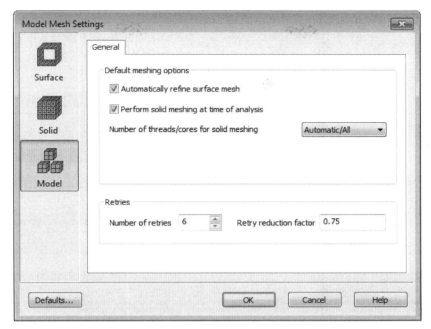

*Figure 5-18 The options displayed in the **Model Mesh Settings** dialog box when the **USE VCAD** tool is not active*

The options available in the **Model Mesh Settings** dialog box are discussed next.

General Tab: If the **Automatically refine surface mesh** check box is selected in the **General** tab then after generating the surface mesh, the refinement points will be added to the meshed model automatically, depending upon its geometry. Note that based on the added refinement points, model will be meshed again and after generating the surface mesh, the solid mesh will be generated. Note that this check box will be available in the **General** tab only if the **Use VCAD** tool is deactivated in the **Mesh** panel of the **Mesh** tab in the **Ribbon** and the legacy mesh engine is used for meshing.

If the **Use automatic geometry-based mesh size function** check box is selected in the **General** tab, the meshing in the curved areas will be automatically refined. Note that this check box will be available only if the **Use VCAD** check box is selected in the **Mesh** panel of the **Mesh** tab in the **Ribbon** and VCAD mesh engine is used for meshing.

On selecting the **Perform solid meshing at time of analysis** check box from the **General** tab, the solid meshing will be generated on the model at the time of analysis. However, if this check box is not selected, the solid mesh will be generated immediately after the surface mesh is created.

On selecting the **Use virtual imprinting** check box from the **General** tab, the connecting surfaces of two parts in a model will split by creating a virtual imprint of one surface on another. This virtual imprint of one surface on another is useful in order to match the meshing of both the contacted parts in an assembly. Note that this check box will be available only if the **Use VCAD** check box is selected and you are working on an assembly file. You can also specify tolerance value in the **Tolerance** edit box of the dialog box for creating virtual imprinting of one surface on another between the contacting parts of an assembly.

Note
*Unlike splitting surfaces, the **Use virtual imprinting** option does not create additional surfaces on the original model at the contacted area.*

Figure 5-19 shows a model meshed when the **Use virtual imprinting** check box is selected and Figure 5-20 shows the same model meshed when the **Use virtual imprinting** check box is cleared.

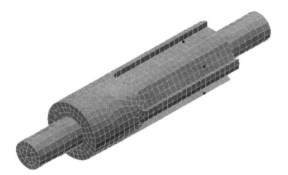 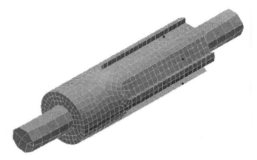

Figure 5-19 *A model meshed with the **Use virtual** imprinting check box selected*   *Figure 5-20* *A model meshed with the **Use virtual imprinting** check box cleared*

The **Perform imprinting within parts** option of the **General** tab functions similar to the **Use virtual imprinting** option. The only difference between these options is that the **Use virtual imprinting** option is used for matching mesh where part to part contacts are available in an assembly by creating virtual imprints on the contacts. However, the **Perform imprinting within parts** option is used for matching mesh where surface to surface contacts are available in a part. Note that this check box will be available only if the **Use VCAD** check box is selected. You can also specify tolerance value in the **Tolerance** edit box of the dialog box for creating virtual imprinting of one surface on another between the contacting surface of a part.

Mesh matching Tab: If the **Fraction of surface mesh size** option is selected in the **On-surface tolerance based on** drop-down list of the **Mesh matching** tab, the range of radius will be defined as the average length of the elements multiplied by the value specified in the **Tolerance value** edit box available on the right of the **On-surface tolerance based on** drop-down list. All the nodes available within the radius range will be combined together automatically. Note that even if two parts do not have proper contact with each other in an assembly but are within the range of radius defined, the meshing of both the parts will match each other.

On selecting the **Absolute length dimension** option from the **On-surface tolerance based on** drop-down list, the absolute radius value can be specified in the **Tolerance value** edit box. Note that the nodes available within the range specified in this edit box will be combined together to match the mesh. The **Fraction of automatic mesh size** option in the **On-surface tolerance based on** drop-down list is similar to the **Fraction of surface mesh size** option. The only difference between these two options is that if the **Fraction of automatic mesh size** option is selected, the average length of the elements that will be multiplied by the value specified in the **Tolerance value** edit box will depend on the size of mesh that will be generated at 100% mesh size.

If the **Do not match the mesh of contact pairs when applicable** check box in the dialog box is selected, the mesh between two contact surfaces will not match forcefully.

After specifying the meshing setting for solid mesh, choose the **OK** button from the **Model Mesh Settings** dialog box; the **Model Mesh Setting** dialog box will be displayed, as shown in Figure 5-21.

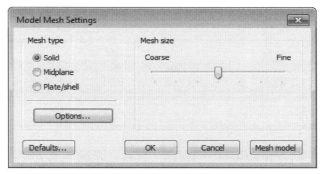

*Figure 5-21 The **Model Mesh Settings** dialog box*

Midplane

The **Midplane** radio button in the **Mesh type** area of the dialog box is used to mesh a model at the midplane location that is formed exactly at the center of its inside and outside surfaces. This type of mesh is required when the geometry of a model is symmetric about its inside and outside surfaces. The midplane mesh is generated by using plate elements instead of brick elements. Therefore, it is recommended to use midplane mesh for thin components that have uniform thickness throughout. Figure 5-22 shows a model before creating midplane mesh and Figure 5-23 shows the same model after creating the midplane mesh.

Figure 5-22 Model before creating the midplane mesh

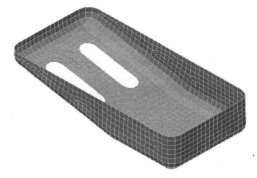

Figure 5-23 Model after creating the midplane mesh

In midplane mesh, the average thickness of the plate elements will be automatically used as the thickness of each surface for analysis. You can control various parameters of the midplane mesh by using the **Options** button of the **Model Mesh Settings** dialog box. To do so, first select the **Midplane** radio button in the **Mesh type** area of the dialog box and then choose the **Options** button; the **Model Mesh Settings** dialog box will be modified, as shown in Figure 5-24. The options available in this modified dialog box are discussed next.

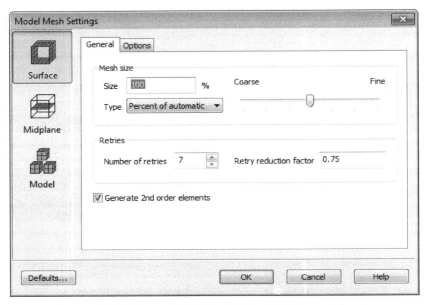

Figure 5-24 *The modified* **Model Mesh Settings** *dialog box*

    Tip. *You can generate midplane mesh on a 3D model which is symmetric about its mid plane.*

Surface

   The **Surface** button is chosen by default. As a result, on the right side of the dialog box, the **General** and **Options** tabs will be displayed, refer to Figure 5-24. The options available in these tabs are same as discussed earlier.

Midplane

   When you choose the **Midplane** button from the **Model Mesh Settings** dialog box, the **General** tab is displayed in the dialog box, refer to Figure 5-25. The options available in this tab are used for specifying the mesh element type, aspect ratio, warp angle, volume-to-length ratio, microholes, properties of tetrahedral meshes, boundary layer meshes, and so on. These options are discussed next.

General Tab: You can specify the element types for the midplane mesh by using the options available in the **Midplane mesh type** area of this tab. In the **Thickness control** area of the **General** tab, the **User-specified maximum thickness** check box is cleared by default. As a result, the minimum thickness of a component to be meshed will be calculated automatically.

Note that the minimum detected thickness will be used as the maximum thickness for midplane mesh. You can also specify the maximum thickness for midplane mesh by selecting the **User-specified maximum thickness** check box.

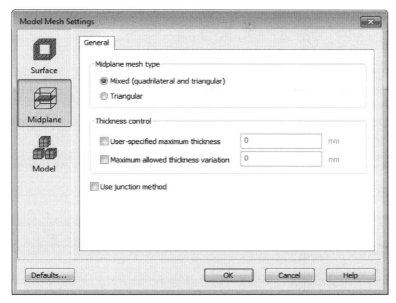

Figure 5-25 *The modified* **Model Mesh Settings** *dialog box displayed on choosing the* **Midplane** *button*

On selecting the **User-specified maximum thickness** check box in the **General** tab, the edit box on the right of this check box will be activated. In this edit box, you can specify maximum thickness for midplane mesh. Note that any area of the component to be meshed that has thickness greater than the specified one will not be midplane meshed.

On selecting the **Maximum allowed thickness variation** check box, the edit box on its right will be enabled. In this edit box, you can specify a value for maximum allowed thickness variation. As a result, the midplane mesh process will generate the midplane mesh on the component only if the difference between the maximum and minimum thickness of the component to be meshed is less than the value specified in this edit box.

The **Use junction method** check box of this tab is used to achieve accurate midplane mesh, if the geometry of the component to be meshed is complex having junctions and intersections. On selecting this check box, a Chordal Axis Transform (CAT) algorithm will be used to generate the midplane mesh. This method eliminates discontinuities at the junctions while meshing in order to achieve accurate midplane mesh.

Model

The options displayed after choosing the **Model** button from the dialog box are same as discussed earlier.

After specifying the meshing parameters for midplane mesh, choose the **OK** button from the **Model Mesh Settings** dialog box; this dialog box will be modified, as shown in Figure 5-26.

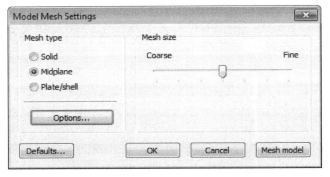

*Figure 5-26 The **Model Mesh Settings** dialog box*

Plate/shell

The **Plate/shell** radio button of the **Mesh type** area of the dialog box is used to mesh a model with plate elements. On selecting this button, all surfaces of the model, inside and outside, will be meshed by using the **Plate/shell** mesh type. However, the thickness of plate elements needs to be defined manually. You will learn more about adding thickness to the plate elements later. This type of mesh is widely used for meshing surface models.

You can control various parameters of the plate/shell mesh by using the **Options** button of the **Model Mesh Settings** dialog box. To do so, first select the **Plate/shell** radio button in the **Mesh type** area of the dialog box as shown in Figure 5-27. Next, choose the **Options** button; the **Model Mesh Settings** dialog box will be modified, as shown in Figure 5-28. The options available in this modified dialog box are discussed next.

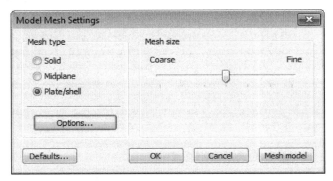

*Figure 5-27 The **Model Mesh Settings** dialog box with **Plate/shell** radio button selected*

Surface

The **Surface** button located on the left in the dialog box is chosen by default. As a result, on the right side of the dialog box, the **General** and **Options** tabs are displayed, refer to Figure 5-27. The options available in these tabs are the same, as discussed earlier.

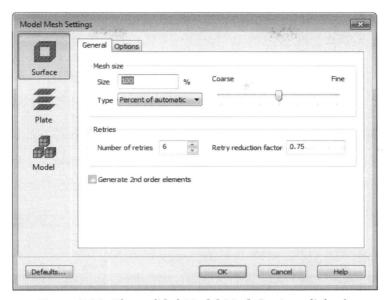

Figure 5-28 *The modified* **Model Mesh Settings** *dialog box*

Plate

On choosing the **Plate** button, the options used for defining the element type for the plate mesh will be displayed in the **General** tab.

Model

On choosing the **Model** button, the options used for defining the element type for the solid mesh will be displayed in the **General** tab.

After specifying the meshing parameters for plate/shell mesh, choose the **OK** button from the **Model Mesh Settings** dialog box; the **Model Mesh Settings** dialog box will be displayed.

The **Defaults** button of the **Model Mesh Settings** dialog box is used to reset the meshing parameters to their default values. The slider available in the **Mesh size** area of the **Model Mesh Setting** dialog box, refer to Figure 5-27, is used to define the mesh size as coarse to fine. Figure 5-29 shows a solid model with coarse mesh and Figure 5-30 shows the same model with fine mesh.

Tip. *If the mesh is too coarse, the results will not be accurate. Similarly, if the mesh is very fine, the results will be accurate but the time taken to obtain them will be more. A fine mesh is required in the region of high stress and strains. In a model, the areas with holes and sharp corners have to be meshed densely to get better results because stress will be more in these areas.*

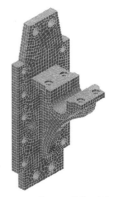

Figure 5-29 *The model with coarse mesh*   ***Figure 5-30*** *The model with fine mesh*

Note

It is evident from Figures 5-29 and 5-30 that the finer the mesh, the more is the number of elements. Since the elements are more in number in a fine mesh, the total deformation obtained in a model after performing analysis is more as compared to the model with coarse mesh. However, more the number of elements, the more computational time will be required for analysis. Therefore, it is important to make proper balance between number of elements in a mesh and computational time taken for analysis. For a complex model, where time is not the primary concern but accuracy is, an FEA engineer would refine the mesh till the analysis shows improved results. However, the time is also an important factor to find a perfect mesh for a model.

Figure 5-31 shows the relation between the number of elements and deformation. In this figure, X coordinate represents the element count and the Y coordinate represents the total deformation evaluated on a particular model. Notice that when the element count is 2200, the value of deformation achieved is 2.45 E-07. As the number of elements increases, the deformation also increases and the maximum value of deformation is achieved where the element count is 11,500. Hence, it is obvious that meshing a component beyond this point is not required. Meshing the model further will result in an increased runtime of the analysis.

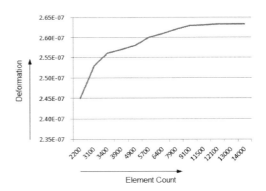

Figure 5-31 *Element count v/s deformation*

After specifying all settings for generating the mesh, choose the **OK** button from the **Model Mesh Settings** dialog box to exit it. If you choose the **Mesh Model** button from this dialog box, the **Meshing Progress** window will be displayed and the process of generating a mesh with the specified settings will start, refer to Figure 5-32.

As discussed earlier, once the meshing process is complete, the **View Mesh Results** window will be displayed with a message **Would you like to view the meshing results at this time?**.

If you choose the **No** button, the **View Mesh Results** window will be closed and the meshed model will be displayed on the screen with the default mesh settings. However, on choosing the **Yes** button from the **View Mesh Results** window, the **Meshing Results** window will be displayed where in you can view the meshing results.

*Figure 5-32 The **Meshing Progress** window*

VIEWING THE MESHING RESULTS

Ribbon: Mesh > Mesh > View Mesh Results

After a mesh has been generated, you can review the mesh results by using the **Meshing Results** dialog box. As discussed earlier, the **Meshing Results** dialog box will be displayed on choosing the **Yes** button from the **View Mesh Results** window displayed on completion of the meshing process. You can also invoke the **Meshing Results** dialog box by choosing the **View Mesh Results** tool from the **Mesh** panel of the **Mesh** tab in the **Ribbon**. Figure 5-33 shows the **Meshing Results** dialog box.

In the **Meshing Results** dialog box, the **Model** button is chosen by default, refer to Figure 5-33. As a result, the dialog box displays total number of parts available in the model that are meshed and the total number of elements generated in them. The **Part** button of the **Meshing Results** dialog box is used to review the mesh results of each part of the model individually. When you choose the **Part** button, by default, the first page of the **Meshing Results** dialog box displays the mesh result of the first part of the model. To review the mesh results of other parts of the model, you need to invoke their respective pages by using the **Part number** spinner that is available at the bottom of the **Meshing Results** dialog box, refer to Figure 5-34.

If any problem occurs while meshing in any of the parts of the model, Autodesk Simulation Mechanical will report the error and the **Problems** button will be displayed in the **Meshing**

Results dialog box. On choosing the **Problems** button, the dialog box displays all the errors and warnings occurred while meshing. Generally, the errors occur while meshing if the input surface does not enclose corresponding surface completely and form a watertight solid. This means such types of errors occur due to availability of unmatched and multi-matched feature lines in the model. Figure 5-35 shows the **Meshing Results** dialog box with the **Problems** button chosen.

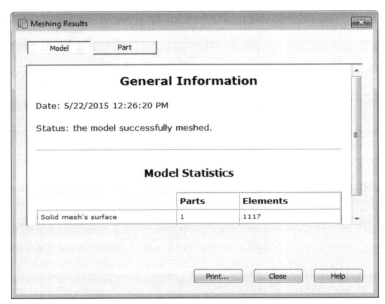

*Figure 5-33 The **Meshing Results** dialog box with the **Model** button chosen*

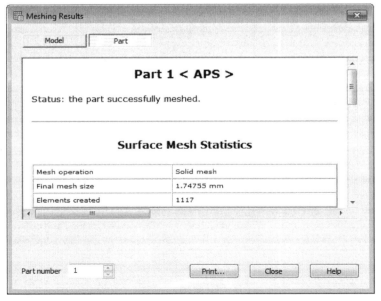

*Figure 5-34 The **Meshing Results** dialog box with the **Part** button chosen*

You can also take the printout of each mesh result by using the **Print** button of the dialog box. To exit the **Meshing Results** dialog box after reviewing the mesh results, choose the **Close** button.

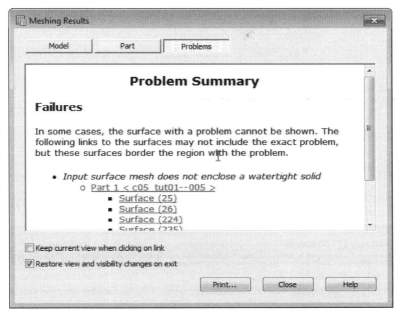

*Figure 5-35 The **Meshing Results** dialog box with the **Problems** button chosen*

UNMATCHED AND MULTI-MATCHED FEATURE LINES

Feature lines of a solid 3D model are the lines that define surfaces in a model. There are two possible configurations of feature lines: Unmatched and Multi-matched feature lines. Due to these configurations, errors can be reported while meshing a model. Both these configurations are discussed next.

Unmatched Feature Lines

Ribbon: View > Appearance > CAD Surfaces

An unmatched feature line is a line that does not knit two surfaces together and forms a gap between them. As a result, the model does not enclose a volume and, therefore, does not allow the solid mesh engine to create a solid mesh. To view the unmatched feature lines of a model, first change the visual style of the model to edges or shaded with edges visual style by using the **Edges** or **Shaded with Edges** tool from the **Visual Style** drop-down. This drop-down is available in the **Appearance** panel of the **View** tab in the **Ribbon**. Next, choose the **CAD Surfaces** tool from the **Appearance** panel of the **View** tab in the **Ribbon**, if not already chosen; the unmatched feature lines will be displayed in the graphics area, in orange color along with other edges of the model, as shown in Figure 5-36.

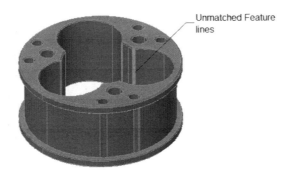

Figure 5-36 Unmatched feature lines

Multi-matched Feature Lines

Ribbon: View > Appearance > CAD Surfaces

A multi-match feature line is formed when more than two surfaces of a part share a single edge. However, to enclose a continuous volume, only two surfaces can share an edge. As a result, the part does not enclose a volume and will, therefore, not allow the solid mesh engine to create a solid mesh. To view the multi-matched feature lines of a model, first change the visual style of a part to edges or shaded with edges visual style by using the **Edges** or **Shaded with Edges** tool from the **Visual Style** drop-down. Next, choose the **CAD Surfaces** tool from the **Appearance** panel of the **View** tab in the **Ribbon**; the multi-match feature lines will be displayed in the graphics area, if any, in orange color along with other edges of the model.

 Tip. *You can also display the unmatched and multi-matched feature lines by choosing the **Only Bad CAD Features** tool from the expanded **Appearance** panel of the **View** tab in the **Ribbon** after changing the visual style of the model to edges or shaded with edges, refer to Figure 5-37.*

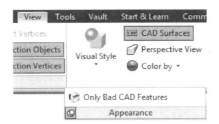

*Figure 5-37 The **Only Bad CAD Features** tool in the **Appearance** panel*

ELIMINATING UNMATCHED AND MULTI-MATCHED FEATURE LINES

Ribbon: Mesh > Mesh > Feature Matching

As discussed earlier, if any unmatched or multi-matched feature line is available in a model then the model does not enclose a volume and will not allow the solid mesh engine to create a solid mesh. You can eliminate the unmatched and multi-matched feature lines from a model to create a solid mesh by using the **Feature Matching** tool. To eliminate these lines, choose the **Feature Matching** tool from the **Mesh** panel of the **Mesh** tab in the **Ribbon**, refer to Figure 5-38; the **Feature Matching** dialog box will be displayed with the default mesh size for matching features in the part, refer to Figure 5-39.

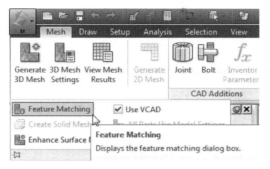

*Figure 5-38 The **Feature Matching** tool in the **Mesh** panel of the **Mesh** tab in the **Ribbon***

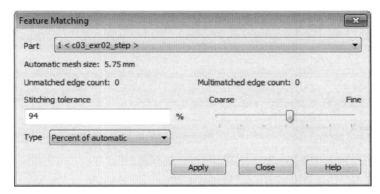

*Figure 5-39 The **Feature Matching** dialog box*

 Note
The default size of a mesh is one sixth of the cube root of the total volume of the model.

The **Feature Matching** dialog box is used to control the mesh size that is responsible for generating unmatched and multi-matched feature lines in a part. If the **Percent of automatic** option is selected in the **Type** drop-down list of this dialog box, you can specify the percentage value in the **Stitching tolerance** edit box and then based on the percentage value entered, the mesh size will automatically be calculated. However, if the **Absolute mesh size** option is selected in the **Type** drop-down list of this dialog box then you can directly enter the mesh size in the **Stitching tolerance** edit box as per the units set for the current session.

You can eliminate the unmatched and multi-matched feature lines from a model by trying different mesh sizes in the **Stitching tolerance** edit box. Note that if you specify bigger mesh size then most unmatched feature lines will be eliminated and if you specify smaller mesh size then the most of the multi-matched feature lines will be eliminated from the model. You can also use the slider available at the right of the **Stitching tolerance** edit box in this dialog box to control the mesh size from coarse to fine. However, this slider will be enabled only when the **Percent of automatic** option is selected in the **Type** drop-down list of the dialog box.

GENERATING 2D MESH

Ribbon: Mesh > Mesh > Generate 2D Mesh

In Autodesk Simulation Mechanical, you can also generate 2D mesh on closed 2D sketches by using the **Generate 2D Mesh** tool that is available in the **Mesh** panel of the **Mesh** tab in the **Ribbon**. The **Generate 2D Mesh** tool is an advanced tool for automatically generating a mesh on any enclosed sketched regions. Figure 5-40 shows a 2D sketch created by using the tools available in the **Draw** panel of the **Draw** tab in the **Ribbon** and Figure 5-41 shows the same sketch after generating 2D mesh on it.

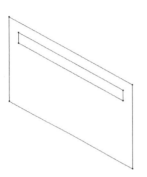   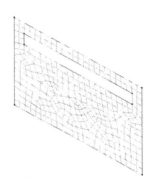

Figure 5-40 *2D sketch before generating mesh*   *Figure 5-41* *2D sketch after generating mesh*

Note
*The **Generate 2D Mesh** tool will be enabled in the **Mesh** panel of the **Mesh** tab in the **Ribbon** only if a 2D sketch is available and selected in the drawing area.*

After creating a sketch to be meshed, select the node corresponding to the sketching plane which was used for creating the sketch from the **Tree View** under the Part heading, refer to Figure 5-42. On doing so, all sketch entities created on the selected sketching plane will be selected and highlighted in the drawing area. As a result, the **Generate 2D Mesh** tool will be enabled in the **Mesh** panel of the **Mesh** tab in the **Ribbon**. Now, you can choose the **Generate 2D Mesh** tool to generate the mesh on the selected sketch. On choosing the **Generate 2D Mesh** tool, the **2-D Mesh Generation** dialog box will be displayed, as shown in Figure 5-43. The options available in this dialog box are used to specify the settings for generating 2D mesh.

Note
The 2D sketch to be meshed must be a closed sketch with no overlapping or open loops.

Specifying 2D Mesh Settings

In Autodesk Simulation Mechanical, you can specify the settings of the mesh to be generated on a 2D sketch by using the **2-D Mesh Generation** dialog box. The options in this dialog box are discussed next.

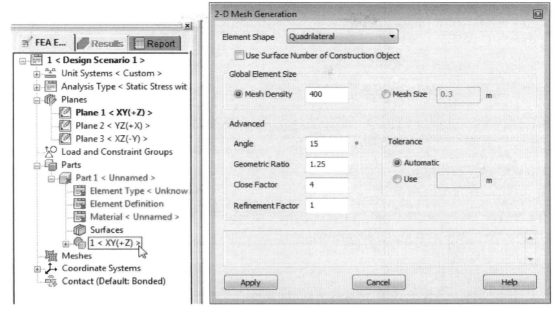

Figure 5-42 The *Tree View*   *Figure 5-43* The *2-D Mesh Generation* dialog box

Element Shape

The **Element Shape** drop-down list is used to select the shape of elements to be used in meshing. You can mesh a 2D sketch with quadrilateral, triangular, or mixed (both quadrilateral and triangular) shaped elements in the meshing. By default, the **Quadrilateral** option is selected in the **Element Shape** drop-down list. If you mesh a 2D sketch with this option selected, the closed region of the sketch will be meshed with quadrilateral elements only, refer to Figure 5-44.

Note

When you mesh a closed region using the quadrilateral elements, each segment of the region will split into minimum two divisions and a mesh of quadrilateral elements will be generated. However, if a segment is very small, then there is possibility of creation of some distorted elements in the area. Note that the possibility of the creation of distorted elements can be overcome by using mixed mesh or adding some refinement points. You will learn more about refinement points later in this textbook.

On selecting the **Triangular** option from the **Element Shape** drop-down list, the closed region of the sketch will be meshed using triangular elements only, refer to Figure 5-45.

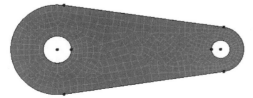   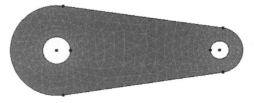

Figure 5-44 A closed region meshed using quadrilateral elements only

Figure 5-45 A closed region meshed using triangular elements only

If you select the **Mixed** option from the **Element Shape** drop-down list, the closed region of the sketch will be meshed with both quadrilateral and triangular elements, refer to Figure 5-46. This type of mesh does not split each segment into two elements and create a uniform mesh in the short segments.

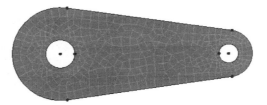

Figure 5-46 *A closed region meshed with both quadrilateral and triangular elements*

Use Surface Number of Construction Object

If the **Use Surface Number of Construction Object** check box is cleared, the surface of each edge of the sketch will be assigned a unique number in the **Tree View**, refer to Figure 5-47. However, on selecting the **Use Surface Number of Construction Object** check box, the surface of each edge of the sketch will merge into one and will be displayed as only one surface in the **Tree View**, refer to Figure 5-48. This option is useful when multiple edges are to be assigned for the same load.

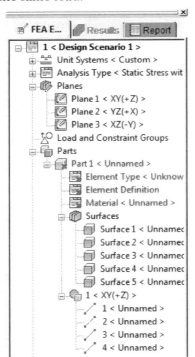

Figure 5-47 *All edge surfaces displayed with unique numbers in the Tree View*

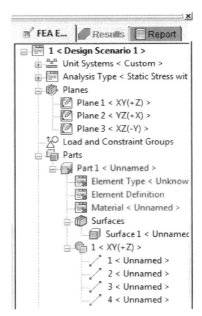

Figure 5-48 *The Tree View with surface number 1 assigned to all edges and construction surfaces of an object*

Global Element Size Area

This area of the **2-D Mesh Generation** dialog box is used to specify or control the size of elements in the mesh. You can specify the size of the elements either by specifying the mesh density or by specifying the mesh size by selecting the respective radio buttons from the **Global Element Size** area. The options available in this area are discussed next.

Mesh Density

The **Mesh Density** radio button is used to specify the number of elements to be generated after meshing an enclosed geometry. By default, this radio button is selected and the value is set to 400 in the **Mesh Density** edit box. Note that holes, arcs, and refinement points affect the actual number of elements generated after meshing. Therefore, the actual generated elements can be bigger or smaller than the value specified in the **Mesh Density** edit box. If the model does not fill the entire enclosed geometry due to holes and arcs in it, the number of generated elements will be lesser than the specified value in the **Mesh Density** edit box. Similarly, if the model contains refinement points, the model will have more generated elements than the value specified in the **Mesh Density** edit box.

Mesh Size

The **Mesh Size** radio button is used to specify the element size. Note that the size of the mesh elements can vary from the value specified in this edit box in order to maintain the smoothness and also for appropriate use of nodes.

Advanced Area

The options in the **Advanced** area of the **2-D Mesh Generation** dialog box are discussed next.

Angle

The **Angle** edit box is used to specify the minimum angle for dividing curved edges made up of arcs, splines, and NURB curves. As a result, the size of mesh elements at the curved edges can be smaller than the specified standard mesh element size in order to provide more elements at the edges where stress concentration is higher.

Geometric Ratio

This edit box is used for defining the approximate size ratio of adjacent elements in a mesh. A value close to 1 will result in slow transition from a region where the elements size is smaller to the region where the elements size is comparability larger. Note that as the value of geometric ratio increases, the transition ratio also becomes faster.

Close Factor

The value specified in the **Close Factor** edit box of the **Advanced** area is used to determine the range within which the 2D mesh generator searches for close nodes. This range will be calculated by multiplying the value specified in the **Close Factor** edit box with the mesh element size. The nodes within this range can be connected to each other for forming an element in the mesh to be generated. By default, the value 4 is entered in this edit box. As a result, if the element size is 0.5 inch, then the range within which 2D mesh generator would search for close nodes would be 2 inch. Here the value 2 inch represents minimum distance between two nodes in the mesh.

Refinement Factor

This edit box is used to quickly change the mesh size, especially the refinement points without doing any editing operation.

Tolerance Area

The options in this area are discussed next.

Automatic

By default, the **Automatic** radio button is selected in the **Tolerance** area of the **2-D Mesh Generation** dialog box. As a result, Autodesk Simulation Mechanical automatically calculates an appropriate tolerance value that defines the maximum distance between two points that can be considered to be coincident to each other.

Use

On selecting the **Use** radio button, the edit box on its right will be enabled. In this edit box, you can specify the tolerance value for the maximum distance between two points that can be coincident to each other.

After specifying all the options in the **2-D Mesh Generation** dialog box, choose the **Apply** button for generating the 2D mesh.

TUTORIALS

To perform the tutorials of this chapter, download the input files from *www.cadcim.com*. The complete path for downloading the file is as follows:

Textbooks > CAE Simulation > Autodesk Simulation Mechanical > Autodesk Simulation Mechanical 2016 for Designers > Input Files > c05_simulation_2016_input.zip

Extract the downloaded **c05_simulation_2016_input** zipped file at the location *C:\Autodesk Simulation Mechanical\c05*.

Tutorial 1

In this tutorial, you will import the IGS file of the model shown in Figure 5-49 to Autodesk Simulation Mechanical. Select the default analysis type that is Static Stress with Linear Material Models. After importing the model, generate fine mesh with 75% mesh size. Also, the solid mesh type elements used for meshing are the combination of bricks and tetrahedra.

(Expected time: 30 min)

The following steps are required to complete this tutorial:

a. Start Autodesk Simulation Mechanical.
b. Import the input file of this tutorial to Autodesk Simulation Mechanical.
c. Display the visibility of unmatched or multi-matched feature lines.
d. Eliminate unmatched or multi-matched feature lines from the model.
e. Generate mesh.
f. Save and close the model.

Figure 5-49 *Model for Tutorial 1*

Importing the IGS File into Autodesk Simulation Mechanical

You need to start Autodesk Simulation Mechanical and import the *c05_tut01.igs* file into it.

1. Start Autodesk Simulation Mechanical by double-clicking on the shortcut icon of Autodesk Simulation Mechanical 2016 available on the desktop of your computer.

2. Choose the **Open** button from the **Launch** panel of the **Start & Learn** tab of the **Ribbon**; the **Open** dialog box is displayed.

   As mentioned in the tutorial description, the file extension of this model is *.igs*.

3. Select the **IGES Solid Files (*.igs; *.ige; *.iges)** file extension from the **Files of type** drop-down list in the **Open** dialog box.

4. Browse to the location *C:\Autodesk Simulation Mechanical\c05\Tut01*; the **c05_tut01.igs** file is displayed in the **Open** dialog box.

5. Select the **c05_tut01.igs** file and then choose the **Open** button from the **Open** dialog box; some window(s) may be displayed. Choose the **Yes** button from these window(s) until the **Choose Analysis Type** dialog box is displayed with the **Static Stress with Linear Material Models** analysis type selected by default.

6. Choose the **OK** button from the **Choose Analysis Type** dialog box; the selected *.igs* file is now opened in the Autodesk Simulation Mechanical, refer to Figure 5-50.

 Note
If the orientation of the model displayed on your system is not the same as shown in Figure 5-50, you need to change the orientation of the model by using the ViewCube.

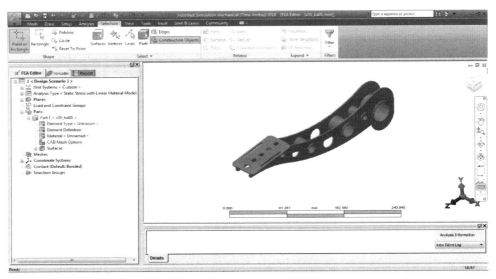

Figure 5-50 *The .igs file imported into Autodesk Simulation Mechanical*

Generating Mesh in the Model

Once the model has been imported in Autodesk Simulation Mechanical, you need to generate mesh in it. However, before you start generating the mesh, you need to first specify mesh settings.

1. Choose the **3D Mesh Settings** tool from the **Mesh** panel of the **Mesh** tab in the **Ribbon**; the **Model Mesh Settings** dialog box is displayed, refer to Figure 5-51.

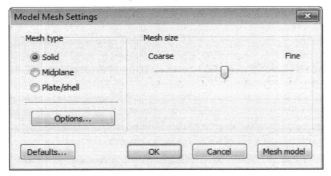

Figure 5-51 *The Model Mesh Settings dialog box*

2. Make sure the **Solid** radio button is selected in the **Model Mesh Settings** dialog box. Next, choose the **Options** button from the dialog box; another **Model Mesh Settings** dialog box is displayed, as shown in Figure 5-52.

   As mentioned in the tutorial description, you need to mesh the model with 75% mesh size and by using the bricks and tetrahedra elements.

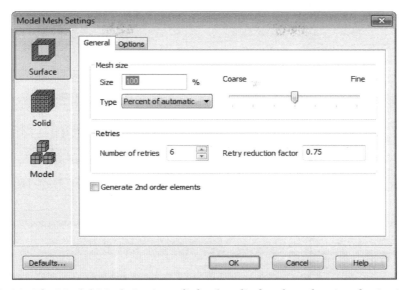

Figure 5-52 *The* **Model Mesh Settings** *dialog box displayed on choosing the* **Options** *button*

3. Make sure that the **Percent of automatic** option is selected in the **Type** drop-down list of the **General** tab in the dialog box. Next, enter **75%** in the **Size** edit box of the **Mesh size** area in this tab.

4. Enter **3** in the **Number of retries** edit box of the dialog box.

5. Choose the **Solid** button from the left side of the dialog box and then make sure that the **Bricks and tetrahedra** radio button is selected in the **General** tab of the dialog box.

6. Choose the **OK** button from the dialog box. Next, choose the **Mesh Model** button from the **Model Mesh Settings** dialog box for generating the mesh on the model; the **Meshing Progress** window is displayed, as shown in Figure 5-53.

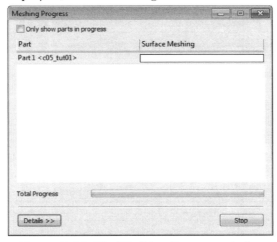

Figure 5-53 *The* **Meshing Progress** *window*

After sometime another **Meshing Progress** window may be displayed with an error message, as shown Figure 5-54. If the error message is displayed then you can perform steps 7 through 13. If the error message is not displayed then the **View Mesh Result** window is displayed. In this case, skip steps 7 through 13.

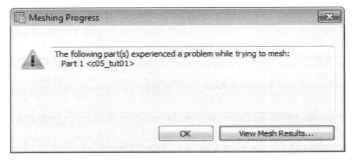

*Figure 5-54 The **Meshing Progress** window with an error message*

7. Choose the **View Mesh Results** button from the **Meshing Progress** window; the **Meshing Results** window is displayed with the **Problems** button chosen. As a result, a page stating the summary of the problem is displayed, refer to Figure 5-55.

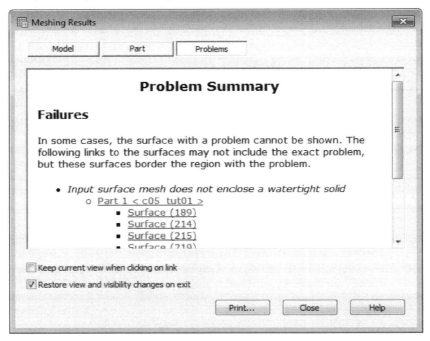

*Figure 5-55 The problem summary displayed in the **Meshing Results** window*

8. As stated in the **Problem Summary** page of the window, this problem occurs because of the unmatched feature lines available in the model which did not allow the model to form a watertight solid. Next, close the window by choosing the **Close** button.

Now, before generating the mesh again on to the model, you need to eliminate the unmatched feature lines to avoid further errors.

9. Choose the **View** tab in the **Ribbon** and then choose the **Shaded with Edges** option from the **Visual Style** drop-down list in the **Appearance** panel of the **View** tab. The visual style of the model changes to the Shaded with Edges visual style.

10. Choose the **CAD Surfaces** tool from the **Appearance** panel of the **View** tab, if not chosen already; all unmatched feature lines of the model are displayed in orange color along with other edges of the model, as shown in Figure 5-56.

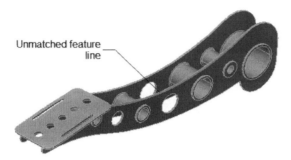

Figure 5-56 Unmatched feature line displayed

11. Choose the **Mesh** tab of the **Ribbon** and then choose the **Feature Matching** tool from the **Mesh** panel of the **Mash** tab in the **Ribbon**; the **Feature Matching** dialog box is displayed, refer to Figure 5-57. Note that you need to expand the **Mesh** panel for choosing the **Feature Matching** tool.

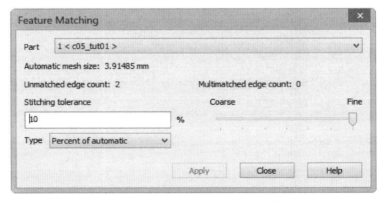

*Figure 5-57 The **Feature Matching** dialog box*

Note that the bigger mesh size will eliminate most unmatched feature lines and the smaller mesh size will eliminate the most multi-matched feature lines from the model. Therefore, you need to specify a larger mesh size to eliminate unmatched feature lines.

12. Make sure that the **Percent of automatic** option is selected in the **Type** drop-down list of the dialog box. Next, enter **120** in the **Stitching tolerance** edit box and then choose the **Apply** button from the dialog box; the unmatched feature lines are eliminated from the model.

 Tip. *You can eliminate the unmatched and multi-matched feature lines from a model by trying different mesh sizes in the* **Stitching tolerance** *edit box of the* **Feature Matching** *dialog box.*

13. Choose the **Close** button to exit the **Feature Matching** dialog box.

After eliminating the unmatched feature lines from the model, you can start meshing process again.

14. Choose the **Yes** button from the **View Mesh Results** window to view the meshing result; the **Meshing Results** window is displayed.

After viewing the meshing result, you need to exit this window.

15. Choose the **Close** button from the **Meshing Results** window to exit it. The model after generating the mesh is displayed in the graphics area, as shown in Figure 5-58.

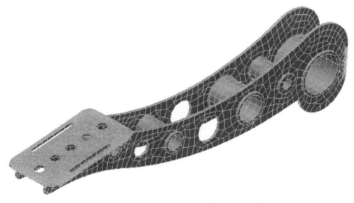

Figure 5-58 The meshed model

Saving the Model

1. Choose the **Save** button from the **Quick Access Toolbar**; the changes made in the FEA model are saved.

2. Choose **Close** from the **Application Menu** to close the file.

Tutorial 2

In this tutorial, you will import the STEP file of the model shown in Figure 5-59 to the Autodesk Simulation Mechanical. Select the unit used in STEP file and the default analysis type which is **Static Stress with Linear Material Models**. After importing the model, generate fine mesh with 1 mm mesh size. Use the combination of bricks and tetrahedra type solid mesh elements for meshing. **(Expected time: 30 min)**

Figure 5-59 *Model for Tutorial 2*

The following steps are required to complete this tutorial:

a. Start Autodesk Simulation Mechanical.
b. Import the input file of this tutorial to Autodesk Simulation Mechanical.
c. Generate mesh.
d. Save and close the model.

Importing the STEP File into Autodesk Simulation Mechanical

You need to start Autodesk Simulation Mechanical and import the *c05_tut02.stp* file into it.

1. Start Autodesk Simulation Mechanical.

2. Choose the **Open** button from the **Launch** panel of the **Start & Learn** tab of the **Ribbon**; the **Open** dialog box is displayed.

    You need to select the **STEP Files (*.stp; *.ste; *.step)** from the **Files of type** drop-down list of the **Open** dialog box.

3. Select **STEP Files (*.stp; *.ste; *.step)** from the **Files of type** drop-down list of the **Open** dialog box.

4. Browse to the location *C:\Autodesk Simulation Mechanical\c05\Tut02*; the **c05_tut02.stp** file is displayed in the **Open** dialog box.

5. Select the **c05_tut02.stp** file and then choose the **Open** button from the **Open** dialog box; some message window(s) may be displayed. Choose the **Yes** button from these window(s) until the **Choose Analysis Type** dialog box is displayed with the **Stress with Linear Material Models** analysis type selected by default.

Note
*If the **Select Length Units** dialog box is displayed on choosing the **Open** button from the **Open** dialog box, select the **Use STEP file units** option from the **Length units to use** drop-down list of the dialog box and then choose the **OK** button.*

6. Choose the **OK** button from the **Choose Analysis Type** dialog box; the selected file is opened in Autodesk Simulation Mechanical.

Generating Mesh on the Model

Once the model is imported to Autodesk Simulation Mechanical, you need to generate mesh in it. However, before you start the process of mesh generation, you need to first specify the mesh settings.

1. Choose the **3D Mesh Settings** tool from the **Mesh** panel of the **Mesh** tab in the **Ribbon**; the **Model Mesh Settings** dialog box is displayed, as shown in Figure 5-60.

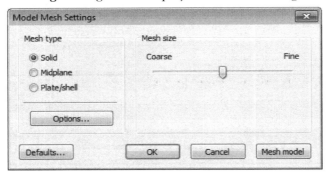

*Figure 5-60 The **Model Mesh Settings** dialog box*

2. Make sure the **Solid** radio button is selected in the **Model Mesh Settings** dialog box. Next, choose the **Options** button from the dialog box; another **Model Mesh Settings** dialog box is displayed, refer to Figure 5-61.

    You need to mesh the model with 1 mm mesh size and by using the bricks and tetrahedra mesh elements.

3. Select the **Absolute mesh size** option from the **Type** drop-down list of the **General** tab of the dialog box. Next, enter **1** in the **Size** edit box of the **Mesh size** area in the **General** tab of the dialog box.

4. Choose the **Solid** button from the left of the dialog box and then make sure that the **Bricks and tetrahedra** radio button is selected in the **General** tab of the dialog box.

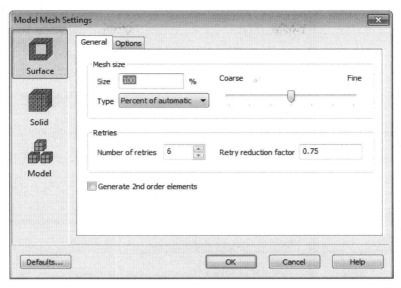

Figure 5-61 *The Model Mesh Settings dialog box displayed on choosing the Options button*

5. Choose the **OK** button from the dialog box. Next, choose the **Mesh Model** button from the previously invoked **Model Mesh Settings** dialog box; the **Meshing Progress** window is displayed. After the mesh is generated, the **View Mesh Results** window is displayed.

6. Choose the **Yes** button from the **View Mesh Results** window to view the meshing result; the **Meshing Results** window is displayed, as shown in Figure 5-62.

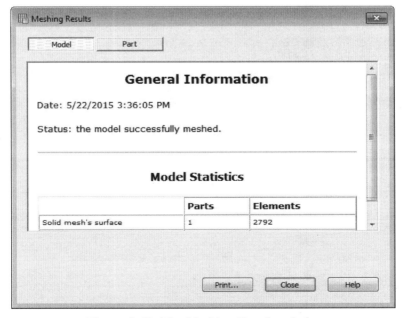

Figure 5-62 *The Meshing Results window*

After viewing the meshing result, you need to exit the **Meshing Results** window.

7. Choose the **Close** button from the **Meshing Results** window to exit it; the model after meshing is displayed in the graphics area, as shown in Figure 5-63.

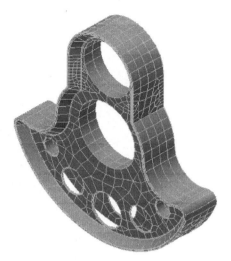

Figure 5-63 *The meshed model*

Saving the Model

1. Choose the **Save** button from the **Quick Access Toolbar**; the FEA model is saved.

2. Choose **Close** from the **Application Menu** to close the assembly.

Self-Evaluation Test

Answer the following questions and then compare them to those given at the end of this chapter:

1. If the _____ check box is cleared in the **General** tab of the **Model Mesh Settings** dialog box, then the solid mesh will be generated immediately after the surface mesh is generated.

2. The **Generate 3D Mesh** tool is used to generate mesh as per the settings specified in the _____ dialog box.

3. The _____ tool is used to review the mesh results.

4. If any problem is found while meshing a model, Autodesk Simulation reports the error and the _____ button appears in the **Meshing Result** dialog box.

5. There are two possible configurations of feature lines _____ and _____ that can cause errors while meshing a model.

6. You can display the unmatched and multi-matched feature lines of the model by first changing the visual style to either edges or shaded with edges visual style and then choosing either the _____ or _____ tool from the **Mesh** panel of the **Mesh** tab in the **Ribbon**.

7. Autodesk Simulation Mechanical needs a finite element model to carry out analysis. (T/F)

8. The **Generate 2D Mesh** tool will be enabled even if no 2D sketch is available in the drawing area. (T/F)

9. In Autodesk Simulation Mechanical, if a valid surface mesh cannot be created with the specified mesh size, the mesh size will be reduced automatically and will generate mesh in the model again. (T/F)

10. In Autodesk Simulation Mechanical, you cannot eliminate errors or warnings caused due to microholes that are very small in size relative to the volume of the part. (T/F)

Review Questions

Answer the following questions:

1. Which of the following dialog boxes is displayed on choosing the **Feature Matching** tool?

    (a) **Match**   (b) **Feature Match**
    (c) **Feature Matching**   (d) None of these

2. Which of the following windows is displayed on choosing the **Generate 3D Mesh** tool?

    (a) **Meshing Progress**   (b) **Generate 3D Mesh**
    (c) **Select Analysis Type**   (d) None of these

3. Which of the following spinners is used to specify the number of repetition performed in order to create a valid surface mesh?

    (a) **Repetition performed**   (b) **Repetition performed**
    (c) **Create**   (d) **Number of retries**

4. The _____ radio button of the **Mesh** type area of the **Model Mesh Settings** dialog box is used to mesh a model with plate elements.

5. On selecting the _____ radio button while generating surface mesh, all surfaces of the model will be meshed using 6-wedge elements and then solid mesh thus created will have 4-node tetrahedral elements.

6. On selecting the _____ check box from the **General** tab of the **Model Mesh Settings** dialog box, the mesh on curved surfaces will be automatically refined.

7. The _____ radio button of the **Mesh type** area of the **Model Mesh Settings** dialog box is used to mesh a model at the midplane location.

8. In Autodesk Simulation Mechanical, you cannot eliminate the unmatched and multi-matched feature lines from a model. (T/F)

9. You can generate 2D mesh on closed or opened 2D sketches. (T/F)

10. On selecting the **All tetrahedra** radio button from the **Solid mesh type** area of the **Model Mesh Settings** dialog box, the solid mesh will be generated with 4-node tetrahedra elements only. (T/F)

EXERCISES

Exercise 1

In this exercise, you will open the FEA model *c03_exr01.fea* created in Exercise 1 of Chapter 3. You will then save this model at the location *C:\Autodesk Simulation Mechanical\c05\Exr01* with the name *c05_exr01* and generate fine solid mesh with 1.5 mm mesh size. Also, for meshing, use only bricks type solid mesh elements. Figure 5-64 shows the final FEA model after meshing.

(Expected time: 10 min)

Figure 5-64 *Final FEA model of Exercise 1*

Exercise 2

In this exercise, you will open the FEA model *c03_exr02.fea* created in Exercise 2 of Chapter 3. You will then save this model at the location *C:\Autodesk Simulation Mechanical\c05\Exr02* with the name *c05_exr02* and generate fine mesh with default mesh settings. Figure 5-65 shows the final FEA model after meshing. **(Expected time: 10 min)**

Figure 5-65 Final FEA model of Exercise 2

Exercise 3

In this exercise, you will open the FEA model *c03_exr03.fea* created in Exercise 3 of Chapter 3. You will then save this model at the location *C:\Autodesk Simulation Mechanical\c05\Exr03* with the name *c05_exr03* and generate fine mesh with 85% mesh size. Also, for meshing, use only the tetrahedra type solid mesh elements. Figure 5-66 shows the final FEA model after meshing.

(**Expected time: 10 min**)

Figure 5-66 Final FEA model of Exercise 3

Answers to Self-Evaluation Test

1. Perform solid meshing at time of analysis, 2. Model Mesh Settings, 3. View Mesh Results, 4. Problem, 5. unmatched, multi-matched, **6. CAD Surfaces, Only Bad CAD Features, 7.** T, **8.** F, **9.** T, **10.** F

Chapter 6

Meshing-II

Learning Objectives

After completing this chapter, you will be able to:
- *Understand the concept of refine mesh*
- *Understand different methods of refining mesh*
- *Add refinement points using different methods*
- *Create refine mesh on a surface*
- *Create refine mesh on an edge*
- *Create refine mesh on a part*
- *Edit or modify the parameters of the existing refinement points*

INTRODUCTION

As discussed earlier, an effective mesh is a mesh that provides you the maximum accuracy in minimum computational time. If the mesh is very fine, the result will be accurate but the time taken to obtain the result will be longer. Therefore, it is very important to maintain a balance between the number of elements in a mesh and the computational time taken for analysis. In Autodesk Simulation Mechanical, you can create a fine mesh only in the areas of high stress and strains by adding the refinement points. You can also refine the mesh of surfaces or parts by specifying fine mesh size for their elements. Figure 6-1 shows a model with mesh generated with default settings and Figure 6-2 shows the same model with mesh generated after adding the refinement points.

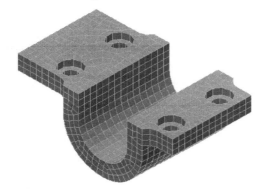

Figure 6-1 *Mesh generated with default settings*   *Figure 6-2* *Mesh generated after adding refinement points*

CREATING REFINE MESH

In complex models, most of the time engineers are more concerned about the accuracy of specific areas of the model. In such cases, creating mesh on the entire model with a fine mesh size, may consume more time for analyzing the fine mesh in regions where the results are not of so much significance. Therefore, you need to create fine mesh only in those areas where you are more concerned about the accuracy. In Autodesk Simulation Mechanical, you can do so by using the methods discussed next.

Refining Mesh by Adding Refinement Points

A refinement point specifies a volume of space in a model in which a finer mesh will be generated. You can add refinement points to a model automatically by using the **Automatic** tool or by specifying the coordinates of the refinement points using the **Specify Nodes** tool. Both these methods of adding refinement points are discussed next.

Adding Refinement Points by Using the Automatic Tool

Ribbon:	Mesh > Refinement > Automatic

In this method, the refinement points will be added automatically to the regions where the fine mesh is required. To do so, choose the **Automatic** tool from the **Refinement** panel of the **Mesh** tab in the **Ribbon**; the **Automatic Refinement Points** dialog box will be displayed, as shown in Figure 6-3.

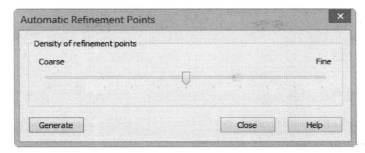

*Figure 6-3 The **Automatic Refinement Points** dialog box*

Note
*You can add refinement points by using the automatic method only to the meshed model. The **Automatic** tool will be enabled, if the mesh model is opened in the graphics area of Autodesk Simulation Mechanical.*

In the **Automatic Refinement Points** dialog box, you can specify the density of refinement points from coarse to fine by using the slider bar. After doing so, choose the **Generate** button; the refinement points will be added to the model automatically depending upon the specified density. Also, the number of refinement points added to the model will be displayed next to the **Generate** button of the dialog box, refer to Figure 6-4. If the **No points created** message is displayed next to the **Generate** button of the dialog box, you need to further refine the density of the refinement point using the slider bar. The finer the density, the more will be the number of refinement points generated. You can increase or decrease the number of refinement points by using the slider bar and then choosing the **Generate** button of the dialog box. Figure 6-5 shows a model with refinement points added automatically and Figure 6-6 shows the same model re-meshed after adding refinement points.

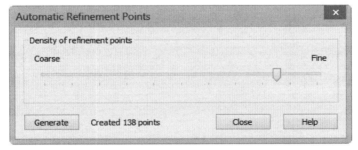

*Figure 6-4 The **Automatic Refinement Points** dialog box with refinement points generated*

Note
*By default, the added refinement points are visible in the model. This is because the **Visibility** button is chosen in the **Refinement Points** panel of the **Mesh** tab. The **Visibility** button is a toggle button. On choosing this button again, all the visible refinement points will disappear from the model.*

Figure 6-5 Mesh model after adding refinement points automatically

Figure 6-6 Model re-meshed after adding the refinement points

Adding Refinement Points by Using the Specify Nodes Tool

Ribbon: Mesh > Refinement > Specify Nodes

    In this method, the refinement points will be added by specifying the coordinates of the points where you want to add refinement points to a model. This method is mostly used where nodes do not exist in a model. For example, the center of a hole in a model where nodes are not available to add refinement points. To add refinement points by specifying the coordinates, choose the **Specify Nodes** tool from the **Refinement** panel of the **Mesh** tab in the **Ribbon**; the **Refinement Point Browser** dialog box will be displayed, as shown in Figure 6-7.

Note

*If a meshed model already contains refinement points, then on choosing the **Specify Nodes** tool, the **Refinement Point Browser** dialog box with a list of all the available refinement points will be displayed, refer to Figure 6-8.*

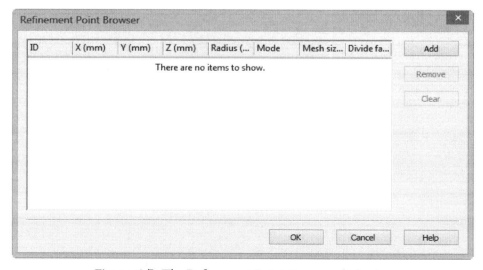

Figure 6-7 The **Refinement Point Browser** dialog box

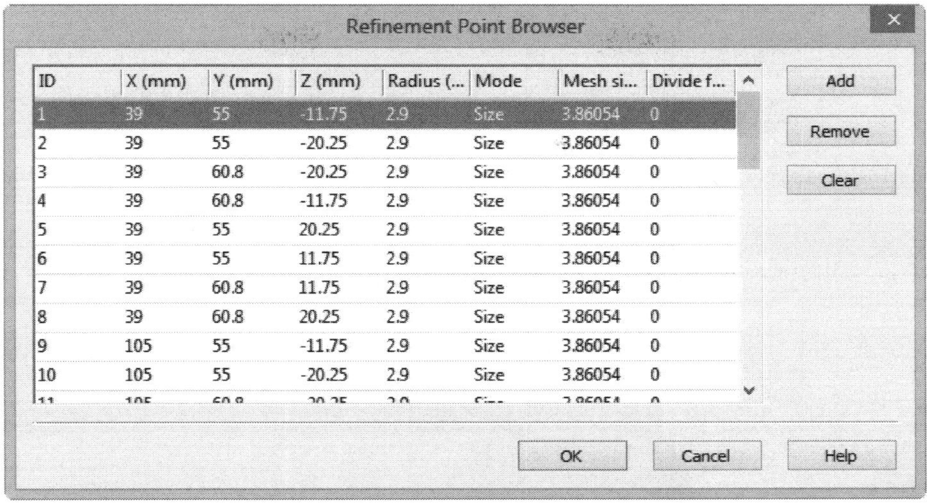

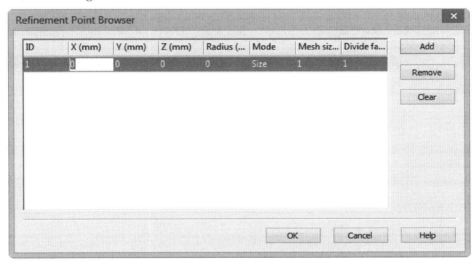

Figure 6-8 The **Refinement Point Browser** *dialog box with added refinement points*

Choose the **Add** button from the **Refinement Point Browser** dialog box; a row will be added to the dialog box. You can define the parameters of the refinement point in their respective fields of the row, refer to Figure 6-9.

Figure 6-9 The **Refinement Point Browser** *dialog box with a new row added to it*

The **X**, **Y**, and **Z** fields of the row are used to specify the X, Y, and Z coordinate values of the point in the model where you want to add a refinement point. To enter a value in a field of a row, you need to first activate its editing mode by double-clicking on it.

The **Radius** field of a row is used to specify the radius which defines the spherical region around the refinement point for creating a refined mesh. For example, if you specify origin (0, 0, 0) as the refinement point, then on entering the value **10** mm in the **Radius** field, the spherical region will be defined by measuring **10** mm radius from the origin as the center of the region.

Note
*The measuring unit will depend on the unit of the current session of Autodesk Simulation Mechanical. You can change the unit of the session as per your requirement by using the **Unit Systems** node of the **Tree View**.*

The **Mode** field of a row is used to select the type of mode for specifying the refined mesh size. To do so, double-click on the **Mode** field of a row, a down arrow will be displayed on its right. Click on the down arrow; a drop-down list will be displayed with the types of modes available in it, refer to Figure 6-10.

On selecting the **Size** option from the **Mode** drop-down list, you can define the size of a mesh element in the refined mesh region by entering the required value in the **Mesh Size** field of the row. Note that the size specified in the **Mesh Size** field will be applied to all the elements starting from the refinement point whose coordinates are specified in the X, Y, and Z fields of the row up to the radius specified in the **Radius** field.

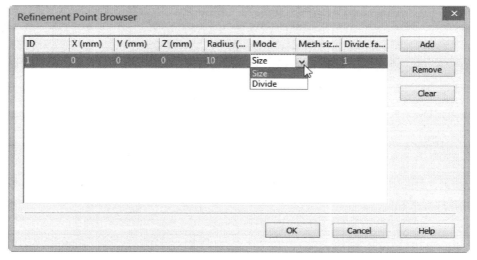

*Figure 6-10 The **Refinement Point Browser** dialog box with the **Mode** drop-down list*

On selecting the **Divide** option from the **Mode** drop-down list, you can define the size of a mesh element in the refine mesh region by entering the mesh division factor in the **Divide factor** field of the row. In this case, the size of a mesh element in the refine region will be equal to the mesh size divided by the divide factor.

Mesh size of an element in the refine region = Mesh Size/Divide factor

After specifying all the parameters for defining the refinement point, choose the **OK** button from the **Refinement Point Browser** dialog box; the refinement point will be added to the model. You can add multiple refinement points in a model by using the **Add** button of the **Refinement Point Browser** dialog box.

Note
*If the added refinement points are not visible in the model then you need to choose the **Visibility** button from the **Refinement Points** panel of the **Mesh** tab in the **Ribbon** to turn on their visibility.*

The **Remove** button of the **Refinement Point Browser** dialog box is used to remove the selected row from the list of rows in the dialog box that defines the refinement points.

To delete or remove all the rows that define the refinement points from the **Refinement Point Browser** dialog box, choose the **Clear** button from the dialog box. Figure 6-11 shows a meshed model with a refinement point added at the center of the hole and Figure 6-12 shows the same model re-meshed after adding the refinement point at the center of the hole.

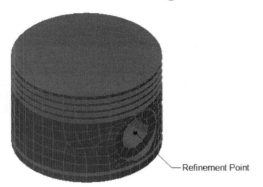

Figure 6-11 A refinement point added at the center of the hole

Figure 6-12 Model remeshed after adding the refinement points at the center of the hole

Note
In Autodesk Simulation Mechanical, alike adding refinement points to 3D models, you can also add refinement points to 2D geometries.

Refining Mesh by Using the Add to Selection Method

In this method, you can refine the mesh of selected geometries by using the **Add to Selection** tool. You can select the vertex, surface, and edge to refine their mesh by using this method.

Refining Mesh by Selecting a Vertex

Ribbon: Mesh > Refinement > Add to Selection

To refine the mesh of a region around a vertex of a model by using the **Add to Selection** tool, select a vertex of the model and then choose the **Add to Selection** tool from the **Refinement** panel of the **Mesh** tab in the **Ribbon**; the **Create Refinement Point** dialog box will be displayed, refer to Figure 6-13.

Note
*To select a vertex of a model, you need to activate the vertex selection mode. To do so, click on the **Selection** tab of the **Ribbon** to display the tools used for selections. Next, choose the **Point or Rectangle** tool from the **Shape** panel and the **Vertices** tool from the **Select** panel of the **Selection** tab in the **Ribbon**.*

Alternatively, to invoke the **Create Refinement Point** dialog box, select a vertex of the model and then right-click in the graphics area; the shortcut menu will be displayed. Next, choose **Add > Refinement Points** from the shortcut menu.

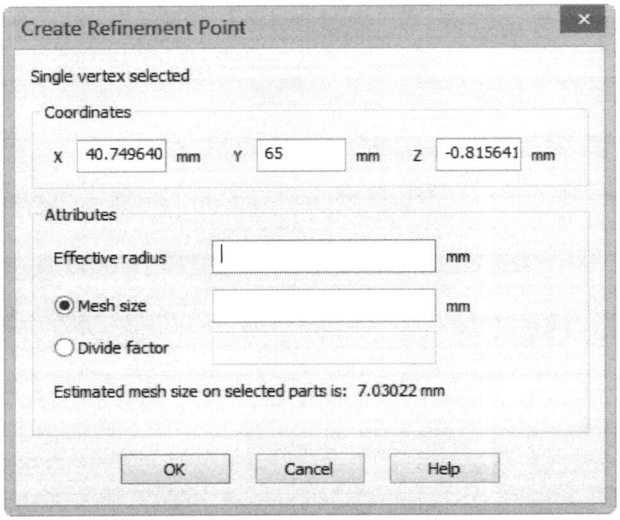

*Figure 6-13 The **Create Refinement Point** dialog box*

In the **Coordinates** area of the **Create Refinement Point** dialog box, the coordinates of the selected vertex will be displayed by default, refer to Figure 6-13. In this dialog box, the **Effective radius** edit box in the **Attributes** area is used to specify the radius which defines the spherical region around the selected vertex that you want to refine. The spherical region within the limit of the specified radius whose center is at the selected vertex will be refined.

You can specify the mesh size for the refine region by using the **Mesh size** edit box. Note that the **Mesh size** edit box will be enabled only on selecting the **Mesh size** radio button. The value entered in this edit box will define the size of the elements in the refine mesh region. The specified refine mesh size will be applied to all the elements that are within the limit of the specified radius defined in the **Effective radius** edit box. Note that the value entered in the **Mesh size** edit box must be less than the average mesh size of the model.

You can also specify the mesh size for the refined region by using the **Divide factor** edit box of the **Attributes** area of the **Create Refinement Point** dialog box. To enable this edit box, select the **Divide factor** radio button. You can enter the desired division factor for the refined mesh size. Note that the mesh size with in the refine radius will be equal to average mesh size of the model divided by the divide factor. The value entered in the **Divide factor** must be greater than 1.

Refine mesh size = Average mesh size of the model / Divide factor

Once you have specified all the parameters for refining the mesh of a particular region, choose the **OK** button from the **Create Refinement Point** dialog box. Next, generate the mesh again by choosing the **Generate 3D Mesh** tool from the **Mesh** panel of the **Mesh** tab in the **Ribbon**. Similarly, you can add multiple refinement points as per the requirement of the design by using the **Add to Selection** tool.

Figure 6-14 shows a mesh model after adding the refinement point with specific parameters and Figure 6-15 shows the same model meshed after adding the refinement point.

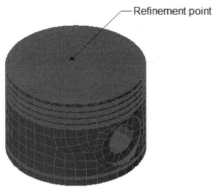

Figure 6-14 *Mesh model after adding the refinement point*

Figure 6-15 *Model remeshed after adding the refinement point*

Refining Mesh by Selecting a Surface

Ribbon: Mesh > Refinement > Add to Selection

 To refine the mesh of a surface of the model by using the **Add to Selection** tool, select the surface and then choose the **Add to Selection** tool from the **Refinement** panel of the **Mesh** tab in the **Ribbon**; the **Creating Surface Refinements** dialog box will be displayed, refer to Figure 6-16.

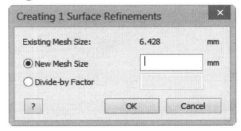

Figure 6-16 *The **Creating Surface Refinements** dialog box*

Alternatively, to invoke the **Creating Surface Refinements** dialog box, select a surface of the model and then right-click in the graphics area; the shortcut menu will be displayed. Next, choose **Add > Surface Refinement** from the shortcut menu.

 Note
*To select a surface of a model, you need to activate the surface selection mode. To do so, click on the **Selection** tab of the **Ribbon** to display the tools used for selections. Next, choose the **Point or Rectangle** tool from the **Shape** panel and the **Surfaces** tool from the **Select** panel of the **Selection** tab in the **Ribbon**.*

In the **Creating Surface Refinements** dialog box, the meshed size of the current elements of the selected surface will be displayed next to the **Existing Mesh Size** text, refer to Figure 6-16. Note that by default, the **New Mesh Size** radio button of the dialog box is selected. As a result, you can enter the new mesh size for the elements of the selected surface in the **New Mesh Size** edit box of the dialog box.

You can also specify the new mesh size for the elements of the selected surface by using the **Divide factor** edit box of the **Creating Surface Refinements** dialog box. To enable this edit box, select the **Divide-by Factor** radio button. You can enter the desired division factor for the refined mesh size. Note that the refined mesh size of the elements of the selected surface will be the average mesh size of the model divided by the divide factor. The value entered in the **Divide factor** edit box must be greater than 1.

Refine mesh size = Average mesh size of the model / Divide factor

Once you have specified the parameters for refined mesh, choose the **OK** button from the **Creating Surface Refinements** dialog box. Note that by using this dialog box, you can also select multiple surfaces for refining their mesh elements size. Figure 6-17 shows a model after refining the elements of the surface.

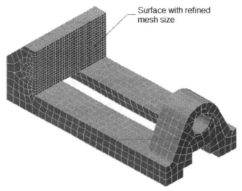

Figure 6-17 *Surface with refined mesh size elements*

Refining Mesh by Selecting an Edge

Ribbon: Mesh > Refinement > Add to Selection

 To refine the mesh of an edge of a model by using the **Add to Selection** tool, select the edge and then choose the **Add to Selection** tool from the **Refinement** panel of the **Mesh** tab in the **Ribbon**; the **Creating Edge Refinements** dialog box will be displayed, refer to Figure 6-18.

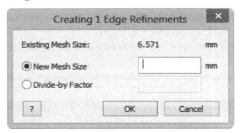

Figure 6-18 *The **Creating Edge Refinements** dialog box*

Alternatively, to invoke the **Creating Edge Refinements** dialog box, select an edge of the model and then right-click in the graphics area; a shortcut menu will be displayed. Next, choose **Add > Edge Refinement** from the shortcut menu.

 Note
To select an edge of a model, you need to activate the edge selection mode. To do so, click on the *Selection tab of the* *Ribbon to display the tools used for selections. Next, choose the* *Point or* *Rectangle tool from the* *Shape panel and then the* *Edges tool from the* *Select panel.*

The options available in this dialog box are same as the options of the **Creating Surface Refinements** dialog box, discussed earlier. You can specify the new mesh size for the elements of the selected edge either by using the **New Mesh Size** edit box or by using the **Divide factor** edit box of the dialog box. Note that by using this dialog box, you can also select multiple edges for refining their mesh elements size. Figure 6-19 shows a model with the refined mesh on an edge.

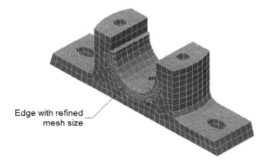

Edge with refined
mesh size

Figure 6-19 Edge with refined mesh size

 Note
If you invoke the *Add to Selection tool without selecting an object or geometry, the* *Create* *Refinement Point dialog box will be displayed, as shown in Figure 6-20. By using this dialog box, you can create a refinement point at the desired location. To do so, enter the X, Y, Z coordinates of the point in the* *X, Y, and* *Z edit boxes, respectively, in the* *Coordinates area of the dialog box. Next, specify the effective radius and the mesh size for the elements in the range of effective radius in their respective edit boxes and then choose the* *OK button; the refinement points with the specified parameters will be created.*

 Tip. *You can select the already refined surfaces and edges again and further refine their element size by following the procedure discussed above.*

In addition to adding refinement points and defining refine mesh size for the elements of a refined mesh on particular area of a component, you can also define refine mesh size for the elements of a part in the assembly model. Creating refine mesh on a part of the assembly model is discussed next.

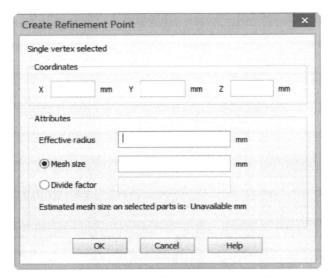

Figure 6-20 The **Create Refinement Point** *dialog box*

Creating Refine Mesh on a Part

The process of creating refine mesh on a part of an assembly is generally used when a part of the model is considered to have high stress and strains. To create the refine mesh on a part, select the part and then right-click in the drawing area; a shortcut menu will be displayed. Next, choose **CAD Mesh Options > Part** from the shortcut menu, refer to Figure 6-21; the **Part Mesh Settings** dialog box will be displayed, as shown in Figure 6-22.

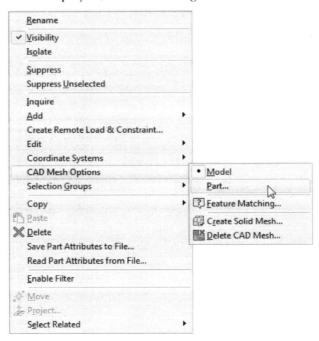

Figure 6-21 The **Part** *option being selected from the shortcut menu*

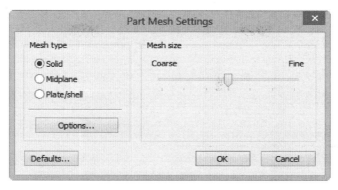

Figure 6-22 The **Part Mesh Settings** *dialog box*

By using the options available in the **Part Mesh Settings** dialog box, you can control the mesh settings of the selected part and define the refined mesh. The options available in this dialog box are same as discussed in Chapter 5. After defining the refined mesh size for the elements of the selected part, exit the dialog box. Next, regenerate the mesh on the model by using the **Generate 3D Mesh** tool for reflecting the results of the mesh settings modified in the **Part Mesh Settings** dialog box. Figure 6-23 shows a model with the refined mesh elements of a part.

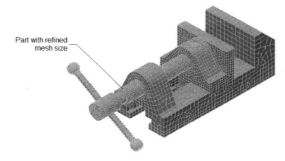

Figure 6-23 Part with refined mesh size elements

EDITING REFINEMENT POINTS

Editing parameters of refinement points is one of the important aspects of the meshing process. In Autodesk Simulation Mechanical, you can edit the parameters of a refinement point either by using the **Refinement Point Browser** or **Modify Refinement Point** dialog box.

Editing Refinement Points by Using the Refinement Point Browser Dialog box

To invoke the **Refinement Point Browser** dialog box for editing the refinement point, choose the **Specify Nodes** tool from the **Refinement** panel of the **Mesh** tab in the **Ribbon**. The **Refinement Point Browser** dialog box displays the list of all the refinement points that are available in the current mesh model. You can double-click on any of the fields of a refinement point you want to edit and then specify the new value in that particular field. For example, to change the effective radius of a refinement point, double click on the **Radius** field of the refinement point in the dialog box; an edit box will be displayed. In this edit box, you can enter the new radius value for the refinement point and choose **OK**.

Editing Refinement Points by Using the Modify Refinement Point Dialog box

As discussed, you can edit the parameters of refinement points by using the **Modify Refinement Point** dialog box. To invoke the **Modify Refinement Point** dialog box, select the refinement point to be edited from the drawing area and then right-click; a shortcut menu will be displayed, as shown in Figure 6-24. Choose the **Edit** option from the shortcut menu; the **Modify Refinement Point** dialog box will be displayed, refer to Figure 6-25. Alternatively, you can invoke the **Modify Refinement Point** dialog box by double-clicking on the refinement point whose parameters are to be edited.

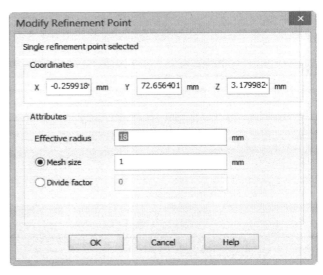

Figure 6-24 *A shortcut menu displayed on right-clicking on a refinement point*

Figure 6-25 *The **Modify Refinement Point** dialog box*

Note
*To select a refinement point of a model, you need to activate the vertex selection mode. To do so, click on the **Selection** tab of the **Ribbon** to display the tools used for selections. Next, choose the **Point** tool from the **Shape** panel and the **Vertex** tool from the **Select** panel of the **Selection** tab in the **Ribbon**.*

The **Modify Refinement Point** dialog box displays the existing parameters of the selected refinement point in the respective edit boxes of the dialog box. You can specify the new values in these edit boxes and choose the **OK** button.

Note
In Autodesk Simulation Mechanical, similar to editing the parameters of refinement points added in 3D models, you can also edit the parameters of refinement points added in 2D geometries.

TUTORIALS

Before you start the tutorials, you need to download the zipped file *c06_simulation_2016_input* from *www.cadcim.com*. The complete path for downloading the file is as follows:

Textbooks > CAE Simulation > Autodesk Simulation Mechanical > Autodesk Simulation Mechanical 2016 for Designers > Input Files

Extract the downloaded *c06_simulation_2016_input* zipped file and save it at the location *C:\Autodesk Simulation Mechanical\c06.*

Tutorial 1

In this tutorial, you will import the Autodesk Inventor model file *c06_Tut01.ipt* into the Autodesk Simulation Mechanical. The model for this tutorial is shown in Figure 6-26. Select the default **Static Stress with Linear Material Models** analysis type. After importing the model, generate fine mesh with the following mesh settings:

Mesh Type: Solid
Mesh Size: 5 mm
Solid Mesh element type: Bricks and tetrahedra

Figure 6-26 *The model for Tutorial 1*

After generating the mesh with the above mesh settings, you need to further refine the mesh by adding the refinement points at the places whose coordinates are given in the following table and the effective radius for refinement points is 5 mm and mesh size is 0.75 mm. Figure 6-27 shows the mesh model after refining the mesh by adding the refinement points.

(Expected time: 30 min)

Table 6-1 Coordinates of refinement points		
X	Y	Z
-10	0	-26.1534

-28	-26.1534	0
-10	0	26.1534
-28	26.1534	0

You need to generate the refine mesh with the effective refine radius of 5 mm and the refine mesh size of 0.75 mm for refinement points.

Figure 6-27 *Refine mesh model*

The following steps are required to complete this tutorial:

a.   Start Autodesk Simulation Mechanical and then import the input file of this tutorial in it.
b.   Generate mesh on the model.
c.   Add refinement points.
d.   Regenerate mesh on the model.
e.   Save and close the model.

Importing the Inventor File into Autodesk Simulation Mechanical

Now, you need to open Autodesk Simulation Mechanical and import the *c06_tut01.ipt* file into it.

1.   Start Autodesk Simulation Mechanical and then invoke the **Open** dialog box.

As mentioned in the tutorial description, the file extension of this model is *.ipt*. Therefore, you need to select **Autodesk Inventor Part (*.ipt)** from the **Files of type** drop-down list of the **Open** dialog box.

2.   Select **Autodesk Inventor Files (*.ipt; *.iam)** from the **Files of type** drop-down list in the **Open** dialog box.

3.   Browse to the location *C:\Autodesk Simulation Mechanical\c06\Tut01*; the **c06_tut01.ipt** file is displayed in the **Open** dialog box.

4. Select the **c06_tut01.ipt** file and then choose the **Open** button from the **Open** dialog box; either the **Import Inventor Work Points, Simulation Mechanical Color Palette,** and **CAD Part Names** window(s) or the **Choose Analysis Type** dialog box may be displayed. If the windows are displayed then choose the **Yes** button from each of these windows to display the **Choose Analysis Type** dialog box. In this dialog box, the Stress with Linear Material Models analysis type is selected by default.

5. Choose the **OK** button from the **Choose Analysis Type** dialog box; the selected Inventor file is opened into the Autodesk Simulation Mechanical, refer to Figure 6-28.

Note

*If the orientation of the model displayed on your system is not same as the one shown in Figure 6-28, then choose the **Home** button from the ViewCube.*

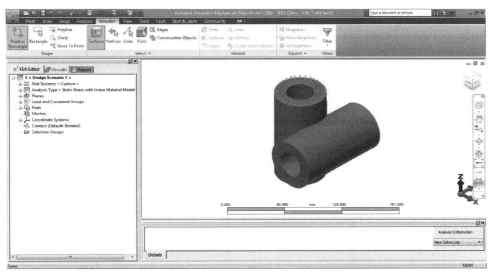

Figure 6-28 The Inventor model imported into Autodesk Simulation Mechanical

Generating Mesh on the Model

Once the model has been imported in the Autodesk Simulation Mechanical, you need to generate mesh on it. However, before starting the meshing process, you need to specify mesh settings.

1. Choose the **3D Mesh Settings** tool from the **Mesh** panel of the **Mesh** tab in the **Ribbon**; the **Model Mesh Settings** dialog box is displayed, as shown in Figure 6-29.

2. In the **Model Mesh Settings** dialog box, specify the mesh setting as mentioned in the tutorial description.

3. After specifying the mesh settings in the dialog box, generate the mesh. The resulting model will be similar to the one shown in Figure 6-30.

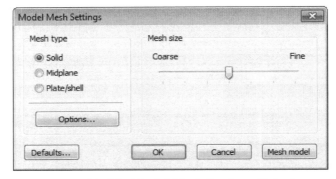

*Figure 6-29 The **Model Mesh Settings** dialog box*

Figure 6-30 The model after generating mesh

Adding Refinement Points

After generating the mesh with the given mesh settings, you further need to refine the mesh by adding refinement points.

1. Choose the **Specify Nodes** tool from the **Refinement** panel of the **Mesh** tab in the **Ribbon**; the **Refinement Point Browser** dialog box is displayed, as shown in Figure 6-31.

2. Choose the **Add** button from the **Refinement Point Browser** dialog box; a row is added in the dialog box.

3. Double-click in the **X** field of the newly added row and enter **-10** in the **X** edit box displayed, refer to Figure 6-32.

4. Make sure that **0** is entered in the **Y** field of the row. Next, double-click in the **Z** field of the row and enter **-26.1534** in the **Z** edit box displayed, refer to Figure 6-32.

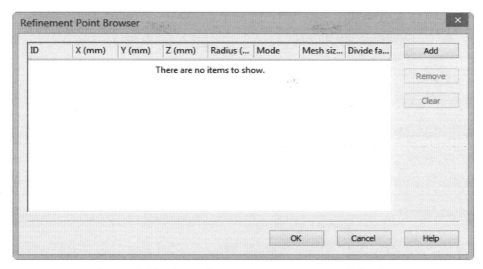

Figure 6-31 The **Refinement Point Browser** *dialog box*

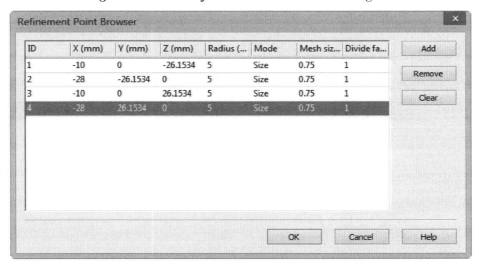

Figure 6-32 The **Refinement Point Browser** *dialog box with refinement points added*

5. Double-click in the **Radius** field of the row and enter **5**, refer to Figure 6-32. Next, double-click in the **Mesh size** field of the row and enter **0.75** as the mesh size for the refined region, defined by the value entered in the **Radius** field of the row, refer to Figure 6-32.

6. Similarly, add the second row by choosing the **Add** button from the **Refinement Point Browser** dialog box and enter **-28**, **-26.1534**, **0**, **5**, and **0.75** in the **X**, **Y**, **Z**, **Radius**, and **Mesh size** fields of the second row, respectively in the dialog box, refer to Figure 6-32.

7. Similarly, add the third row by choosing the **Add** button from the **Refinement Point Browser** dialog box and enter **-10**, **0**, **26.1534**, **5**, and **0.75** in the **X**, **Y**, **Z**, **Radius**, and **Mesh size** fields of the third row, respectively, refer to Figure 6-32.

8. Similarly, add the fourth row and enter **-28**, **26.1534**, **0**, **5**, and **0.75** in the **X**, **Y**, **Z**, **Radius**, and **Mesh size** fields, respectively of the fourth row, refer to Figure 6-32.

9. After specifying the parameters of all the refinement points in the dialog box, choose the **OK** button; the refinement points are added to the model with respect to the coordinates of the points specified. Figure 6-33 shows the model after adding refinement points.

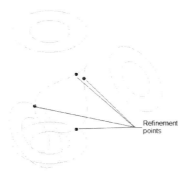

Figure 6-33 The model after adding refinement points

Note
*If the refinement points are not visible in the model, you need to turn on the visibility of the refinement points. To do so, choose the **Visibility** tool from the **Refinement** panel of the **Mesh** tab in the **Ribbon**.*

*In Figure 6-33, for a better visibility, the display style of the model has been changed to Edges display style. To do so, choose the **Edges** tool from the **Visual Style** drop-down of the **Appearance** panel of the **View** tab in the **Ribbon**.*

Regenerating Mesh On the Model

After defining the refined regions by adding the refinement points in the model, you need to regenerate the mesh.

1. Choose the **Generate 3D Mesh** tool from the **Mesh** panel of the **Mesh** tab in the **Ribbon**; the process of meshing starts and the **Meshing Progress** window is displayed. Once the process of generating mesh is done successfully, the **View Mesh Results** window is displayed. Also, the meshed model is displayed at the background of this window.

2. Choose the **No** button from the **View Mesh Results** window if you do not wish to view the meshing result now; the model with refined mesh on the areas defined by the refinement points is displayed, as shown in Figure 6-34.

Saving the Model

After performing all the meshing operations, you need to save the model.

1. Choose the **Save** button from the **Quick Access Toolbar** to save the model.

2. Choose **Close** from the **Application Menu** to close the file.

Figure 6-34 *The final meshed model*

Tutorial 2

In this tutorial, you will import the IGS file *c06_Tut02.igs* into Autodesk Simulation Mechanical. The model for this tutorial is shown in Figure 6-35. Select the default analysis type **Static Stress with Linear Material Models**. After importing the model, generate fine mesh with the following mesh settings:

Mesh Type: Solid
Mesh Size: 2 inch
Solid Mesh element type: Bricks and tetrahedra

After generating the mesh with the above mesh settings, you need to further refine the mesh by adding the refinement points using the automatic method.

Once the refinement points are added by using the automatic method, you need to edit the effective radius to 0.1 inch and refine mesh size to 0.35 inch for all the automatically added refinement points. Figure 6-36 shows the mesh model after refining the mesh by adding the refinement points. **(Expected time: 30 min)**

The following steps are required to complete this tutorial:

a. Start Autodesk Simulation Mechanical and then import the input file of this tutorial.
b. Generate mesh on the model.
c. Add refinement points.
d. Edit the parameters of the refinement points.
e. Regenerate the mesh on the model.
f. Save and close the model.

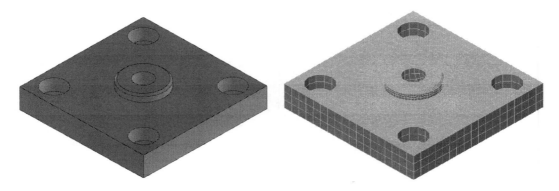

Figure 6-35 *The model for Tutorial 2*   **Figure 6-36** *Refined mesh model*

Importing the IGS File into Autodesk Simulation Mechanical

Now, you need to open Autodesk Simulation Mechanical and import the *c06_tut02.igs* file into it.

1. Start Autodesk Simulation Mechanical and then choose the **Open** button from the **Launch panel** of the **Start & Learn** tab of the **Ribbon** to invoke the **Open** dialog box.

    As mentioned in the tutorial description, the file used in this tutorial is IGES file. Therefore, you need to select the **IGES Solid Files (*.igs; *.ige; *.iges)** file extension from the **Files of type** drop-down list of the **Open** dialog box.

2. Select the **IGES Solid Files (*.igs; *.ige; *.iges)** file extension from the **Files of type** drop-down list of the **Open** dialog box.

3. Browse to the location *C:\Autodesk Simulation Mechanical\c06\Tut02*; the *c06_tut02.igs* file is displayed in the **Open** dialog box.

4. Select the **c06_tut02.igs** file and then choose the **Open** button from the **Open** dialog box; the **Choose Analysis Type** dialog box is displayed with the **Stress with Linear Material Models** analysis type selected, by default.

5. Choose the **OK** button from this dialog box; the selected file is opened into the Autodesk Simulation Mechanical, refer to Figure 6-37.

Note

1. If the orientation of the model displayed on your system is not same as the one shown in Figure 6-37 then use the ViewCube to change the orientation of the model.

*2. If the **Import Inventor Work Points, Simulation Mechanical Color Palette, or CAD Part Names** window(s) are displayed. Choose the **Yes** button from these windows; the **Choose Analysis Type** dialog box is displayed with the **Stress with Linear Material Models** analysis type selected by default.*

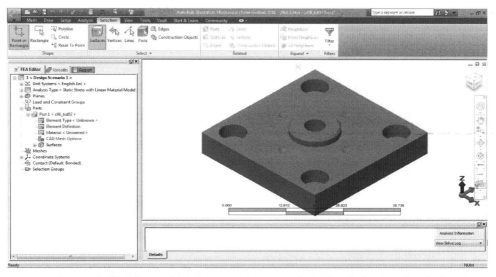

Figure 6-37 *The model is imported into Autodesk Simulation Mechanical*

Generating Mesh On the Model

Once the model has been imported in the Autodesk Simulation Mechanical, you need to generate mesh on it. However, before starting the meshing process, you need to first specify the mesh settings.

1. Choose the **3D Mesh Settings** tool from the **Mesh** panel of the **Mesh** tab in the **Ribbon**; the **Model Mesh Settings** dialog box is displayed, as shown in Figure 6-38.

2. In the **Model Mesh Settings** dialog box, specify the mesh settings as given in the tutorial description.

3. After specifying the mesh settings in the dialog box, generate the mesh. The model after generating the mesh is displayed, refer to Figure 6-39.

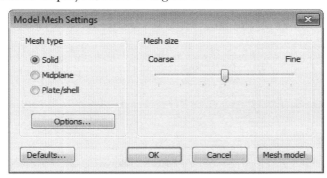

Figure 6-38 *The **Model Mesh Settings** dialog box*

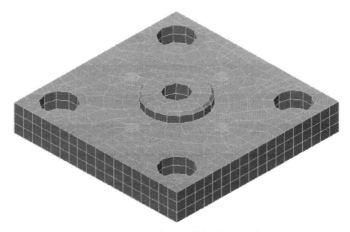

Figure 6-39 The model after meshing

Adding Refinement Points

After generating the mesh with the specified mesh settings, you need to further refine the mesh by adding refinement points using the automatic method.

1. Choose the **Automatic** tool from the **Refinement** panel of the **Mesh** tab in the **Ribbon**; the **Automatic Refinement Points** window is displayed, as shown in Figure 6-40.

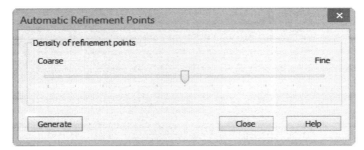

Figure 6-40 The **Automatic Refinement Points** window

2. Set the density for the refinement points from coarse to fine by using the slider and then choose the **Generate** button; the refinement points are added to the model and the information about the number of generated refinement points is displayed next to the **Generate** button in the **Automatic Refinement Points** window, refer to Figure 6-41.

3. Again set the density for the refinement points from coarse to fine by using the slider and then choose the **Generate** button. Repeat the procedure till about 85 refinement points are generated, refer to Figure 6-41.

4. Choose the **Close** button from the **Automatic Refinement Points** window to exit from it; the refinement points are added to the mesh model, refer to Figure 6-42.

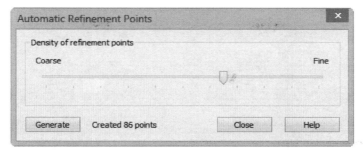

Figure 6-41 The **Automatic Refinement Points** *window displaying the number of refinement points generated*

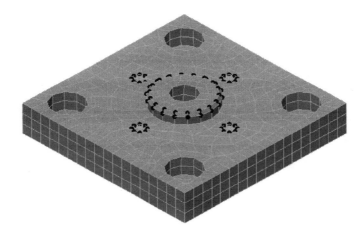

Figure 6-42 The meshed model after adding refinement points

 Note

*If the refinement points are not visible in the model, you need to turn on their visibility. To do so, choose the **Visibility** tool from the **Refinement** panel of the **Mesh** tab in the **Ribbon**.*

Editing the Default Parameters of the Refinement Points

After adding the refinement points, you need to edit their default parameters.

1. Choose the **Selection** tab in the **Ribbon**; the selection tools are displayed.

2. Choose the **Rectangle** tool from the **Shape** panel and the **Vertices** tool from the **Select** panel of the **Selection** tab in the **Ribbon**.

3. Select the entire meshed model by creating a rectangular window around it, refer to Figure 6-43; all the vertices of the mesh model are selected, refer to Figure 6-44.

4. Right-click in the drawing area; a shortcut menu is displayed, refer to Figure 6-45.

5. Choose the **Edit** option from the shortcut menu; the **Modify Multiple Refinement Points** dialog box is displayed, as shown in Figure 6-46.

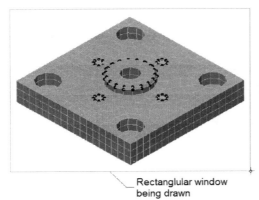

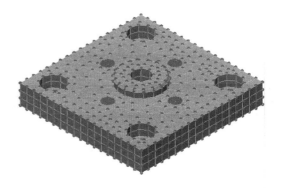

Figure 6-43 Creating a rectangular window

Figure 6-44 All the vertices of the meshed model selected

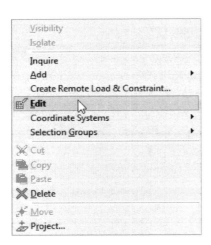

Figure 6-45 The shortcut menu displayed

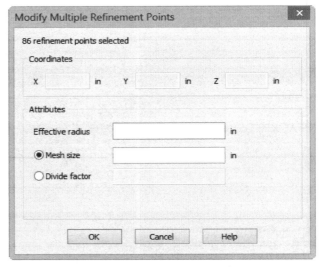

Figure 6-46 The Modify Multiple Refinement Points dialog box

6. Enter **0.1** in the **Effective radius** edit box and **0.35** in the **Mesh size** edit box of the dialog box.

7. Choose the **OK** button from the dialog box. Now, click anywhere in the drawing area to exit the current selection set.

Regenerating Mesh On the Model

After defining the refined regions by adding the refinement points in the model and editing their parameters as per the requirement, you need to regenerate the mesh.

1. Choose the **Generate 3D Mesh** tool from the **Mesh** panel of the **Mesh** tab in the **Ribbon**; the process of meshing starts and the **Meshing Progress** window is displayed. Once the

process of generating mesh is completed successfully, the **View Mesh Results** window is displayed. Also, the meshed model is displayed in the background of this window.

2. Choose the **No** button from the **View Mesh Results** window if you do not wish to view the meshing result now; the model with refined mesh generated on the areas defined by the refinement points is displayed, as shown in Figure 6-47.

Note
In the Figure 6-47, the visibility of refinement points has been turned off for clarity.

Saving the Model

1. Choose the **Save** button from the **Quick Access Toolbar**; the FEA model is saved.

2. Choose **Close** from the **Application Menu** to close the assembly.

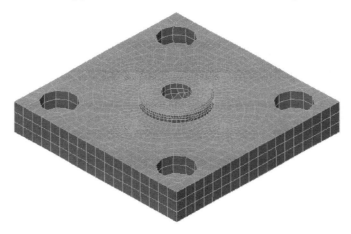

Figure 6-47 The final meshed model

Tutorial 3

In this tutorial, you will import the STEP file *c06_Tut03.STEP* into Autodesk Simulation Mechanical. The model of this tutorial is shown in Figure 6-48. Select the **Static Stress with Linear Material Models** analysis type. After importing the model, generate mesh with the default mesh settings.

After generating the mesh with the default mesh settings, you need to refine the mesh of the top curved face of mesh size 2 mm. Figure 6-49 shows the mesh model after refining the mesh of the top curved face of the model.   (**Expected time: 30 min**)

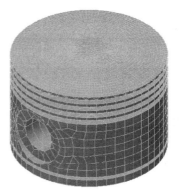

Figure 6-48 The model for Tutorial 3   *Figure 6-49* Refined mesh model

The following steps are required to complete this tutorial:

a.   Start Autodesk Simulation Mechanical and then import the input file of this tutorial.
b.   Generate mesh on the model.
c.   Refine the top curved face of the model.
d.   Regenerate the mesh on the model.
e.   Save and close the model.

Importing the STEP File into Autodesk Simulation Mechanical

Now, you need to open Autodesk Simulation Mechanical and import the *c06_tut03.STEP* file into it.

1.   Start Autodesk Simulation Mechanical and then invoke the **Open** dialog box.

2.   Select the **STEP Files (*.stp; *.ste; *.step)** from the **Files of type** drop-down list of the **Open** dialog box.

3.   Browse to the location *C:\Autodesk Simulation Mechanical\c06\Tut03*; the *c06_tut03.STEP* file is displayed in the **Open** dialog box.

4.   Select the **c06_tut03.STEP** file and then choose the **Open** button from the **Open** dialog box; the **Choose Analysis Type** dialog box is displayed with the **Stress with Linear Material Models** analysis type selected by default.

5.   Choose the **OK** button from the **Choose Analysis Type** dialog box; the selected file is imported into the Autodesk Simulation Mechanical, refer to Figure 6-50.

Note
If the orientation of the model displayed on your system is not same as shown in Figure 6-50, then adjust the orientation by using the ViewCube.

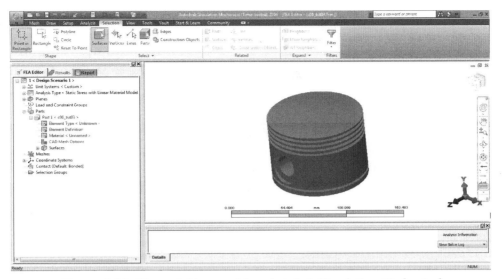

Figure 6-50 *The model imported into Autodesk Simulation Mechanical*

Generating Mesh On the Model

Once the model has been imported in the Autodesk Simulation Mechanical, you need to generate mesh with default mesh settings.

1. Choose the **Generate 3D Mesh** tool from the **Mesh** panel of the **Mesh** tab in the **Ribbon**; the process of meshing starts and the **Meshing Progress** window is displayed. Once the process of generating mesh is completed successfully, the **View Mesh Results** window is displayed. Also, the meshed model is displayed in the background of this window.

2. Choose the **No** button from the **View Mesh Results** window for not viewing the meshing result at this moment; the model with default mesh settings is shown in Figure 6-51.

Refining the Top Curved Face

After generating the mesh with the default mesh settings, you need to further refine the mesh of the top curved face of the model.

1. Select the top curved face of the model, refer to the Figure 6-52.

Note
*To select a surface of a model, you need to activate the surface selection mode. To do so, choose the **Selection** tab of the **Ribbon** to display the tools used for selections. Next, choose the **Point or Rectangle** tool from the **Shape** panel and the **Surfaces** tool from the **Select** panel of the **Selection** tab in the **Ribbon**.*

2. Choose the **Add to Selection** tool from the **Refinement** panel of the **Mesh** tab in the **Ribbon**; the **Creating Surface Refinements** dialog box is displayed, refer to Figure 6-53.

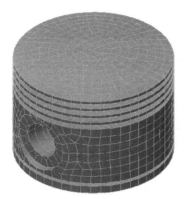

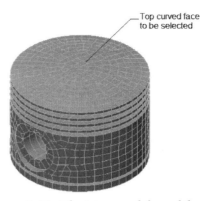

Top curved face
to be selected

Figure 6-51 *The model meshed with default mesh settings*

Figure 6-52 *The top curved face of the model to be selected*

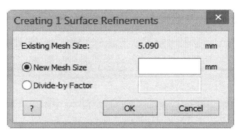

Figure 6-53 *The **Creating Surface Refinements** dialog box*

3. By default, the **New Mesh Size** radio button is selected in the dialog box. As a result, the **New Mesh Size** edit box is enabled. Enter **2** in the **New Mesh Size** edit box of the dialog box. Next, choose the **OK** button.

    After defining the new mesh size for the elements of the top curved surface of the model, you need to regenerate the mesh.

4. Choose the **Generate 3D Mesh** tool from the **Mesh** panel of the **Mesh** tab in the **Ribbon**; the process of meshing starts and the **Meshing Progress** window is displayed. Once the process of generating mesh is completed successfully, the **View Mesh Results** window is displayed. Also, the meshed model is displayed in the background of this window.

5. Choose the **No** button from the **View Mesh Results** window if you do not wish to view the meshing result now; the final meshed model with refined mesh on the top curved surface is displayed, as shown in Figure 6-54.

Saving the Model

1. Choose the **Save** button from the **Quick Access Toolbar**; the FEA model is saved.

2. Choose **Close** from the **Application Menu** to close the assembly.

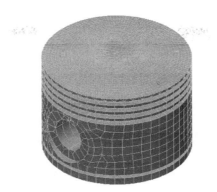

Figure 6-54 *The final meshed model*

Self-Evaluation Test

Answer the following questions and then compare them to those given at the end of this chapter:

1. The _____ tool is used to add refinement points automatically to the regions of the model where the fine mesh is required.

2. On choosing the **Automatic** tool from the **Refinement Points** panel of the **Mesh** tab in the **Ribbon**, the _____ dialog box will be displayed.

3. You can turn on or off the visibility of the refinement points available in the model by using the _____ tool.

4. The _____ button of the **Refinement Point Browser** dialog box is used to add refinement points.

5. In Autodesk Simulation Mechanical, you can edit the parameters of a refinement point either by using the _____ or _____ dialog box.

6. In Autodesk Simulation Mechanical, you can create fine mesh on specific areas of a model by adding refinement points. (T/F)

7. A refinement point specifies the volume of space in a model in which a finer mesh will be generated. (T/F)

8. In Autodesk Simulation Mechanical, you cannot add refinement points by specifying the coordinates of the points where you want to add refinement points to a model. (T/F)

9. You cannot add a refinement point with user defined parameters at a specific point, vertex, or node in the model. (T/F)

10. Once the refinement point is added to the model, you cannot delete it. (T/F)

Review Questions

Answer the following questions:

1. Which of the following dialog boxes is displayed on choosing the **Add to Selected Nodes** tool?

   (a) **Refinement Point**   (b) **Create Refinement Point**
   (c) **Add Refinement Point**   (d) None of these

2. Which of the following dialog boxes is displayed on choosing the **Specify** tool?

   (a) **Refinement Point Browser**   (b) **Add Refinement Point**
   (c) **Create Refinement Point**   (d) None of these

3. Which of the following buttons of the **Refinement Point Browser** dialog box is used to delete or remove all the existing refinement points from the model?

   (a) **Delete**   (b) **Remove**
   (c) **Clear**   (d) None of these

4. Which of the following dialog boxes is displayed when you double-click on a refinement point in the drawing area?

   (a) **Edit Refinement Point**   (b) **Modify Refinement Point**
   (c) **Modify**   (d) None of these

5. The _____ edit box of the **Create Refinement Point** dialog box is used to specify the radius which defines the spherical region around the selected vertex to be refined.

6. To add refinement points by specifying the coordinates, choose the _____ tool from the **Refinement Points** panel of the **Mesh** tab in the **Ribbon**.

7. You can specify the density of refinement points by using the slider bar from coarse to fine in the **Automatic Refinement Points** dialog box. (T/F)

8. The **Automatic** tool will be enabled only if a meshed model is opened in the graphics area of the Autodesk Simulation Mechanical. (T/F)

9. In Autodesk Simulation Mechanical, you cannot edit the parameters of the refinement points after adding them. (T/F)

10. In Autodesk Simulation Mechanical, the procedure of adding refinement points in 2D geometries is same as adding refinement points in 3D models. (T/F)

EXERCISE

Exercise 1

In this exercise, you will open the FEA model *c03_tut02.fea* created in Tutorial 2 of the Chapter 3, refer to Figure 6-55. You will then save this model at the location *C:\Autodesk Simulation Mechanical\c06\Exr01* with the name *c06_exr01* and generate mesh with default mesh settings.

Figure 6-55 *The model for Exercise 1*

After generating the mesh with the default mesh settings, you need to refine the mesh by adding a refinement point at the point whose X, Y, and Z coordinates are -72.6181, 32.1889, and 22.451, respectively. Also, specify the effective radius to 50 mm and the refine mesh size to 1.5 mm for the refinement point.

Figure 6-56 shows the mesh model after refining the mesh by adding refinement points.

(Expected time: 30 min)

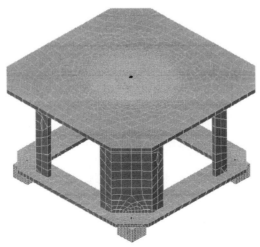

Figure 6-56 *Refined mesh model*

Answers to Self-Evaluation Test

1. Automatic, **2.** Automatic Refinement Points, **3.** Visibility, **4.** Add, **5.** Refinement Point Browser, Modify Refinement Point, **6.** T, **7.** T, **8.** T, **9.** F, **10.** F

Chapter 7

Working with Joints
and Contacts

Learning Objectives

After completing this chapter, you will be able to:
- *Understand the concept of creating joints in Autodesk Simulation Mechanical*
- *Create pin and universal joints*
- *Create bolted connections between contacted parts*
- *Understand the concept of applying contacts*
- *Differentiate between different contact types*
- *Apply contact between parts*
- *Override the existing contact type*
- *Specify friction coefficients for surface/edge contact*
- *Rename applied contacts*

INTRODUCTION

In this chapter, you will learn about joints, bolted fitting connections, and different types of contacts used between the parts. You must already be aware that in some of the real world models, it is necessary to simulate joinery connections for rotational purposes. But creating the actual joinery fitting components for simulating the joints in a model results in spending unnecessary time for analyzing. Also, in most of the cases, the effects on these components are not much significant. Therefore, it is better to avoid modeling such components, and instead create truss elements that will represent the joints in a model. The truss elements do not resist the rotation and allow parts to rotate about the axis of joint. It is also recommended to create truss elements or joints to get a better result in less computational time. Figure 7-1 shows a model with joinery fitting components and Figure 7-2 shows a meshed model with the pin joint connection created instead of actually creating the fitting components.

Figure 7-1 *A model with joinery fitting components*

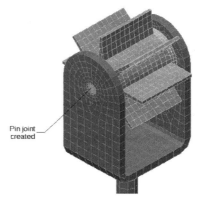

Figure 7-2 *The meshed model with pin joint connection created*

Similarly, you can represent the bolted connection instead of actually creating it in a model in order to avoid wasting unnecessary time for analyzing. Figure 7-3 shows a model with bolted connection and Figure 7-4 shows the same model after creating representation of the bolted connection.

Figure 7-3 *Model with actual bolted connection*

Figure 7-4 *Model with the representation of bolted connection*

CREATING JOINTS

Ribbon: Mesh > CAD Additions > Joint

In Autodesk Simulation Mechanical, while working with 3D CAD assembly, you can create joints between two parts to simulate pinned or universal connections by using the **Joint** tool. To do so, choose the **Joint** tool from the **CAD Additions** panel of the **Mesh** tab in the **Ribbon**; the **Joint Mesh Setup** dialog box will be displayed, as shown in Figure 7-5.

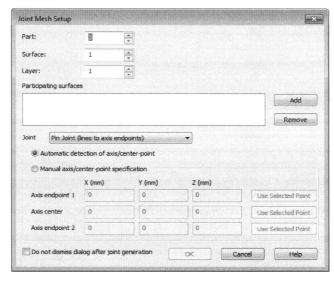

*Figure 7-5 The **Joint Mesh Setup** dialog box*

In the graphics area, select the surfaces of the model that define the joint and form an axis. You can use different selection methods for selecting the surfaces. To select multiple surfaces one by one, click the left mouse button and press the CTRL key. After selecting the surfaces that define the joint, choose the **Add** button from the **Joint Mesh Setup** dialog box; all the selected surfaces will be listed in the **Participating surfaces** selection area of the dialog box.

Note
*You can also select surfaces for defining a joint before invoking the **Joint Mesh Setup** dialog box.*

To remove a surface from the list of the **Participating surfaces** selection area of the **Joint Mesh Setup** dialog box, select it and then choose the **Remove** button from the **Joint Mesh Setup** dialog box.

After defining the surfaces for creating the joint, you need to specify the type of joint you need to create. In Autodesk Simulation Mechanical, you can use the **Joint** drop-down list in the **Joint Mesh Setup** dialog box to create two types of joints, namely pin joint and universal joint. Select the required type of joint from the **Joint** drop-down list. You can also specify the part, surface, and layer numbers for the joint being created in their respective edit boxes. Next, choose the **OK** button; the required joint will be created.

Note

*1. The joint created by using the **Joint Mesh Setup** dialog box will be visible in the meshed model only.*

2. You will learn more about defining part, surface, and layer numbers later in this chapter.

On creating the pin joint, an imaginary axis will be created based on the geometry of the surfaces selected for creating the joint. As a result, the connecting parts will rotate about the imaginary axis created. The imaginary axis will be defined by two opposite points. Figure 7-6 shows a model with pin joint created.

On creating the universal joint, an imaginary point will be created at the volumetric center of the surfaces selected for creating the joint. As a result, the connecting parts will rotate about any axis that passes through the center point. Figure 7-7 shows a model with universal joint created.

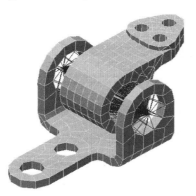

Figure 7-6 A pin joint created between parts

Figure 7-7 A universal joint created between parts

The other options of the **Joint Mesh Setup** dialog box are discussed next.

Automatic detection of axis/center point

The **Automatic detection of axis/center point** radio button in the **Joint Mesh Setup** dialog box is selected by default. As a result, the axis/center point for the pin/universal joint will be automatically detected based on the geometry of the surfaces selected for creating the joint.

Manual axis/center-point specification

The **Manual axis/center-point specification** radio button is used to specify the axis/center point for the joint as per the requirement of design. You can do so by entering their coordinates in the respective edit boxes of the **Joint Mesh Setup** dialog box. These edit boxes will be enabled as soon as you select this radio button.

If the **Pin Joint (lines to axis endpoints)** option is selected in the **Joint** drop-down list of the dialog box for creating the pin joint, then on selecting the **Manual axis/center-point specification** radio button, the **X**, **Y**, and **Z** edit boxes of the **Axis endpoint 1** and **Axis endpoint 2** fields will be enabled in the dialog box. As a result, you can enter the coordinates of two end points which will define the axis of rotation of the pin joint.

If the **Universal Joint (lines to axis midpoints)** option is selected in the **Joint** drop-down list of the dialog box for creating the universal joint, then on selecting the **Manual axis/center-point specification** radio button, the **X**, **Y**, and **Z** edit boxes of the **Axis center** field will be enabled. As a result, you can enter the coordinates of a point which will define the center point to swivel the connected parts of the joint.

CREATING BOLTED CONNECTIONS

Ribbon: Mesh > CAD Additions > Bolt

 As discussed earlier in Autodesk Simulation Mechanical, you can also create representation for bolted connection instead of actually creating it in a model in order to save the computational time required for analysis, refer to Figures 7-3 and 7-4.

To create the representation of the bolted connection in Autodesk Simulation Mechanical, choose the **Bolt** tool from the **CAD Additions** panel of the **Mesh** tab in the **Ribbon**; the **Bolted Mesh Setup** dialog box will be displayed, refer to Figure 7-8.

The options of the **Bolted Mesh Setup** dialog box are discussed next.

Part Number
The **Part Number** edit box is used to specify the part number for the bolted connection to be created.

Bolt diameter
The **Bolt diameter** edit box is used to specify the diameter of the bolt.

Type of Bolt
The **Type of Bolt** drop-down list is used to select the type of bolted connection to be created. You can create bolt with nut, bolt without nut, and grounded bolt connections by using the options available in this drop-down list. Figure 7-9 shows the options available in the **Type of Bolt** drop-down list.

Number of spokes
The **Number of spokes** edit box is used to specify the number of spokes or lines to represent the bolt head and nut.

Bolt head
The **Bolt head** area is used to define the contact surface of the bolted head and the head diameter. To define the contact surface for the head, select the surface from the drawing area and then choose the **Add** button from the **Bolt head** area; the selected surface will be selected as the contact surface for the bolted head. Also, the name of the selected surface will be listed in the **Contact surface(s)** selection area. To remove the selected surface from the **Contact surface(s)** selection area, select the surface to be removed and then choose the **Remove** button.

The **Head diameter** edit box in the **Bolt head** area of the dialog box is used to specify the diameter of the bolted head.

Figure 7-8 The **Bolted Mesh Setup** *dialog box*

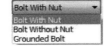

Figure 7-9 Options in the
Type of Bolt drop-down list

Interior hole surface(s) for one hole

The **Interior hole surface(s) for one hole** area is used to select the interior surfaces of the hole to define the axis of hole. To do so, select the surfaces from the drawing area and then choose the **Add** button; all the selected surfaces will be listed in the **Interior hole surface(s) for one hole** selection area. Also the **Tight Fit** check boxes will be displayed beside each of the listed surfaces. By default, these check boxes are cleared. As a result, the bolted connection to be created will have a clearance between the connected surfaces. If the **Tight Fit** check boxes are selected, then the bolted connection will have no clearance between the connected surfaces.

Nut

The **Nut** area is used to define the contact surface for the nut and the nut diameter. To define the contact surface for the nut, select the surface from the drawing area by clicking the left mouse button and then choose the **Add** button of the **Nut** area; the selected surface will be selected as the contact surface for the nut. Also, the name of the selected surface will be listed in the **Contact surface(s)** selection area. To remove the selected surface from the **Contact surface(s)** selection area, select the surface to be removed and then choose the **Remove** button.

The **Nut diameter** edit box of the **Nut** area is used to specify the diameter of the nut.

Preload magnitude

The **Preload magnitude** area is used to specify the magnitude of the axial force acting on the joint. By default, the **Axial Force** radio button of the **Preload magnitude** area is selected. As a result, the **Magnitude** edit box available at front of this radio button gets enabled. You can enter the magnitude of the axial force in this edit box.

On selecting the **Torque** radio button, the **Magnitude** and **Torque coefficient (K)** edit boxes beside this radio button will be enabled. These edit boxes are used to enter the torque magnitude and friction factor. Note that if you have entered the torque magnitude and the friction factor, then the axial force will be calculated based on the formulas.

For calculating the axial force when the bolted connection has a nut, the formula $F = T / K*D$ is used. And, when the bolted connection has no nut, the formula $F = T / 1.2*K*D$ will be used for calculating the axial force.

In the above formulas, T is the torque magnitude, K is the torque coefficient, and D is the bolt diameter.

After specifying all parameters to define the bolted connection, choose the **OK** button from the **Bolted Mesh Setup** dialog box; the bolted connection will be created and the dialog box will be closed. However, if you select the **Do not dismiss dialog after bolt generation** check box and choose the **OK** button, the dialog box will not be closed and you can continue creating the next bolted connection by using the options in the dialog box.

WORKING WITH CONTACTS

In general, a contact is formed between two surfaces or a surface and a point in contact. In a model, load or heat can transfer from an element to the adjacent element only if they are in proper contact with each other. In most of the real world assemblies, you will find that some of the parts have some motion related to their adjacent parts. To analyze such models, you need to have proper contact between the parts to represent the correct conditions or behavior of the model for analysis. In Autodesk Simulation Mechanical, you can define different types of contacts between the parts of the model.

Types of Contacts

The different types of contacts are discussed next.

Bonded Contact

The Bonded contact forces all nodes of the contact faces to be in perfect contact throughout the analysis process. In case of bonded contact, when the nodes of the first contact are deflected, the nodes on the adjoining faces will also get deflected by the same amount and in the same direction. When you import an assembly model, the bonded contact will automatically be assigned to the contact faces of the parts in a model. You will learn more about defining contact between two faces later in this chapter.

Welded Contact

The Welded contact forces the nodes of the contact edges of the faces to be in perfect contact throughout the analysis process and they act in the same way as the bonded contact type. However, the interior surfaces inside the contact edges will be allowed to have free movement.

Free/No Contact

In the Free/No contact, the contact nodes of the faces that are in the contact pair will not merge together, even if the mesh is matched. In this type of contact, as the two contact nodes are not collapsed into one node, there will be no transmission of load between the parts and the parts are free to move independently.

Surface Contact

In the Surface contact, the contact nodes of the faces that are in the contact pair will merge with each other like the bonded contact. The only difference between the Surface contact and the Bonded contact is that the Surface contact allows the contact parts to move with respect to each other. Here the nodes of the contacted surfaces get collapsed into one, therefore when the nodes move with respect to each other, the stiffness will be applied to restrict their motion.

Sliding/No Separation Contact

The Sliding/No Separation contact forces the nodes of the contact edges of the surfaces to be in contact with each other throughout the analysis process. At the same time, this contact type allows the parts to slide with respect to each other without losing the contact.

Separation/No Sliding Contact

The Separation/No Sliding contact allows the contact surfaces to move away from each other without any sliding motion.

Edge Contact

The Edge contact forces the nodes of the contact edges of the surfaces to be in contact with each other by merging the contacted edges nodes together. As the nodes of the contact edges are collapsed to one node, when nodes move toward each other, stiffness will be applied to restrict the motion. Also, in edge contact, the interior surfaces inside the contact edges will have relative motion with each other.

Shrink Fit/Sliding Contact

The Shrink Fit/Sliding contact works same as the Surface contact and forces the nodes of the contact surfaces that have interference fit to match the mesh with each other. At the same time, it allows parts to slide with respect to each other without losing the contact.

Figure 7-10 shows a meshed model before applying the Shrink Fit/Sliding contact. Note that in this figure, due to the interference between the parts, the resultant mesh on both the parts do not match each other. On the other hand, Figure 7-11 shows the same model meshed after applying the Shrink Fit/Sliding contact. In this figure, the resultant mesh of both the parts match each other.

Shrink Fit/No Sliding Contact

The Shrink Fit/No Sliding contact works same as Separation/No Sliding contact and forces the nodes of the contact surfaces having interference fit to match the mesh with each other. At the same time, it restricts the sliding motion between the contacted surfaces.

 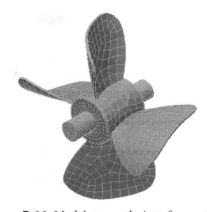

Figure 7-10 *Mesh between the interference parts before applying the Shrink Fit/Sliding contact*   *Figure 7-11* *Mesh between the interference parts after applying the Shrink Fit/Sliding contact*

Applying Contact between Parts

In Autodesk Simulation Mechanical, you can define contacts between the parts before or after meshing. To define a contact, select a pair of parts or surfaces from the drawing area or from the **Tree View** by pressing the CTRL key and then right-click; a shortcut menu will be displayed. Choose **Contact** from the shortcut menu; a cascading menu will be displayed with the list of contacts, refer to Figure 7-12. Select the required contact type from the cascading menu; the

selected contact will be applied between the selected parts or surfaces. Also, the name of the applied contact will be displayed under the heading **Contact (Default:** *name of the default contact***)** in the **Tree View**.

Note that when you import a CAD model in Autodesk Simulation Mechanical, the default contact type will automatically be applied between the contact surfaces of the imported model. This is because the **Automatically generate contact pairs** option is set to **Yes** in the **Global CAD Import Options** dialog box. If the **Automatically generate contact pairs** option is set to **No** then on importing the CAD model, the contacts between the contacted pairs will not be applied automatically.

Figure 7-12 *The cascading menu containing the list of contacts*

To set the **Automatically generate contact pairs** option to **Yes** or **No**, choose the **Application Options** button from the **Options** panel of the **Tools** tab in the **Ribbon**; the **Options** dialog box will be displayed. Next, choose the **CAD Import** tab from the **Options** dialog box; the options related to importing a CAD geometry will be displayed in the dialog box. Next, choose the **Global CAD Import Options** button; the **Global CAD Import Options** dialog box will be displayed, as shown in Figure 7-13. In this dialog box, you can set the **Automatically generate contact pairs** option to either **Yes** or **No** as per the requirement.

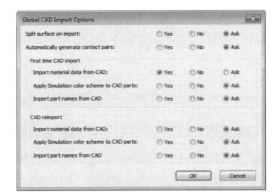

Figure 7-13 The **Global CAD Import Options** *dialog box*

Overriding Existing Contact Type

As discussed earlier, all the applied contacts between the contact surfaces or parts of the model will be listed under the heading **Contact (Default:** *name of the default contact*) in the **Tree View**. Therefore, to override any of the existing contact type with the new one, select the required contact to be overridden from the **Tree View** and then right-click; a shortcut menu will be displayed, refer to Figure 7-14. Now you can select the required contact type from the shortcut menu displayed. The existing contact will be overridden with the newly selected contact type.

Specifying Friction Coefficients, Direction, and Interference for Contact

After applying the contact between the contacting surfaces of a model, you can further modify their default settings such as friction coefficients between the contact pairs, direction of contact pairs, and interference between the contact pairs. The procedure to modify these settings is discussed next.

Setting Friction Coefficients for Contact

As discussed earlier in the Surface Contact, Edge Contact, and Shrink Fit contact, the contacted nodes of the contact surfaces/edges collapse into one node and when nodes move towards each other, stiffness will be applied to restrict the motion. Therefore, when Surface Contact, Edge contact, or Shrink fit contact is applied between a contact pair, then you may need to specify friction. To specify friction, expand the **Contact (Default:** *name of the default contact*) heading in the **Tree View**; the list of all the applied contacts will be displayed. Select the Surface Contact, Edge Contact, or Shrink Fit contact from the list of contacts displayed in the **Tree View** and right-click; a shortcut menu will be displayed, refer to Figure 7-15.

Choose the **Settings** options from the shortcut menu displayed; the **Contact Options** dialog box will be displayed, refer to Figure 7-16.

Note

If you invoke a shortcut menu by selecting a Surface Contact from the **Tree View***, then on choosing the* **Settings** *option from it, the* **Contact Options** *dialog box will be displayed, as shown in Figure 7-16. If you invoke the shortcut menu by selecting an Edge contact from the* **Tree View***, then after choosing the* **Settings** *option, the* **Contact Options** *dialog box will be displayed, as shown* *in Figure 7-17. On the other hand, if you invoke the shortcut menu by selecting the Shrink Fit/Sliding or Shrink Fit/No Sliding contact, then after choosing the **Settings** option, the **Contact Options** dialog box will be displayed, as shown in Figure 7-18.*

Figure 7-14 *A shortcut menu containing the list of contacts*

Figure 7-15 *A shortcut menu displayed after right clicking on a Surface Contact in the **Tree View***

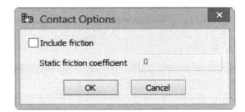

Figure 7-16 *The **Contact Options** dialog box displayed in case of the Surface Contact type*

Figure 7-17 *The **Contact Options** dialog box displayed in case of the Edge Contact type*

Figure 7-18 *The **Contact Options** dialog box displayed in case of Shrink Fit contact type*

The **Include friction** check box of the **Contact Options** dialog box is used to include the friction coefficient between the contact pairs. By default, this check box is cleared. As a result, there is no friction specified between the contact. On selecting the **Include friction** check box, the **Static friction coefficient** edit box will be enabled in the **Contact Options** dialog box. The **Static friction coefficient** edit box is used to specify the static friction coefficient for the selected contact. Note that greater the value entered in this edit box, larger is the force required to create movement between the contacted faces.

Specifying Surface Contact Direction

When the contact is defined as Surface contact then the surface contact direction for each contacted element in the pair will be calculated by using the option selected in the **Surface contact direction** drop-down list. By default, the **Calculate by matching directions** option is selected in the **Surface contact direction** drop-down list of the **Contact Options** dialog box. As a result, the surface contact direction will be calculated by taking the average of the normal direction of individual element of the contacted faces. This is the recommended method for calculating the contact direction. However, if some of the elements of the contacted faces are non flat then the default method for calculating contact direction will not be applicable. This is because, the normal direction of the elements of both the contacted faces may not be match each other. In such a case, you may need to select the other option for calculating the contact direction from the **Surface contact direction** drop-down list.

If one of the contacted faces contains flat elements and other contains non flat elements, then you can select either the **Normal to the first part/surface** or **Normal to the second part/surface** option from the **Surface contact direction** drop-down list for calculating the contact direction. You can also specify the tolerance for the direction angle in the **Direction tolerance angle** edit box of the **Contact Options** dialog box.

Note

The options for specifying the surface contact direction and tolerance will not be available for Edge Contact.

Specifying Interference for Shrink Fit Contact Types

In the Shrink Fit/Sliding and Shrink Fit/No Sliding contacts, the nodes of the interference surfaces collapse into one node and the mesh automatically gets matched. This is because the **Automatic** check box is selected by default in the **Contact Options** dialog box. Note that when the **Automatic** check box is selected, the interference between the geometry will be automatically calculated by analyzing the effect of various intensities of fit due to the interference.

On clearing the **Automatic** check box, the **Interference** edit box will be enabled in the **Contact Options** dialog box. In this edit box, you can manually enter the value of interference between the geometries. This method is most widely used when the 3D geometry does not have any interference but you want to analyze the effect of various intensities of fit due to the interference in actual part.

Note

The options for specifying the interference between the contacted parts will be available only for Shrink Fit/Sliding and Shrink Fit/No Sliding contacts.

Renaming Contacts

By default, the names of the applied contacts are dependent upon the type of contact selected and the names of the contacted faces. For example, **Bonded < 3 with 4 >** is the default name of a contact where **Bonded** is the type of contact applied, and **3** and **4** are the default names of the contacted faces of the contact.

In Autodesk Simulation Mechanical, you can rename the default nomenclature of the contact faces between which the contact is applied as per your requirement. To do so, select the name of the contact to be renamed from the **Contact (Default:** *name of the default contact*) node of the **Tree View** and right-click; a shortcut menu will be displayed, refer to Figure 7-19. Now, choose the **Rename** option; the **Rename** window will be displayed, as shown in Figure 7-20.

Figure 7-19 *A shortcut menu displayed*   *Figure 7-20* *The **Rename** window*

In the **Description** edit box of the **Rename** window, you can enter a new name or description for the contact faces of the selected contact and then choose the **OK** button; the name of the selected contact will be renamed and displayed under the heading **Contact (Default:** *name of the default contact*) in the **Tree View**.

Deleting Contacts

In Autodesk Simulation Mechanical, you can delete the applied contacts of a model. To do so, select the contact to be deleted from the **Contact (Default:** *name of the default contact*) node of the **Tree View** and right-click to display the shortcut menu. Next, choose the **Delete** option from the shortcut menu; the selected contact will be deleted from the model.

TUTORIALS

To perform the tutorials, download the input files of this chapter from *www.cadcim.com*. The complete path for downloading the file is as follows:

Textbooks > CAE Simulation > Autodesk Simulation Mechanical > Autodesk Simulation Mechanical 2016 for Designers > Input Files > c07_simulation_2016_input.zip

Extract the downloaded *c07_simulation_2016_input* zipped file and save it at the location *C:\Autodesk Simulation Mechanical\c07*.

Tutorial 1

In this tutorial, you will import the model shown in Figure 7-21 into Autodesk Simulation Mechanical. You can select the default **Static Stress with Linear Material Models** analysis type. After importing the model, generate fine mesh with the mesh settings given next.

Mesh type: Solid
Mesh size: 2 mm
Solid mesh element type: Bricks and tetrahedra

After generating the mesh with the above mesh settings, you need to add pin joint.
                                                                     (Expected time: 30 min)

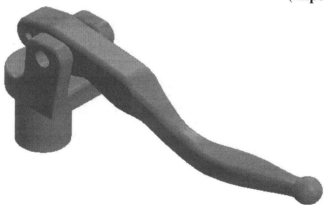

Figure 7-21 The model for Tutorial 1

The following steps are required to complete this tutorial:

a.   Start Autodesk Simulation Mechanical.
b.   Import the input file of this tutorial into Autodesk Simulation Mechanical.
c.   Generate mesh on the model.
d.   Add a pin joint.
e.   Save and close the model.

Importing the Model

You need to open Autodesk Simulation Mechanical and import the *c07_tut01.STEP* file to it.

1. Start Autodesk Simulation Mechanical and then invoke the **Open** dialog box.

2. Select **STEP Files (*.stp; *.ste; *.step)** from the **Files of type** drop-down list of the **Open** dialog box.

3. Browse to the location *C:\Autodesk Simulation Mechanical\c07\Tut01*; the *c07_tut01.STEP* file is displayed in the **Open** dialog box.

4. Select the **c07_tut01.STEP** file and then choose the **Open** button from the **Open** dialog box; the **Choose Analysis Type** dialog box is displayed with the **Static Stress with Linear Material Models** analysis type selected by default.

5. Choose the **OK** button from the **Choose Analysis Type** dialog box; the selected file is opened in Autodesk Simulation Mechanical, refer to Figure 7-22.

Note
1. If the orientation of the model displayed on your system is not the same as the one shown in Figure 7-22, you can change it by using the ViewCube.

2. Sometimes when you open a file, one or more message windows may be displayed. Choose the **Yes** *button from these windows.*

Figure 7-22 The model imported into Autodesk Simulation Mechanical

Generating Mesh on the Model

Once the model is imported into Autodesk Simulation Mechanical, you need to specify the mesh settings and generate mesh on it.

1. Invoke the **Model Mesh Settings** dialog box and specify the mesh setting as follows.

   Mesh type: Solid
   Mesh size: 2 mm
   Solid mesh element type: Bricks and tetrahedra

2. After specifying the mesh settings, generate the mesh. The model after generating the mesh is shown in Figure 7-23.

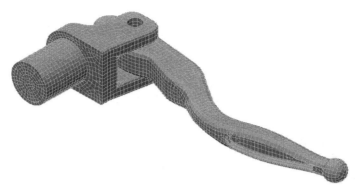

Figure 7-23 *The model displayed after generating the mesh*

Adding the Pin Joint

Now you need to add the pin joint between the large holes of two parts so that the parts can rotate about the center of the holes.

1. Choose the **Mesh** tab in the **Ribbon** and then choose the **Joint** tool from the **CAD Additions** panel of the **Mesh** tab; the **Joint Mesh Setup** dialog box is displayed, as shown in Figure 7-24.

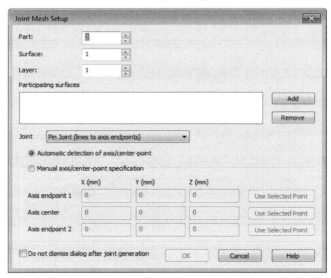

Figure 7-24 *The **Joint Mesh Setup** dialog box*

Now, you need to select the surfaces of the model that will define the axis of the hole.

2. Change the view of the model by using the ViewCube such that you can view the hole from its front side by using the ViewCube, refer to Figure 7-25.

3. Choose the **Selection** tab in the **Ribbon**; the selection tools are displayed.

4. Choose the **Circle** tool from the **Shape** panel and then choose the **Surfaces** tool from the **Select** panel of the **Selection** tab in the **Ribbon**.

Figure 7-25 The model displayed after changing its view

5. Draw a circle around the large hole, refer to Figure 7-26. All the surfaces that are completely enclosed inside the drawn circle are selected in the drawing area, refer to Figure 7-27.

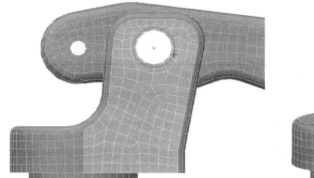

Figure 7-26 A circle being drawn around the large hole of the model

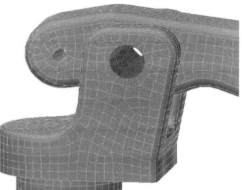

Figure 7-27 Surfaces completely enclosed inside the circle are selected

6. Choose the **Add** button from the **Joint Mesh Setup** dialog box; all the selected surfaces are listed in the **Participating surfaces** selection area of the dialog box.

7. Make sure that the **Pin Joint (lines to axis endpoints)** option is selected in the **Joint** drop-down list and the **Automatic detection of axis/center-point** radio button is selected in the **Joint Mesh Setup** dialog box. Next, choose the **OK** button; the pin joint is created, as shown in Figure 7-28. Figure 7-29 shows the complete model after adding the pin joint.

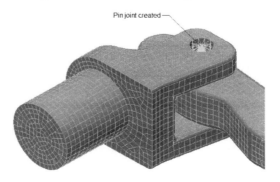

Pin joint created

Figure 7-28 Model after adding pin joint

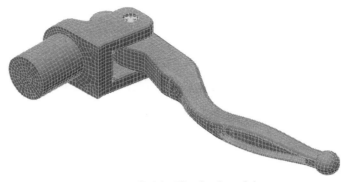

Figure 7-29 *The final model*

Saving and Closing the Model

1. Choose the **Save** button from the **Quick Access Toolbar** to save the model.

2. Choose **Close** from the **Application Menu** to close the file.

Tutorial 2

In this tutorial, you will import the model shown in Figure 7-30 into Autodesk Simulation Mechanical. You can select the default **Static Stress with Linear Material Models** analysis type. After importing the model, generate mesh with the following mesh settings:

Mesh type: Solid
Mesh size: 6 mm
Solid mesh element type: Bricks and tetrahedra

After generating the mesh with the settings mentioned above, you need to add the pin joints between the contacting pairs 1 through 6 of the assembly, refer to Figure 7-31. Also, apply the surface contact between the contacting pairs 1 through 7 of the model, refer to Figure 7-31.

(Expected time: 30 min)

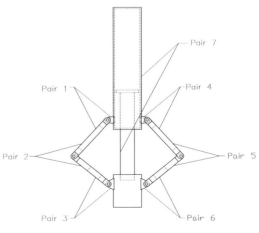

Figure 7-30 *The model for Tutorial 2*   *Figure 7-31* *Contacting pairs of the assembly*

The following steps are required to complete this tutorial:

a. Start Autodesk Simulation Mechanical.
b. Import the input file of this tutorial into Autodesk Simulation Mechanical.
c. Generate mesh on the model.
d. Add pin joints.
e. Apply surface contacts.
f. Save and close the model.

Importing the Model

Now, you need to open Autodesk Simulation Mechanical and import the *c07_tut02.STEP* file to it.

1. Start Autodesk Simulation Mechanical and then invoke the **Open** dialog box.

2. Select **STEP Files (*.stp; *.ste; *.step)** from the **Files of type** drop-down list in the **Open** dialog box.

3. Browse to the location *C:\Autodesk Simulation Mechanical\c07\Tut02*; the *c07_tut02.STEP* file is displayed in the **Open** dialog box.

4. Select the **c07_tut02.STEP** file and then choose the **Open** button from the **Open** dialog box; the **Choose Analysis Type** dialog box is displayed with the **Static Stress with Linear Material Models** analysis type selected by default.

5. Choose the **OK** button from the **Choose Analysis Type** dialog box; the selected file is opened in Autodesk Simulation Mechanical, refer to Figure 7-32.

Note
1. If the orientation of the model displayed on your system is not the same as the one shown in Figure 7-32, you can change it by using the ViewCube.

*2. Sometimes when you open a file, one or more message windows may be displayed. Choose the **Yes** button from these windows and proceed.*

Generating Mesh on the Model

Once the model has been imported into the Autodesk Simulation Mechanical, you need to specify mesh settings and generate mesh.

1. Invoke the **Model Mesh Settings** dialog box and specify the mesh setting as follows.

   Mesh type: Solid
   Mesh size: 6 mm
   Solid mesh element type: Bricks and tetrahedra

2. After specifying the mesh settings, generate the mesh. The model after generating the mesh is shown in Figure 7-33.

Figure 7-32 The model imported into
Autodesk Simulation Mechanical

Figure 7-33 The model after generating
the mesh

Adding Pin Joints

Now, you need to add pin joints between the contacting pairs.

1. Choose the **Mesh** tab in the **Ribbon** and then choose the **Joint** tool from the **CAD Additions** panel of the this tab; the **Joint Mesh Setup** dialog box is displayed, as shown in Figure 7-34.

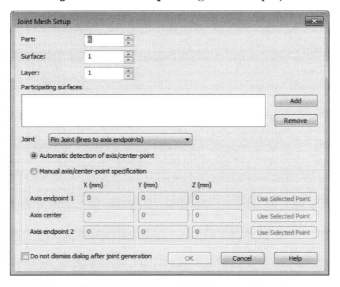

Figure 7-34 The **Joint Mesh Setup** dialog box

Now you need to select the surfaces of the model that will define the axis of hole.

2. Change the view of the model such that it appears as shown in Figure 7-35.

3. Choose the **Selection** tab in the **Ribbon**; the selection tools are displayed in the **Ribbon**.

4. Choose the **Circle** tool from the **Shape** panel and then choose the **Surfaces** tool from the **Select** panel of the **Selection** tab in the **Ribbon**.

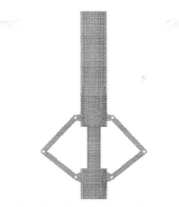

Figure 7-35 *The model after changing its orientation for adding pin joints*

5. Draw a circle around the hole of the contact pair 1, refer to Figures 7-31 and 7-36. All the surfaces that are completely enclosed inside the circle are selected in the drawing area, refer to Figure 7-37.

6. Choose the **Add** button from the **Joint Mesh Setup** dialog box; all the selected surfaces are listed in the **Participating surfaces** selection area of the dialog box.

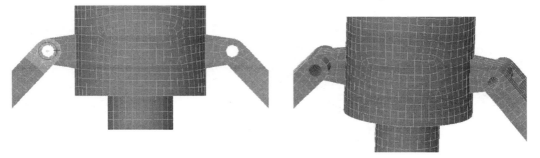

Figure 7-36 *A circle being drawn around the large hole of the model*

Figure 7-37 *Surfaces completely enclosed inside the circle drawn are selected*

7. Make sure that the **Pin Joint (lines to axis endpoints)** option is selected in the **Joint** drop-down list and the **Automatic detection of axis/center-point** radio button is selected in the **Joint Mesh Setup** dialog box. Next, choose the **OK** button; the pin joint is created, as shown in Figure 7-38.

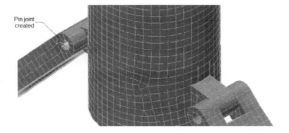

Figure 7-38 *The pin joint created*

8. Similarly, create pin joints between other contacting pairs of the model. Refer to Figure 7-31 for contact pairs number. The partial view of the model after adding all pin joints is shown in Figure 7-39.

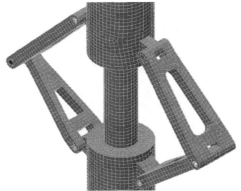

Figure 7-39 *Partial view of the model after adding all pin joints*

Applying Surface Contacts

Now, you need to apply surface contact between the contacting parts.

1. Select the contacting parts of the pair 1, refer to Figure 7-31, from the **Tree View** by using the CTRL key. Next, right-click; a shortcut menu is displayed.

Note
*To select parts, you need to activate the part selection model. To do so, choose the **Selection** tab of the **Ribbon** to display various tools used for selection. Next, choose the **Point or Rectangle** tool from the **Shape** panel and the **Parts** tool from the **Select** panel.*

2. Choose **Contact > Surface Contact** from the shortcut menu and then click anywhere in the drawing area; the surface contact is applied between the selected parts and its name is displayed in the **Tree View** under the heading **Contact (Default: Bonded)**.

3. Similarly, apply the surface contact between the contacting parts of all other pairs, refer to Figure 7-31. The final model after meshing, adding pin joints, and applying surface contacts is shown in Figure 7-40.

Saving and Closing the Model

1. Choose the **Save** button from the **Quick Access Toolbar** to save the model.

2. Choose **Close** from the **Application Menu** to close the file.

Figure 7-40 *The final model*

Tutorial 3

In this tutorial, you will import the model shown in Figure 7-41 into the Autodesk Simulation Mechanical. You can select the default **Static Stress with Linear Material Models** analysis type. After importing the model, generate mesh with the following mesh settings:

Mesh type: Solid
Mesh size: 45 %
Solid mesh element type: Bricks and tetrahedra

After generating the mesh with the above mesh settings, apply Free/No contact between the contacting parts of the model. Also, you will add the Bolt connection at the holes to hold the parts together. **(Expected time: 30 min)**

Figure 7-41 The model for Tutorial 3

The following steps are required to complete this tutorial:

a. Start Autodesk Simulation Mechanical.
b. Import the input file of this tutorial into Autodesk Simulation Mechanical.
c. Generate mesh on the model.
d. Add bolt connections.
e. Apply Free/No contact.
f. Save and close the model.

Importing the Model

Now, you need to open Autodesk Simulation Mechanical and import the *c07_tut03.STEP* file into it.

1. Start Autodesk Simulation Mechanical and then invoke the **Open** dialog box.

2. Select **STEP Files (*.stp; *.ste; *.step)** from the **Files of type** drop-down list in the **Open** dialog box.

3. Browse to the location *C:\Autodesk Simulation Mechanical\c07\Tut03*; the *c07_tut03.STEP* file is displayed in the **Open** dialog box.

4. Select the **c07_tut03.STEP** file and then choose the **Open** button from the **Open** dialog box; the **Choose Analysis Type** dialog box is displayed with the **Static Stress with Linear Material Models** analysis type selected by default.

5. Choose the **OK** button from the **Choose Analysis Type** dialog box; the selected file is opened in Autodesk Simulation Mechanical, refer to Figure 7-42.

Note
If the orientation of the model displayed on your system is not the same as shown in Figure 7-42, you can change it by using the ViewCube.

Generating the Mesh on the Model

Once the model has been imported into the Autodesk Simulation Mechanical, you need to specify mesh settings and generate mesh on it

1. Invoke the **Model Mesh Settings** dialog box and specify the mesh setting as follows.

   Mesh type: Solid
   Mesh size: 45 %
   Solid mesh element type: Bricks and tetrahedra

2. After specifying the mesh settings, generate the mesh. The model after generating the mesh is shown in Figure 7-43.

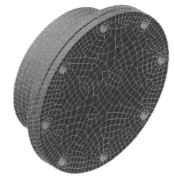

Figure 7-42 The model imported into Autodesk Simulation Mechanical

Figure 7-43 The model after generating the mesh

Adding Bolt Connections

Now, you need to add bolt connections between the contacting pairs.

1. Choose the **Mesh** tab in the **Ribbon** and then choose the **Bolt** tool from the **CAD Additions** panel of the **Mesh** tab; the **Bolted Mesh Setup** dialog box is displayed, as shown in Figure 7-44.

*Figure 7-44 The **Bolted Mesh Setup** dialog box*

Now, you need to specify the part number, bolt diameter, type of bolt connection, number of spokes, bolt head connected surface, bolt head diameter, nut connected surface, nut diameter, interior hole surface, and preload magnitude for adding the bolted connections.

2. Set the value in the **Part Number** spinner to **4** and select the **Bolt With Nut** option from the **Type of Bolt** drop-down list of the dialog box, if not already selected.

3. Enter **2.75** inch in the **Bolt diameter** edit box of the dialog box. Next, make sure that the value in the **Number of spokes** edit box is set to **12**.

Now, you need to select the surface that is connected with the bolt head surface.

4. Choose the **Selection** tab in the **Ribbon** and then choose the **Point or Rectangle** tool from the **Shape** panel. Next, choose the **Surfaces** tool from the **Select** panel of the **Selection** tab in the **Ribbon**.

5. Select the front planar surface of the model from the graphics area as the connecting surface for bolted head, refer to Figure 7-45.

6. Choose the **Add** button from the **Bolt head** area of the **Bolted Mesh Setup** dialog box; the name of the selected surface is displayed in the **Contact surface(s)** selection area.

7. Enter **3.5** inch in the **Head diameter** edit box of the **Bolt head** area.

   Now you need to select the surface that is connected with the nut base surface.

8. Change the orientation of the model such that you can view the model from back side and then select the planar surface of the model, refer to Figure 7-46.

Figure 7-45 *The surface to be selected as the connecting surface for bolt head*

Figure 7-46 *The surface to be selected as the connecting surface for nut*

9. Choose the **Add** button from the **Nut** area of the **Bolted Mesh Setup** dialog box; the name of the selected surface is displayed in the **Contact surface(s)** selection area.

10. Enter **3.5** inch in the **Nut diameter** edit box of the **Nut** area.

    Now you need to select inner hole surfaces to define the hole axis.

11. Change the orientation of the model such that the front face of the model becomes normal to the viewing direction.

12. Choose the **Selection** tab in the **Ribbon** and then choose the **Circle** tool from the **Shape** panel. Next, choose the **Surfaces** tool from the **Select** panel of the **Selection** tab in the **Ribbon**.

13. Draw a circle around a hole to select the interior surfaces, refer to Figure 7-47. All the surfaces that are completely enclosed inside the circle are selected in the drawing area, refer to Figure 7-48.

14. Choose the **Add** button from the **Interior hole surface(s) for one hole** area of the **Bolted Mesh Setup** dialog box; the names of the selected surfaces are displayed in the selection area.

15. Select the **Do not dismiss dialog after bolt generation** check box.

16. Choose the **OK** button; the bolted connection is created and the **Bolted Mesh Setup** dialog box is still displayed in the graphics area. The model after creating the bolted connection on a hole is shown in Figure 7-49.

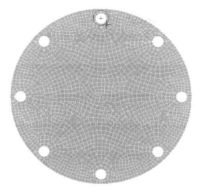

Figure 7-47 *A circle being drawn around a hole of the model*

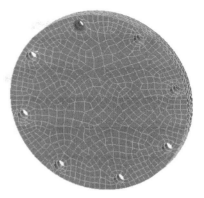

Figure 7-48 *Surfaces selected*

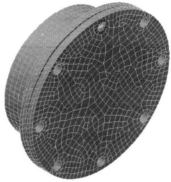

Figure 7-49 *The bolted connection created*

Now you need to add the other bolted connections. As the **Bolted Mesh Setup** dialog box is still invoked and the parameters of the bolt created are still mentioned, you only need to define the interior surfaces for creating other bolts.

17. Change the orientation of the model such that the front face of the model becomes normal to the viewing direction.

18. Draw a circle around the + hole to select the interior surfaces, refer to Figure 7-50. All the surfaces that are completely enclosed inside the circle are selected in the drawing area.

19. Choose the **Add** button from the **Interior hole surface(s) for one hole** area of the **Bolted Mesh Setup** dialog box; the names of the selected surfaces are displayed in the selection area.

20. Choose the **OK** button from the dialog box; the bolted connection is created and the **Bolted Mesh Setup** dialog box is still displayed in the graphics area. The model after creating the bolt connection on other hole is shown in Figure 7-51.

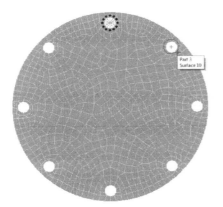 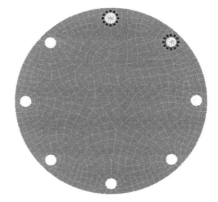

Figure 7-50 *A circle being drawn around a hole of the model*   *Figure 7-51* *The bolt connection created*

21. Similarly, create other bolted connections. The model after creating all the bolts is shown in Figure 7-52. Next, exit the dialog box.

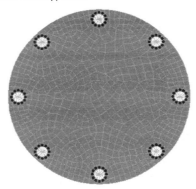

Figure 7-52 *The model after adding all bolts*

Applying Free/No Contact between Contacting Parts

Now you need to apply Free/No contact between the contacting parts of pairs.

1. Select **Part 1** and **Part 2** from the **Tree View** by pressing the CTRL key and then right-click; a shortcut menu is displayed.

2. Select **Contact > Free/No Contact** from the shortcut menu and then click anywhere in the drawing area; the **Surface Contact** is applied between the selected parts and its name is displayed in the **Tree View** under the heading **Contact (Default: Bonded)**.

3. Similarly, apply Free/No contact between Part 2 and Part 3. The final model after meshing, adding bolt connection, and Free/No contact is shown in Figure 7-53.

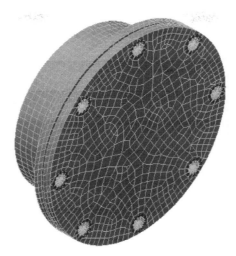

Figure 7-53 *The final model*

Saving and Closing the Model

1. Choose the **Save** button from the **Quick Access Toolbar** to save the model.

2. Choose **Close** from the **Application Menu** to close the file.

Self-Evaluation Test

Answer the following questions and then compare them to those given at the end of this chapter:

1. On choosing the **Bolt** tool from the **CAD Additions** panel of the **Mesh** tab in the **Ribbon**, the _____ dialog box will be displayed.

2. The _____ edit box of the **Bolted Mesh Setup** dialog box is used to specify the diameter of the bolt.

3. You can create _____ , _____ , and _____ types of bolted connections by using the options available in the **Type of Bolt** drop-down list of the **Bolted Mesh Setup** dialog box.

4. If the _____ radio button of the **Joint Mesh Setup** dialog box is selected, the center point for the universal joint will be automatically detected based on the geometry of the surfaces selected for creating the joint.

5. The _____ contact type forces the nodes of the contact edges of the faces to be in perfect contact throughout the analysis process. However, the interior surfaces inside the contact edges can move with respect to each other.

6. In Autodesk Simulation Mechanical, you cannot create joints between two parts to simulate connections. (T/F)

7. In general, a contact is formed when two surfaces or a surface and a point meet together. (T/F)

8. The Bonded contact forces all the nodes of the contacting faces of parts to be in perfect contact throughout the analysis process in a model. (T/F)

9. In Autodesk Simulation Mechanical, you cannot rename the default nomenclature of the applied contact. (T/F)

10. On creating the pin joint, an imaginary axis will be created based on the geometry of the surfaces selected for creating the joint. (T/F)

Review Questions

Answer the following questions:

1. Which of the following dialog boxes is displayed on choosing the **Joint** tool?

   (a) **Pin/Universal Joint**   (b) **Joint Mesh Setup**
   (c) **Joint**   (d) None of these

2. Which of the following dialog boxes is displayed on choosing the **Global CAD Import Options** button from the **CAD Import** tab of the **Options** dialog box?

   (a) **Global CAD Import Options**   (b) **Options**
   (c) **CAD Import Options**   (d) None of these

3. Which of the following buttons of the **Joint Mesh Setup** dialog box is used to remove the selected surface from the **Participating surfaces** selection area of the dialog box?

   (a) **Delete**   (b) **Remove**
   (c) **Clear**   (d) None of these

4. Which of the following dialog boxes is displayed after choosing the **Settings** option from the shortcut menu displayed after right-clicking on the Surface Contact?

   (a) **Settings**   (b) **Settings Options**
   (c) **Contact Options**   (d) None of these

5. Which of the following contact types is used to force the nodes of the contacting surfaces having interference fit to match the mesh with each other and allows the contacting parts to slide with respect to each other without losing the contact?

   (a) **Shrink Fit/Sliding**   (b) **Shrink Fit/No Sliding**
   (c) **Fit/Sliding**   (d) None of these

6. The _____ edit box of the **Nut** area in the **Bolted Mesh Setup** dialog box is used to specify the diameter of the nut.

7. The _____ area of the **Bolted Mesh Setup** dialog box is used to specify the magnitude of the axial force acting on the joint.

8. In Autodesk Simulation Mechanical, you cannot override any of the existing contact types. (T/F)

9. On creating the universal joint, an imaginary point is created at the volumetric center of the surfaces selected for creating the joint. (T/F)

10. In Autodesk Simulation Mechanical, you cannot delete applied contacts for the model. (T/F)

EXERCISES

The input files used in the Exercises are also available in their respective chapter folders where the input files of the Tutorials are available. You can download the input files of the Exercises used in this chapter from *www.cadcim.com*, if not already downloaded. The complete path for downloading the file is given below:

Textbooks > CAE Simulation > Autodesk Simulation Mechanical > Autodesk Simulation Mechanical 2016 for Designers > Input Files > c07_simulation_2016_input.zip

Exercise 1

In this exercise, you will import the model *c07_exr01*, shown in Figure 7-54 into the Autodesk Simulation Mechanical. You can select the default **Static Stress with Linear Material Models** analysis type. After importing the model, generate fine mesh with the default mesh settings.

After generating the mesh with default mesh settings, apply the Surface contact between the contacting parts of the model. Also, add the pin joint connection to hold the parts together.

(Expected time: 30 min)

Figure 7-54 The model for Exercise 1

Exercise 2

In this exercise, you will import the model *c07_exr02*, shown in Figure 7-55, into the Autodesk Simulation Mechanical. You can select the default **Static Stress with Linear Material Models** analysis type. After importing the model, generate fine mesh with default mesh settings.

After generating the mesh with default mesh settings, apply Surface contact between the contacting parts of the model. Also, add bolt connection at the holes to hold the parts together. Figure 7-56 shows the model with bolted connections for your reference.

(**Expected time: 30 min**)

Figure 7-55 *The model for Exercise 2*   *Figure 7-56* *The model with bolted connections*

Answers to Self-Evaluation Test
1. Bolted Mesh Setup, **2. Bolt diameter**, **3.** Bolt with Nut, Bolt without Nut, and Grounded Bolt, **4. Manual axis/center-point specification**, **5.** Edge **6.** F, **7.** T, **8.** T, **9.** F, **10.** T

Chapter 8

Defining Materials and Boundary Conditions

After completing this chapter, you will be able to:
- *Define materials and boundary conditions in FEA models*
- *Assign material to FEA models*
- *Manage material libraries in Autodesk Simulation Mechanical*
- *Apply boundary conditions*
- *Understand different types of constraints*
- *Apply constraints to an FEA model*
- *Understand different types of loads*
- *Define loading conditions for an FEA Model*

INTRODUCTION

Defining material and boundary conditions of an FEA model is one of the most important steps in an analysis process. In Autodesk Simulation Mechanical, you are provided with Autodesk Simulation Material Library and Autodesk Simulation Plastics Library that contain almost all the standard materials. You can select the required material from these libraries and assign it to the selected part of the model. You can also edit some of the properties of the standard materials available in the libraries such as mass density, modulus of elasticity, poisson's ratio, thermal coefficient of expansion, and shear modulus of elasticity as per the project requirement. Additionally, in Autodesk Simulation Mechanical, you can also create a new library with user-defined materials.

ASSIGNING MATERIAL

As discussed earlier, material properties must be assigned for each part in the model on which the analysis process is to be carried out. When you assign a material to a part, all physical properties of the selected material will be assigned to the part. To assign a material to a part of the model, right-click on the **Material <Unnamed>** option available under the **Part** node in the **Tree View** and then choose the **Edit Material** option from the shortcut menu displayed, refer to Figure 8-1; the **Element Material Selection** dialog box will be displayed, as shown in Figure 8-2.

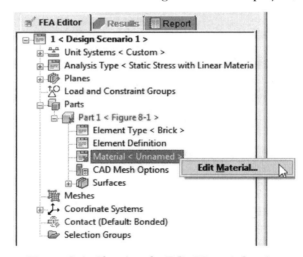

Figure 8-1 *Choosing the* ***Edit Material*** *option*

Note

The ***Edit Material*** *option is enabled in the shortcut menu only when the part is meshed.*

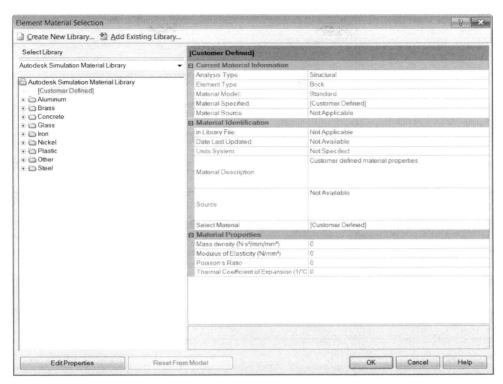

*Figure 8-2 The **Element Material Selection** dialog box*

By default, **Autodesk Simulation Material Library** is selected in the **Select Library** drop-down list of the **Element Material Selection** dialog box. As a result, a number of material families that are available in this library are listed in the left area of the dialog box. Click on the +sign located on the left of the desired material family to display all materials under that family. Figure 8-3 shows the **Element Material Selection** dialog box with the **Aluminum** material family expanded. Select the required material from the expanded material family from the left area of the dialog box; all the properties of the selected material will be displayed on the right of the dialog box. Next, choose the **OK** button; the material properties of the selected material will be assigned to the selected part.

In Autodesk Simulation Mechanical, you can also change or modify the material properties of the selected material. To do so, select the required material from the left of the **Element Material Selection** dialog box. Next, choose the **Edit Properties** button available on the bottom of the dialog box; the **Element Material Specification** dialog box will be displayed, refer to Figure 8-4. In this dialog box, you can specify new mass density, modulus of elasticity, poisson's ratio, and thermal coefficient of expansion for the selected material. After specifying the new material properties for the selected material in the dialog box, choose the **OK** button; the material properties of the selected material will be modified accordingly.

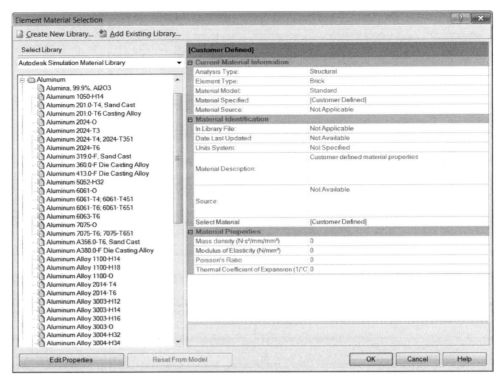

Figure 8-3 *The expanded* **Aluminum** *category in the* **Element Material Selection** *dialog box*

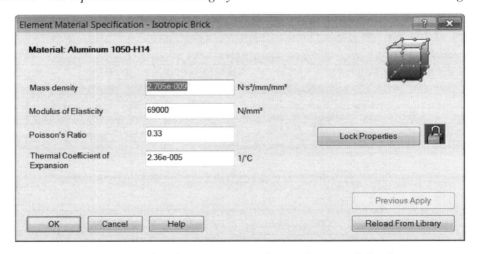

Figure 8-4 *The* **Element Material Specification** *dialog box*

Note
The changes made in the material properties of the selected material by using the **Element Material Specification** *dialog box will be saved in the model database only. In other words, there will be no modification made in the material library of Autodesk Simulation Mechanical. As a result, when you invoke the same material next time, its default or standard properties will be displayed.*

MANAGING MATERIAL LIBRARIES

Autodesk Simulation Mechanical allows you to manage material libraries by creating new material libraries, material categories, user-defined materials, setting selected library as default library, and so on. The procedure to perform all these operations for managing material libraries is discussed next.

Creating a New Material Library

By default, in Autodesk Simulation Mechanical, you are provided with **Autodesk Simulation Material Library** and **Autodesk Simulation Plastics Library**. In addition to the default material libraries in Autodesk Simulation Mechanical, you can also create new material libraries. To do so, choose the **Tools** tab in the **Ribbon** to display all the tools available in this tab. Next, choose the **Manage Material Library** tool from the **Options** panel of the **Tools** tab; the **Autodesk Material Library Manager** dialog box will be displayed, as shown in Figure 8-5.

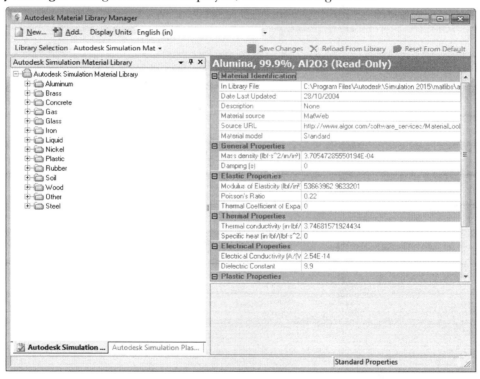

Figure 8-5 The Autodesk Material Library Manager dialog box

Choose the **New** button from the dialog box; the **Create Material Library** dialog box will be displayed. Browse to the location where you want to save the material library file and then specify a name for it in the **File name** edit box of the dialog box. Next, choose the **Save** button; the **Create Library** window will be displayed, refer to Figure 8-6. By default, the name entered for the previously created material library file will be displayed in the edit box of the **Create Library** window. You can enter a unique name for the material library in this edit box. Note that the material library created will be displayed in the list of available material libraries with the specified name. Enter the name for the material library and then choose the **OK** button; a new material library of the specified name will be created and selected in the **Library Selection** drop-down list of the **Autodesk Material Library Manager** dialog box. Figure 8-7 shows the **Autodesk Material Library Manager** dialog box with the new material library created.

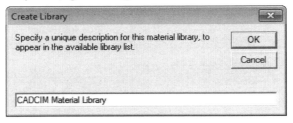

*Figure 8-6 The **Create Library** window*

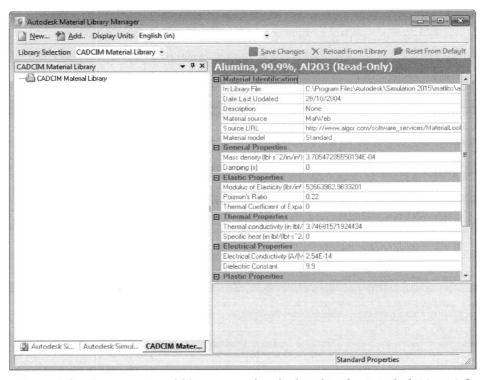

*Figure 8-7 The new material library created and selected in the **Autodesk Material Library Manager** dialog box*

Adding Material Categories

In a library, you can group materials of same family in a category. To create a category inside the user-defined library, invoke the **Autodesk Material Library Manager** dialog box by choosing the **Manage Material Library** tool from the **Options** panel of the **Tools** tab. Next, select the required library from the left of the **Autodesk Material Library Manager** dialog box and right-click; a shortcut menu will be displayed, refer to Figure 8-8.

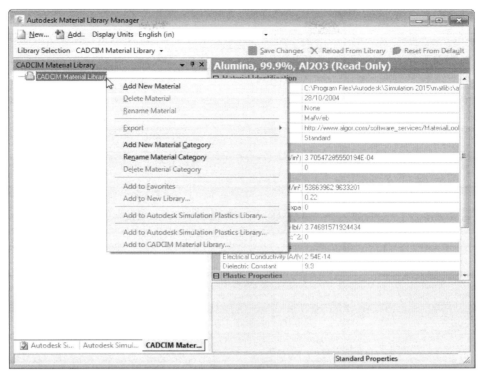

Figure 8-8 *A shortcut menu displayed in the* ***Autodesk Material Library Manager*** *dialog box*

Choose the **Add New Material Category** option from the shortcut menu displayed; the **Material Category Name** window will be displayed, as shown in Figure 8-9. Enter the name of the category in the edit box of the window and then choose the **OK** button; the category of the specified name is created. Similarly, you can create multiple categories in the library. Figure 8-10 shows the **Autodesk Material Library Manager** dialog box with categories **AB**, **GA**, and **KV** created in the **CADCIM Material Library** library.

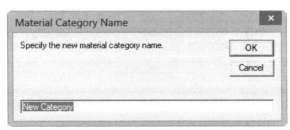

Figure 8-9 *The* ***Material Category Name*** *window*

Adding Materials Under a Material Category
To add material under a category of a user-defined library, invoke the **Autodesk Material Library Manager** dialog box and then select the category for which you want to add the material. Next, right-click and choose the **Add New Material** option from the shortcut menu displayed; the **New Material** dialog box will be displayed, refer to Figure 8-11.

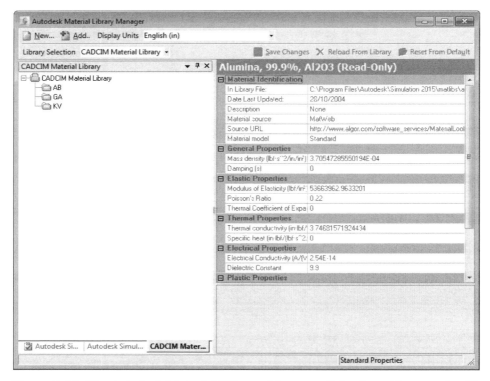

Figure 8-10 The *Autodesk Material Library Manager* dialog box with categories *AB*, *GA*, and *KV* created in the *CADCIM Material Library*

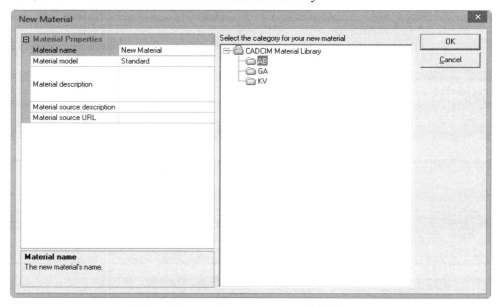

Figure 8-11 The *New Material* dialog box

Specify the name of the material, type of material, material description, material source description (if any), and material source URL in their respective fields available on the left of the **New Material** dialog box. After specifying all the details, choose the **OK** button; the material with the specified details will be created and listed under the selected category in the **Autodesk Material Library Manager** dialog box, refer to Figure 8-12.

Now, you need to specify the material properties for the newly created material. To do so, select the newly created material from the left of the **Autodesk Material Library Manager** dialog box; all the default material properties of the selected material will be displayed on the right of the dialog box. Specify the mass density, modulus of elasticity, poisson's ratio, shear modulus of elasticity, thermal conductivity, and so on of the material in their respective fields available on the right of the dialog box. After specifying all the material properties for the added material, choose the **Save Changes** button available at the top of the dialog box; the specified material properties will be saved in the material library file. Similarly, you can add a number of materials with user-defined material properties as per the requirement.

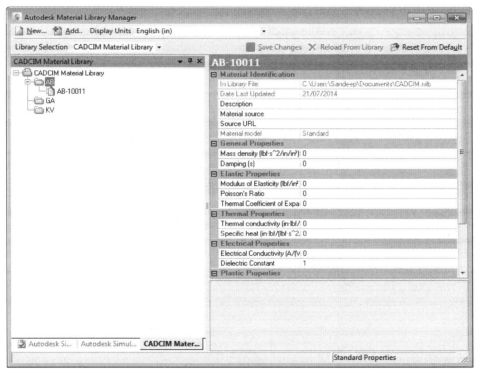

Figure 8-12 The new material AB-10011 added under the AB category of the library

Deleting Material

In Autodesk Simulation Mechanical, you can delete a material from the user-defined material library. To do so, invoke the **Autodesk Material Library Manager** dialog box and then select the material library from the **Library Selection** drop-down list of the dialog box whose material is to be deleted. Next, select the material to be deleted from the material library list and right-click to display a shortcut menu. Choose the **Delete Material** option from the shortcut menu displayed; the **Delete Material** window will be displayed prompting you to confirm whether you want to delete the selected material and its properties or not. Choose the **Yes** button from the window; the selected material will be deleted from the list of materials available in the library.

Deleting Material Category

Similar to deleting material from the library, you can also delete a material category from the user-defined material library. To do so, invoke the **Autodesk Material Library Manager** dialog box and then select the material library from the **Library Selection** drop-down list of the dialog box whose material category is to be deleted. Next, select the material category to be deleted from the material library list and right-click to display a shortcut menu. Choose the **Delete Material Category** option from the shortcut menu; the **Delete Category and Materials** window will be displayed prompting you to confirm whether you want to delete the selected category and all its materials or not. Choose the **Yes** button from the window; the selected material category and all the materials available under it will be deleted from the library.

Loading Material Library File

In Autodesk Simulation Mechanical, the material library file is saved in *.mlb* format. The *.mlb* is the file extension for material library files. If you have a material library file (*.mlb*) which is currently not loaded in Autodesk Simulation Mechanical, you can load it by using the **Add** button available in the **Autodesk Material Library Manager** dialog box. To load a material library file, invoke the **Autodesk Material Library Manager** dialog box and choose the **Add** button from it; the **Add Library to Available Library List** dialog box will be displayed. Browse to the location where the library file (*.mlb*) is saved. Next, select the library file to be loaded and then choose the **Open** button from the dialog box; the **Add Library** dialog box will be displayed. Choose the **OK** button from the dialog box; the selected material file will be uploaded and a list of all the materials available in library file will be displayed in the left of the dialog box.

Setting Material Library as Default Library

By default, the **Autodesk Simulation Material Library** library is set as the default library. As a result, whenever you invoke the **Element Material Selection** dialog box to assign material to a model, the **Autodesk Simulation Material Library** is selected in the **Library Selection** drop-down list and displays all the available material categories on the left in the dialog box. You can also set any other material library as the default library as per the requirement. To do so, invoke the **Autodesk Material Library Manager** dialog box by choosing the **Manage Material Library** tool from the **Options** panel of the **Tools** tab in the **Ribbon**. Next, select the material library that you want to set as the default library from the **Library Selection** drop-down list of the dialog box. Click on the down arrow available on the right of the selected library title in the dialog box to display a menu, refer to Figure 8-13. Select the **Set as Default** option from the menu displayed; the selected library will be set as the default library.

Removing/Deleting Library

In Autodesk Simulation Mechanical, you can remove user-defined material libraries from the list of libraries that are no longer required. To do so, invoke the **Autodesk Material Library Manager** dialog box and then select the material library that you want to remove from the list of libraries. Next, click on the down arrow available on the right of the selected library title in the dialog box to display a menu, refer to Figure 8-13. Choose the **Remove Library** option from the menu displayed; the **Database Removal** window will be displayed prompting you to specify whether you want to remove the library or not, refer to Figure 8-14.

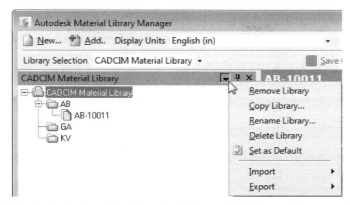

Figure 8-13 *Partial view of the **Autodesk Material Library Manager** dialog box with a menu displayed by clicking on the down arrow*

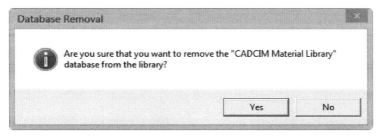

Figure 8-14 *The **Database Removal** window*

Choose the **Yes** button from the **Database Removal** window; the selected library will be removed. However, its data files will still be available in your system hard disk. As a result, you can load the removal material library again by loading its data files as discussed earlier in the **Loading Material Library File** section.

To permanently remove the library, select the **Delete Library** option from the menu displayed instead of selecting the **Remove Library** option. On selecting this option, the **Database Deletion** window will be displayed prompting you to specify whether you want to remove the library and its database files from the disk or not, refer to Figure 8-15. Choose the **Yes** button; the selected library will be removed from the list of libraries. Also, its data files will be deleted from the system hard disk.

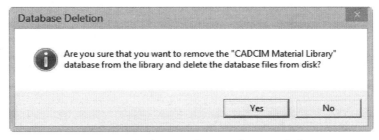

Figure 8-15 *The **Database Deletion** window*

Copying Material Library

In Autodesk Simulation Mechanical, you can create a copy of a material library to create a new material library by editing the material properties of the materials available in the copied library. To create a copy of an existing library, invoke the **Autodesk Material Library Manager** dialog box and then select the material library to be copied from the **Library Selection** drop-down list. Next, click on the down arrow available on the right of the selected library title in the dialog box to display a menu, refer to Figure 8-16. Select the **Copy Library** option from the menu displayed; the **Copy Library To** dialog box will be displayed. Browse to the location where you want to copy the library file and then specify a name for the new library file in the **File name** edit box of the dialog box. Next, choose the **Save** button; the **Copy Library** window will be displayed. By default, the previously entered name appears in the edit box of this window. You can specify a unique name for the new material library in this edit box. Note that the material library created will be displayed in the list of available material libraries with the name entered in this edit box. Enter the name for the material library and then choose the **OK** button; a new material library with the specified name will be created and selected in the **Library Selection** drop-down list of the **Autodesk Material Library Manager** dialog box. Note that in the material library created, all the materials and their material properties are same as that of the parent material library. You can change the material properties as discussed earlier and then choose the **Save Changes** button.

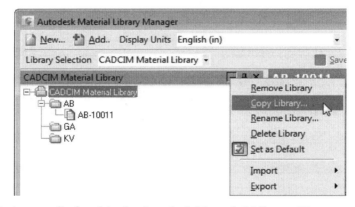

*Figure 8-16 A menu displayed in the **Autodesk Material Library Manager** dialog box*

BOUNDARY CONDITIONS

Boundary conditions are defined on the boundary of an FEA model to represent the effect of the surrounding environment. It includes forces, pressures, displacement, and so on. In Autodesk Simulation Mechanical, the boundary conditions are divided into two categories: loads and constraints. Both are discussed next.

Constraints

Constraints are defined as the boundary conditions that prevent the model from translation or rotational movement. In other words, the constraints are used to restrict the degree of freedom of a model. The applied constraints act as connection between the model and reference. You can select nodes, edges, and surfaces of a model to apply constraints. The tools used for applying constraints is discussed next.

General Constraint

Ribbon: Setup > Constraints > General Constraint

The **General Constraint** tool is used to constrain the degree of freedom of nodes. You can select nodes, edges, or surfaces of a model to apply constraint by using the **General Constraint** tool. Note that depending upon the geometry selected for applying the constraint, the degree of freedom of the nodes of the selected geometry will be constrained. The procedure of applying general constraint on nodes, edges, and surfaces of the model are discussed next.

Applying Nodal General Constraint

General constraint applied on a node is called nodal general constraint. To apply a nodal general constraint, select a node of the model on which you want to apply the constraint and then choose the **General Constraint** tool from the **Constraints** panel of the **Setup** tab in the **Ribbon**; the **Creating Nodal General Constraint Object** dialog box will be displayed, as shown in Figure 8-17. Alternatively, you can invoke this dialog box by selecting a node and right-clicking in the drawing area to display a shortcut menu. Next, choose **Add > Nodal General Constraint** from the shortcut menu displayed. The options available in this dialog box are discussed next.

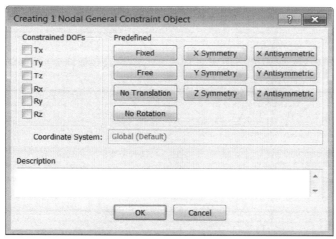

*Figure 8-17 The **Creating Nodal General Constraint Object** dialog box*

 Tip. *You can also select multiple nodes of a model to constrain by using the CTRL key. On selecting nodes, the name of the dialog box invoked on choosing the **General Constraint** tool will be changed to **Creating** (number of nodes selected) **Nodal General Constraint Objects**.*

Constrained DOFs Area

The **Constrained DOFs** area of the dialog box is used to control or specify the constraints for the selected node. By default, all the check boxes available in this area are cleared. As a result, all degrees of freedom of the selected node are free. On selecting the **Tx** check box, the translation degree of freedom along the X-axis of the selected node will be constrained. Similarly, on selecting the **Ty** and **Tz** check boxes, the translation degree of freedom along the Y and Z axes will be constrained.

To constrain the rotational degree of freedom along the X, Y, and Z axes, you can select the **Rx**, **Ry**, and **Rz** check boxes, respectively, from the **Constrained DOFs** area of the dialog box.

Predefined Area

The buttons available in the **Predefined** area of the dialog box are used to select predefined sets of constraints to be applied to the selected nodes. On choosing the **Fixed** button from the **Predefined** area, the degree of freedom of the selected nodes will be fixed and all the check boxes available in the **Constrained DOFs** area of the dialog box will be selected.

On choosing the **Free** button from the **Predefined** area, all the selected check boxes of the **Constrained DOFs** area will be cleared and the degree of freedom of the selected nodes will be free. If you choose the **No Translation** button, the translation degree of freedom along X, Y, and Z axes of the selected nodes will be constrained and the rotational degree of freedom along the X, Y, and Z axes will be free to rotate. On selecting the **No Rotation** button, the rotational degree of freedom of the selected nodes along X, Y, and Z axes will be constrained and the translation degree of freedom along X, Y, and Z axes will be free.

Similarly, you can choose the **X Symmetry**, **Y Symmetry**, **Z Symmetry**, **X Antisymmetric**, **Y Antisymmetric**, and **Z Antisymmetric** buttons from the **Predefined** area of the dialog box to apply the predefined sets of symmetric or antisymmetric constraints to the selected nodes.

Description

The **Description** area of the dialog box is used to enter a comment or description about the constraint being applied.

After specifying the required set of constraints to be applied to the selected nodes, choose the **OK** button from the dialog box; the constraint will be applied. Figure 8-18 shows a meshed model with fixed constraint applied on some of the nodes.

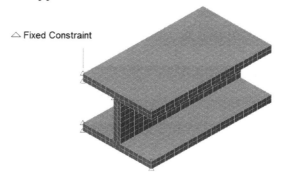

△ Fixed Constraint

Figure 8-18 *The fixed constraint applied to some of the nodes of the model*

Applying Surface General Constraint

The constraint applied on a surface is called surface general constraint. To apply surface general constraint, select the surface of the model on which you want to apply constraint and then choose the **General Constraint** tool from the **Constraints** panel of the **Setup** tab in the **Ribbon**; the **Creating Surface General Constraint Object** dialog box will be displayed, as shown in Figure 8-19. Alternatively, you can invoke this dialog box by selecting a surface and right-clicking in the drawing area to display a shortcut menu. Next, choose **Add > Surface General Constraint** from the shortcut menu displayed. The options available in this dialog box are same as discussed earlier. Figure 8-20 shows a meshed model with the fixed constraint applied on a surface.

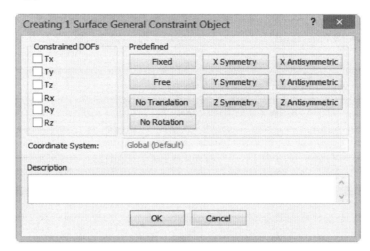

Figure 8-19 The **Creating Surface General Constraint Object** *dialog box*

Applying Edge Constraint

The method of applying constraints on edges is similar to applying constraints on surfaces and nodes. Figure 8-21 shows a meshed model with the fixed constraint applied on an edge.

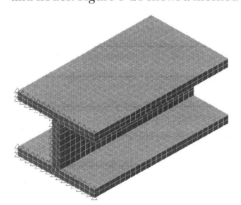 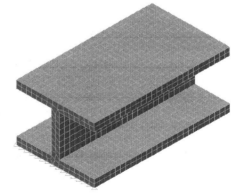

Figure 8-20 The fixed constraint applied on the surface of the model

Figure 8-21 The fixed constraint applied on an edge of the model

Note
*To select a node/surface/edge of a model, choose the **Selection** tab in the **Ribbon**. Next, choose the **Point or Rectangle** tool from the **Shape** panel and the **Vertices/Surfaces/Edges** tool from the **Select** panel of the **Selection** tab in the **Ribbon**.*

Tip. *You can select nodes, surfaces, or edges for applying constraint either before or after invoking the dialog box.*

Pin Constraint

Ribbon:	Setup > Constraints > Pin Constraint

 The **Pin Constraint** tool is used to constrain the radial, tangential, and axial degrees of freedom of the selected cylindrical surface. This constraint is widely used to simulate pin connection. You can apply pin constraint only to a cylindrical surface. To apply the pin constraint, select a cylindrical surface of the model on which you want to apply the pin constraint and then choose the **Pin Constraint** tool from the **Constraints** panel of the **Setup** tab in the **Ribbon**; the **Creating Pin Constraint Object** dialog box will be displayed, as shown in Figure 8-22. Alternatively, you can invoke this dialog box by selecting a cylindrical surface and right-clicking in the drawing area to display a shortcut menu. Next, choose **Add > Surface Pin Constraint** from the shortcut menu displayed. The options available in this dialog box are discussed next.

*Figure 8-22   The **Creating Pin Constraint Object** dialog box*

Constraints Area
The check boxes available in this area are used to constrain the radial, tangential, and axial degrees of freedom of the selected surface. These check boxes are discussed next.

Fix Radial
On selecting the **Fix Radial** check box, the selected cylindrical surface will be restricted to move or deform radially.

Fix Tangential
On selecting the **Fix Tangential** check box, the selected cylindrical surface will be restricted to rotate tangentially around the circumference of the cylindrical surface selected for applying the constraint.

Fix Axial
On selecting the **Fix Axial** check box, the selected cylindrical surface will be restricted to move axially along the axis of the cylindrical surface selected for applying the constraint.

Description Area
The **Description** area of the dialog box is used to enter a comment or description about the applied constraint.

After fixing the required degree of freedom of the selected cylindrical surface to simulate the pin connection, choose the **OK** button from the dialog box; the pin constraint will be applied. Figure 8-23 shows a model with the pin constraint applied.

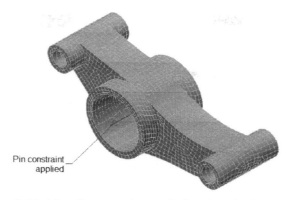

Figure 8-23 *The pin constraint applied to the cylindrical surface*

1D Spring Support

Ribbon: Setup > Constraints > 1D Spring Support

The **1D Spring Support** tool is used to apply stiffness to nodes, edges, or surfaces of a model to resist their translational or rotational degree of freedom along or about any specified direction vector. Note that the deformation that occurs after applying the 1D spring support constraint to a node, edge, or a surface is calculated based on the spring equation F = KX. The greater the stiffness value, the lower will be the deformation. The procedure of applying 1D Spring support constraint on nodes, edges, and surfaces of a model are discussed next.

Applying Nodal 1D Spring Support

The 1D Spring support applied on a node is called nodal 1D Spring support. To apply nodal 1D Spring support, select a node of the model and then choose the **1D Spring Support** tool from the **Constraints** panel of the **Setup** tab in the **Ribbon**; the **Creating Nodal 1D Spring Support Element** dialog box will be displayed, as shown in Figure 8-24. Alternatively, you can invoke this dialog box by selecting a node and right-clicking in the drawing area to display a shortcut menu. Next, choose **Add > Nodal 1D Spring Support** from the shortcut menu displayed. You can also select multiple nodes by pressing the CTRL key. The options available in the **Creating Nodal 1D Spring Support Element** dialog box are discussed next.

Type Area
On selecting the **Translation** radio button in the **Type** area of the dialog box, you can restrict the translation movement of the selected node or nodes. To restrict the rotational movement of the selected node or nodes, select the **Rotation** radio button.

Direction Area
The **Direction** area of the dialog box allows you to select the required direction vector along/about which you want to restrict the translational/rotational movement. You can select the **X**, **Y**, or **Z** radio button to restrict the translational/rotational movement along/about X, Y, or Z axis. You can also select the **Custom** radio button to specify the direction vector other than the X, Y, and Z axes directions of the coordinate system.

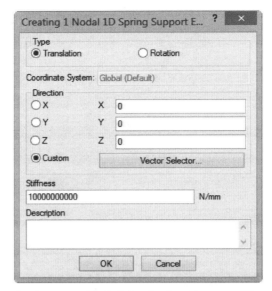

*Figure 8-24 The **Creating Nodal 1D Spring Support Element** dialog box*

To specify the direction vector other than X, Y, and Z directions, select the **Custom** radio button; the **X**, **Y**, and **Z** edit boxes of the dialog box will be enabled. In these edit boxes, you can specify the X, Y, and Z coordinates of the direction vector.

Moreover, you can specify the direction vector by selecting two points. To do so, choose the **Vector Selector** button available on the right of the **Custom** radio button in the dialog box; the dialog box will disappear and you will be prompted to specify the direction vector. Specify the start point and then the endpoint toward which the direction will point. As soon as you specify the endpoint for the direction vector, the **Creating Nodal 1D Spring Support Element** dialog box will be displayed again with the coordinate value of the specifying direction vector displayed in the X, Y, and Z edit boxes of the dialog box.

Stiffness Area
The **Stiffness** edit box is used to specify the stiffness of the elements for resisting the movement in the specified direction.

Description Area
The **Description** area of the dialog box is used to enter a comment or description about the applied constraint.

After specifying all the parameters to define 1D Spring support in a particular direction, choose the **OK** button from the dialog box; the 1D Spring support will be applied.

Applying Surface 1D Spring Support
The 1D Spring support applied on a surface is called surface 1D Spring support. To apply Surface 1D Spring support, select a surface of the model and then choose the **1D Spring Support** tool from the **Constraints** panel of the **Setup** tab in the **Ribbon**; the **Creating**

Surface 1D Spring Support Element dialog box will be displayed, as shown in Figure 8-25. Alternatively, you can invoke this dialog box by selecting a surface and right-clicking in the drawing area to display a shortcut menu. Next, choose **Add > Surface 1D Spring Support** from the shortcut menu. You can also select multiple surfaces by using the CTRL key. The options available in the **Creating Surface 1D Spring Support Element** dialog box are same as discussed earlier.

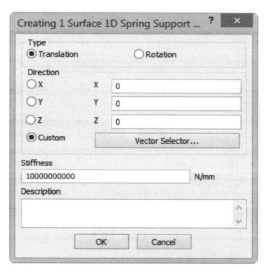

*Figure 8-25 The **Creating Surface 1D Spring Support Element** dialog box*

Applying Edge 1D Spring Support

The 1D Spring support applied on an edge is called edge 1D Spring support. To apply edge 1D Spring support, select an edge of the model and then choose the **1D Spring Support** tool; the **Creating Edge 1D Spring Support Element** dialog box will be displayed. Alternatively, you can invoke this dialog box by selecting an edge and right-clicking on the drawing area to display a shortcut menu. Next, choose **Add > Edge 1D Spring Support** from the shortcut menu. The options available in this dialog box are same as discussed earlier.

Figure 8-26 shows a model with 1D Spring support applied on the bottom front edge of the top plate.

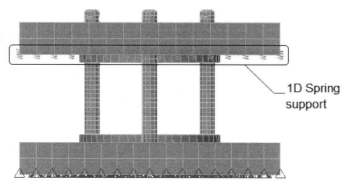

Figure 8-26 The edge 1D Spring support applied

3D Spring Support

Ribbon:	Setup > Constraints > 3D Spring Support

The **3D Spring Support** tool is used to apply stiffness to nodes, edges, or surfaces of a model to restrict their translational/rotational degree of freedom along/about the X, Y, or Z global direction. You can also restrict the translational/rotational degree of freedom along/about more than one (X, Y, and Z) global directions by using this tool. Note that the deformation that occurs after applying the 3D spring support constraint to a node, edge, or a surface is based on the spring equation $F = KX$. The greater the stiffness value, the lower will be the deformation.

The procedure of applying 3D Spring support constraint on nodes, edges, and surfaces of a model is same as discussed for applying 1D spring support. The only difference in 3D Spring support is that you can restrict the translational or rotational movements in all directions globally.

To apply 3D Spring support, select a node/edge/surface and then choose the **3D Spring Support** tool from the **Constraints** panel of the **Setup** tab in the **Ribbon**; the **Creating Nodal 3D Spring Support Element/Creating Edge 3D Spring Support Element/Creating Surface 3D Spring Support Element** dialog box will be displayed. Figure 8-27 shows the **Creating Edge 3D Spring Support Element** dialog box displayed on choosing the **3D Spring Support** tool after selecting an edge.

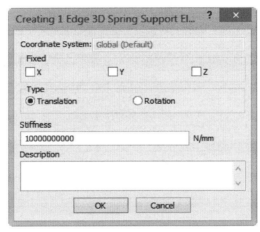

*Figure 8-27 The **Creating Edge 3D Spring Support Element** dialog box*

The options available in this dialog box are same as discussed while applying 1D Spring support, except the options in the **Fixed** area. The **Fixed** area and its options are discussed next.

Fixed Area

The **Fixed** area of the dialog box is used to select the required directions to be resisted for translational or rotational movements along or about the X, Y, and Z directions. Select the required check boxes from the **Fixed** area of the dialog box along or about which you want to restrict the translational or rotational movements.

Figure 8-28 shows a model with 3D Spring support applied on the bottom front edge of the top plate.

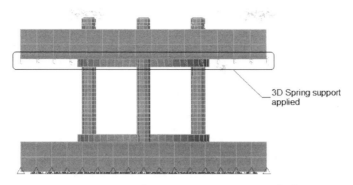

Figure 8-28 *The 3D spring support applied*

Prescribed Displacement Constraint

Ribbon: Setup > Constraints > Prescribed Displacement

⟜ Prescribed Displacement   A prescribed displacement allows nodes to translate or rotate through a specified distance in a specified vector direction. You can select nodes, edges, or surfaces of a model to apply prescribed displacement by using the **Prescribed Displacement** tool. Note that on selecting an edge or a surface, the prescribed displacement will be applied to all the nodes of the selected edge or surface.

To apply prescribed displacement to a surface, select the surface of a model and then choose the **Prescribed Displacement** tool from the **Constraints** panel of the **Setup** tab in the **Ribbon**; the **Creating Surface Prescribed Displacement Element** dialog box will be displayed, refer to Figure 8-29. The options available in this dialog box are discussed next.

Figure 8-29 *The **Creating Surface Prescribed Displacement Element** dialog box*

Note
*The name of the dialog box displayed on choosing the **Prescribed Displacement** tool depends upon the geometry selected. If you select a node and choose the **Prescribed Displacement** tool, the **Creating Nodal Prescribed Displacement Element** dialog box will be displayed and if you select an edge, the **Creating Edge Prescribed Displacement Element** dialog box will be displayed. However, the options available in the dialog boxes are same.*

Type Area
The **Translation** radio button of the **Type** area in the dialog box is used to allow the translation movement upto the specified magnitude in a specified direction. To allow the rotational movement of the selected geometry upto the specified magnitude about an axis, select the **Rotation** radio button from the **Type** area of the dialog box.

Note
The rotational movement can only be possible for the elements that have rotational degrees of freedom such as shell and beam elements.

Magnitude Edit Box
The **Magnitude** area of the dialog box is used to specify the magnitude for the translation or rotational movements.

Direction Area
The **Direction** area of the dialog box is used to select the required direction vector along or about which you want to allow the translational or rotational movements. You can select the **X**, **Y**, or **Z** radio button to allow the translational or rotational movement along or about X, Y, or Z axis. You can also specify the direction vector other than along the X, Y, and Z axes by using the **Custom** radio button and the **Vector Selector** button of the dialog box, as discussed earlier.

Stiffness Edit Box
The **Stiffness** edit box in this area is used to specify the stiffness of the elements to allow the movement in the specified direction. Note that the translational or rotational movement of the nodes of the model will also depend upon the stiffness value specified in this edit box. The higher the stiffness value specified, the greater will be the displacement.

Description Area
The **Description** area of the dialog box is used to enter a comment or description about the applied prescribed displacement.

Similarly, you can select nodes or edges to apply prescribed displacement. Figure 8-30 shows a model with prescribed displacement applied on the curved surface.

Frictionless Constraint

Ribbon:	Setup > Constraints > Frictionless

The frictionless constraint is used to restrict the deformation of the selected surface along its normal direction due to the force applied on the model. You can select planar and cylindrical surfaces of a model to apply frictionless constraint by using the **Frictionless** tool.

To apply frictionless constraint to a surface of the model, select a surface and then choose the **Frictionless** tool from the expanded **Constraints** panel of the **Setup** tab in the **Ribbon**; the **Creating Frictionless Constraint Object** dialog box will be displayed, refer to Figure 8-31. In this dialog box, you can enter a comment or description about the frictionless constraint being applied in the **Description** edit box. Next, choose the **OK** button; the frictionless constraint will be applied to the selected surface.

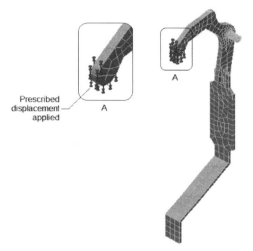

Figure 8-30 *The prescribed displacement applied on the curved surface*

Figure 8-31 *The* ***Creating Frictionless Constraint Object*** *dialog box*

Figure 8-32 shows a model with frictionless constraint applied to the bottom planar surface of a structure member in FEA editor environment and Figure 8-33 shows the model in the Result environment of Autodesk Simulation Mechanical after performing the analysis. Note that the frictionless constraint does not allow the selected surface to deform along its normal direction, refer to Figure 8-33.

 Note
You will learn more about Result environment of Autodesk Simulation Mechanical in later chapters.

Loads

Load is defined as the external force acting on the body. It includes forces, pressures, movement, and so on. The tools that are used for applying different loading conditions are discussed next.

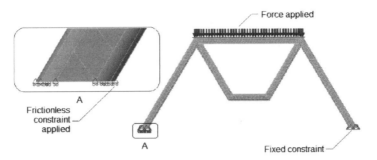

Figure 8-32 *The frictionless constraint applied to the bottom planar surface of a structure member*

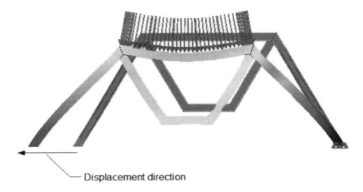

Figure 8-33 *The model with displacement along the parallel direction after applying the frictionless constraint*

Force

Ribbon:	Setup > Loads > Force

A force is an influence that causes an object to undergo a certain change, either concerning its movement, direction, or geometrical construction. In Autodesk Simulation Mechanical, you can apply force to a node, edge, and surface.

Applying Nodal Force

The force which is applied on a node is called as nodal force. To apply nodal force, select the node of the model on which you want to apply the force and then choose the **Force** tool from the **Loads** panel of the **Setup** tab in the **Ribbon**; the **Creating Nodal Force Object** dialog box will be displayed, as shown in Figure 8-34. Alternatively, select a node and right-click to display a shortcut menu. Next, select **Add > Nodal Force** from the shortcut menu displayed, refer to Figure 8-35; the **Creating Nodal Force Object** dialog box will be displayed.

Note
*To select a node of a model, choose the **Selection** tab in the **Ribbon**. Next, choose the **Point or Rectangle** tool from the **Shape** panel and the **Vertices** tool from the **Select** panel of the **Selection** tab in the **Ribbon**.*

Tip. *If you invoke the **Force** tool to apply force while multiple nodes are selected, the name of the dialog box displayed will be **Creating (number of nodes selected) Nodal Force Objects**.*

Figure 8-34 The **Creating Nodal Force Object** *dialog box*

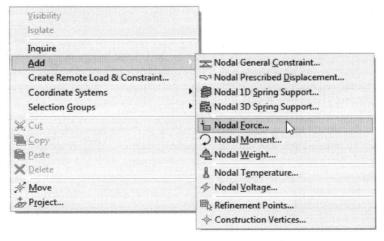

Figure 8-35 *Choosing the **Nodal Force** option from the shortcut menu*

The options available in the **Creating Nodal Force Object** dialog box are discussed next.

Magnitude
The **Magnitude** edit box is used to specify the magnitude of the force to be applied on the selected object.

Direction Area

The **Direction** area of the dialog box is used to specify the direction of force. You can specify the direction of force normal to the X, Y, or Z direction by selecting the respective radio button from this area. For example, on selecting the **X** radio button, the direction of force will be toward the X direction with respect to the coordinate system displayed in the **Coordinate System** display area of the dialog box. The Global coordinate system is the default coordinate system.

You can also specify the direction of force other than the X, Y, and Z directions by using the **Custom** radio button. To do so, select the **Custom** radio button; the **X**, **Y**, and **Z** edit boxes of the dialog box will be enabled. In these edit boxes, you can specify the coordinates of the force direction.

Moreover, you can also specify the direction of force by specifying the direction vector. To do so, choose the **Vector Selector** button in the dialog box; the dialog box will disappear and you will be prompted to specify the direction vector for the force to be applied. Specify the start point and then the endpoint toward which the direction should point, refer to Figure 8-36. As soon as you specify the endpoint for the direction vector, the **Creating Nodal Force Object** dialog box will be displayed again with the coordinate value of the specified direction vector entered in the **X**, **Y**, and **Z** edit boxes.

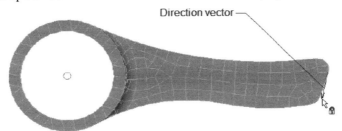

Direction vector

Figure 8-36 The direction vector for specifying the direction of force in a model

Load Case/Load Curve Spinner

The **Load Case/Load Curve** spinner of the dialog box is used to select the required load case/load curve to associate with the force being applied. You will learn more about load case and load curve later in this chapter.

Note
*The **Load Case/Load Curve** spinner will be available only when you apply nodal and edge forces in their respective dialog boxes.*

*Also, the **Curve** button, which is available on the right of the **Load Case / Load Curve** spinner in the dialog box, will be enabled only at the time of performing MES with nonlinear material analysis, the static stress with nonlinear material analysis and Transient stress with linear material. In case of Transient stress the **Curve** button will be available only enable for surface force. This button is used to define the load curve to be followed by the force being applied.*

Description

The **Description** area of the dialog box is used to enter a comment or description about the force being applied.

After specifying all the parameters for applying the nodal force, choose the **OK** button from the dialog box; the force will be applied on the selected node and an arrow representing the force will be displayed in the drawing area.

Figure 8-37 shows a model with nodal force applied along the direction vector shown in Figure 8-36.

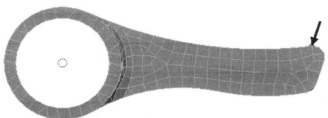

Figure 8-37 *Arrow representing the applied nodal force*

Note
You can select multiple nodes to apply force by using the CTRL key. However, in that case, the total magnitude of the force applied on the model will be equal to the number of nodes selected multiplied by the value of magnitude specified. For example, if you select 4 nodes to apply force of 100 N, then the total magnitude of the force to be applied on the model will be 400 N (4 X 100).

Applying Edge Force
The force applied on an edge is called edge force. To apply the edge force, select the edge of the model on which you want to apply the force and then choose the **Force** tool from the **Loads** panel of the **Setup** tab in the **Ribbon**; the **Creating Edge Force Object** dialog box will be displayed, as shown in Figure 8-38. Alternatively, select the edge and right-click to display a shortcut menu. Next, choose **Add > Edge Force** from the shortcut menu to invoke this dialog box. The options available in the **Creating Edge Force Object** dialog box are same as discussed earlier.

Note
*To select an edge of a model, choose the **Selection** tab in the **Ribbon**. Next, choose the **Point or Rectangle** tool from the **Shape** panel and the **Edges** tool from the **Select** panel of the **Selection** tab in the **Ribbon**.*

Applying Surface Force
The force applied on a surface is called surface force. To apply the surface force, select the surface of the model on which you want to apply the force and then choose the **Force** tool from the **Loads** panel of the **Setup** tab in the **Ribbon**; the **Creating Surface Force Object** dialog box will be displayed, as shown in Figure 8-39. Alternatively, select the surface and right-click to display a shortcut menu. Next, choose **Add > Surface Force** from the shortcut menu to invoke this dialog box.

*Figure 8-38 The **Creating Edge Force Object** dialog box*

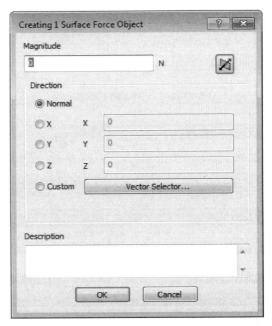

*Figure 8-39 The **Creating Surface Force Object** dialog box*

The options available in the **Creating Surface Force Object** dialog box are the same as those in the **Creating Nodal Force Object** and **Creating Edge Force Object** dialog boxes except the **Normal** radio button.

Note

*To select a surface of a model, choose the **Selection** tab in the **Ribbon**. Next, choose the **Point or Rectangle** tool from the **Shape** panel and the **Surface** tool from the **Select** panel of the **Selection** tab in the **Ribbon**.*

On selecting the **Normal** radio button available in the **Direction** area of the **Creating Surface Force Object** dialog box, you can specify the direction of force normal to the selected surface. Note that after applying the force normal to a surface, the direction arrows representing the direction of force may not be displayed normal to the surface in the FEA Editor environment. However, in the Result environment, the arrows representing the direction of force will be displayed normal to the applied surface.

Figure 8-40 shows the direction of arrows in the FEA Editor environment on applying the surface force normal to the curved surface of the model. Figure 8-41 shows the same model in Result environment with the direction of arrows normal to the curved surface.

Note

You will learn more about the Result environment of Autodesk Simulation Mechanical in later chapters.

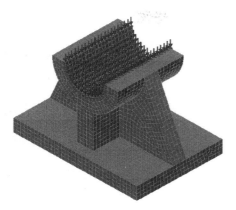

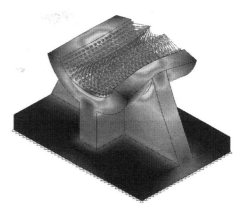

Figure 8-40 *Direction of arrows displayed in the FEA Editor environment on applying surface force normal to the curved surface of the model*

Figure 8-41 *Direction of arrows displayed in the Result environment on applying surface force normal to the curved surface of the model*

 Tip. *You can select nodes, surfaces, or edges for applying force either before or after invoking the dialog box.*

Pressure

Ribbon: Setup > Loads > Pressure

 Pressure is defined as the force per unit area. In Autodesk Simulation Mechanical, you can apply normal/traction pressure to the selected surface by using the options available in the **Creating Surface Pressure/Traction Object** dialog box. A normal pressure refers to the pressure that is applied normal to the selected surface whereas a traction pressure is applied in a specific direction (tangentially).

To apply normal/traction surface pressure, select the surface of the model on which you want to apply pressure and then choose the **Pressure** tool from the **Loads** panel of the **Setup** tab in the **Ribbon**; the **Creating Surface Pressure/Traction Object** dialog box will be displayed, as shown in Figure 8-42. Alternatively, select the surface and right-click to display a shortcut menu. Next, choose **Add > Surface Pressure/Traction** from the shortcut menu to invoke this dialog box.

The options available in the **Creating Surface Pressure/Traction Object** dialog box are discussed next.

Pressure
The **Pressure** radio button is used to apply pressure normal to the selected surface. To apply the normal pressure, select the **Pressure** radio button and then enter the magnitude of the pressure in the **Magnitude** edit box available below this radio button.

Traction
The **Traction** radio button is used to apply a pressure that is pointed along a specific direction. To apply the traction pressure, select the **Traction** radio button; the **X Magnitude**,

Y Magnitude, and **Z Magnitude** edit boxes will be enabled. In these edit boxes, you can specify the magnitude of the pressure in each of the global directions.

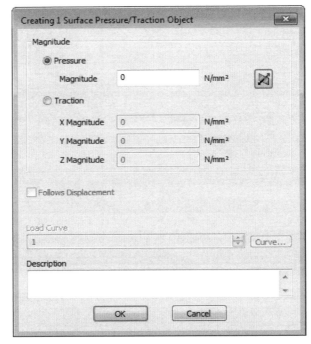

*Figure 8-42 The **Creating Surface Pressure/Traction Object** dialog box*

Load Curve

The **Load Curve** spinner is used to set the load case to be used for the pressure being applied. After setting the load case to be used, you can modify the load curve of the respective load case by using the **Curve** button available next to the **Load Curve** spinner. On choosing the **Curve** button, the **Multiplier Table Editor** dialog box will be displayed, as shown in Figure 8-43.

Note

*The **Load Curve** spinner and the **Curve** button in the **Creating Surface Pressure/Traction Object** dialog box will be enabled only while performing MES with nonlinear material analysis and the static stress with nonlinear material analysis.*

In the **Multiplier Table Editor** dialog box, the load curve is defined as the relationship between the time and multiplier. By default, two data points are defined for the load curve. You can define multiple data points for the load curve. To do so, choose the **Add Row** button from the dialog box; a new row will be added to the left of the dialog box with the default time and multiplier values. Also, a new point will be added in the load curve with the default values. To specify new values for time and multiplier in a particular row, click on the field corresponding to the **Time (s)** / **Multiplier** column of that row to activate its exit mode and then enter the required value in it. Next, press ENTER. If you want to delete a row, select the row to be deleted from the left side of the dialog box and then choose the **Delete Row(s)** button from the **Multiplier Table Editor** dialog box.

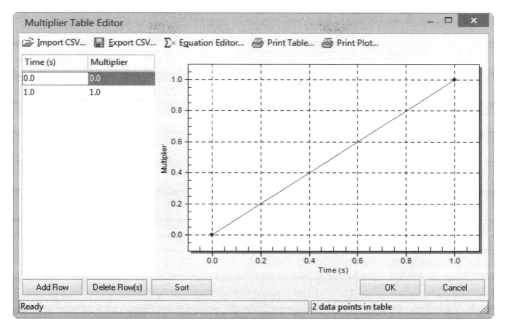

Figure 8-43 *The **Multiplier Table Editor** dialog box*

You can import the data of an existing load curve, which is saved in the *.csv* file format, using the **Import CSV** button of the dialog box. Similarly, you can export the current load curve data to a *.csv* file format using the **Export CSV** button of the dialog box. You can also print the table containing the time and multiplier values of the data points by using the **Print Table** and **Print Plot** buttons in the dialog box. After specifying all the data points for defining the load curve, exit from the dialog box by choosing the **OK** button.

Description Area
The **Description** area of the **Creating Surface Pressure/Traction Object** dialog box is used to enter a comment or description about the pressure being applied.

Figure 8-44 shows a mesh model with pressure applied on its inner surfaces. Figure 8-45 shows the same model after performing analysis with normal pressure applied and Figure 8-46 shows the model with traction pressure of same magnitude applied towards the Y axis direction.

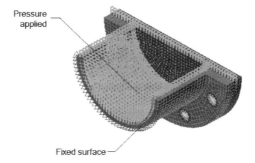

Figure 8-44 *The pressure applied on the inner surfaces of the model*

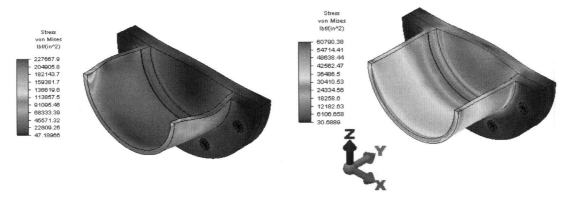

Figure 8-45 Model after performing analysis with normal pressure applied on all its inner surfaces

Figure 8-46 Model after performing analysis with traction pressure applied in the direction of Y axis

Variable Pressure

Ribbon:	Setup > Loads > Variable Pressure

The variable pressure is applied by specifying different pressures at uniform intervals along the direction of the selected surface. To apply variable pressure to a surface, select the surface of the model on which you want to apply variable pressure. Next, choose the **Variable Pressure** tool from the expanded **Loads** panel of the **Setup** tab in the **Ribbon**, refer to Figure 8-47; the **Creating Surface Variable Pressure Object** dialog box will be displayed, as shown in Figure 8-48.

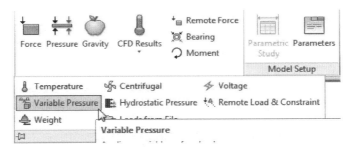

Figure 8-47 The **Variable Pressure** tool in the expanded **Loads** panel

Alternatively, select the surface and right-click to display a shortcut menu. Next, choose **Add > Surface Variable Pressure** from the shortcut menu displayed to invoke the **Creating Surface Variable Pressure Object** dialog box. The options available in the **Creating Surface Variable Pressure Object** dialog box are discussed next.

Coordinate system

The **Coordinate system** drop-down list displays the list of all coordinate systems defined in the current FEA Editor environment in addition to the global coordinate system. You can select the required coordinate system from this drop-down list to be followed by the variable loads. Note that if you want the starting magnitude of the variable pressure to be zero and start from the starting node of an edge of the surface then you need to create a coordinate system that has the origin at the starting node of the edge.

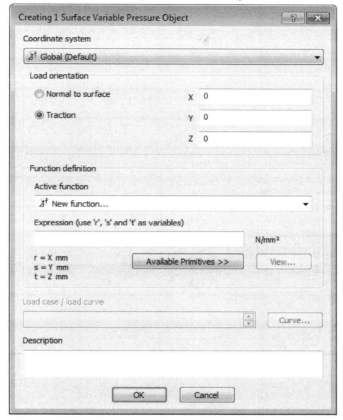

*Figure 8-48 The **Creating Surface Variable Pressure Object** dialog box*

Load orientation Area

The options in the **Load orientation** area are used to specify the direction of pressure. On selecting the **Normal to surface** radio button, the direction of pressure being applied will be perpendicular to the surface selected for applying the pressure. On selecting the **Traction** radio button, the **X**, **Y**, and **Z** edit boxes on the right of this radio button will be enabled. By using these edit boxes, you can specify the direction vector for the variable pressure to be followed.

Function definition Area

The options in the **Function definition** area are used to specify a function that defines the variable load. The **Active function** drop-down list of this area displays the list of all the existing functions that are already specified in the current session of Autodesk Simulation Mechanical. From this drop-down list, you can select an existing function to be used for the variable load being applied. By default, the **New function** option is selected in this drop-down list. You can also enter a new name in the edit field of the **Active function** drop-down list to specify a new function for the variable load.

The **Expression** field in the **Function definition** area is used to specify the expression for the function selected in the **Active function** drop-down list. You can write the equation in the **Expression** field for representing the pressure to be applied to the selected surface. You can use the variables r, s, and t in the expression, where r represents the distance in the X direction, s represents the distance in the Y direction, and t represents the distance in the Z direction. Also, you can use the basic operators such as +, -, * ,/, (,), and ^ in the expression. For example, if **25*(r)** is the expression entered in the **Expression** field and the unit system is set to **English (in)**, then the pressure at the distance of 50 inches along the X direction is equal to 1250 psi (25 x 50). In other words, as per the expression **25*(r)**, the pressure 25 psi is increasing per inch distance toward the X direction. If the **25*(r^2)** will be the expression entered in the **Expression** field, then the pressure at the distance of 50 inches along the X direction will be equal to 62500 psi [25 x (50x50)]. Similarly, 250000 psi pressure at the distance of 100 inches along the X direction.

You can also use some commonly used functions for defining the variable pressure. To do so, click on the **Available Primitives** button of the dialog box; a flyout will be displayed. You can select the required functions from the flyout displayed.

To view the graphical representation of the variable pressure as per the expression entered in the **Expression** field along a specific direction, choose the **View** button from the dialog box. As soon as you choose the **View** button, the **Variable Load Viewer** window will be displayed, refer to Figure 8-49.

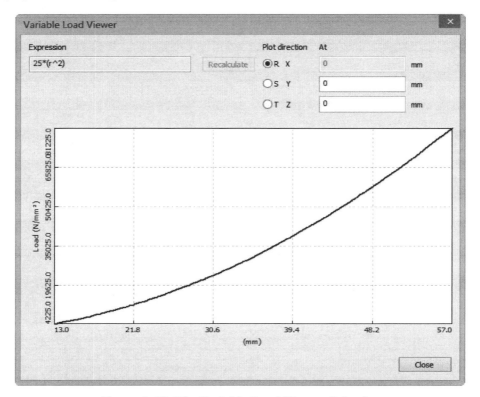

Figure 8-49 *The **Variable Load Viewer** dialog box*

Load case/load curve Spinner

The **Load case/load curve** spinner is used to set the load case to be used for the pressure being applied. After setting the load case to be used, you can modify the load curve of the respective load case by using the **Curve** button available next to the **Load Curve** spinner. On choosing the **Curve** button; the **Multiplier Table Editor** dialog box will be displayed. The options in this dialog box area are same as discussed earlier.

Note

*The **Load case/load curve** spinner and the **Curve** button of the **Surface Variable Load Object** dialog box will be enabled only when performing MES with nonlinear material analysis and the static stress with nonlinear material analysis.*

Description Area

The **Description** area of the dialog box is used to enter a comment or description about the pressure being applied.

Figure 8-50 shows the front view of a model in the Result environment with pressure that varies along the x axis by using the expression $25*(r \char`\^ 2)$. Note that the origin of the coordinate system used for the variable pressure applied in the model shown in Figure 8-50 is at the starting node of the left side of the top edge. You will learn more about the Result environment in the later chapters.

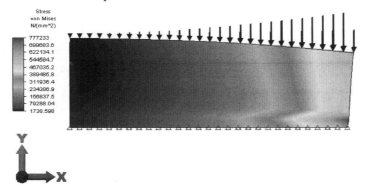

Figure 8-50 Variable pressure applied on the top surface of the model

Hydrostatic Pressure

Ribbon: Setup > Loads > Hydrostatic Pressure

Hydrostatic pressure is the pressure that is exerted by a fluid at equilibrium state. It varies linearly from the level of the fluid in the direction of increasing depth of the fluid. The magnitude of the hydrostatic pressure is directly proportional to the depth of the fluid and to the density of the fluid.

To apply hydrostatic pressure, select the surface of the model. You can also select multiple surfaces by pressing the CTRL key. After selecting the surface, choose the **Hydrostatic Pressure** tool from the expanded **Loads** panel of the **Setup** tab in the **Ribbon**; the **Creating Surface Hydrostatic Pressure Object** dialog box will be displayed, refer to Figure 8-51. Alternatively, select the surface and right-click to display a shortcut menu. Next, select **Add > Surface Hydrostatic**

Pressure from the shortcut menu displayed to invoke this dialog box. The options available in this dialog box are discussed next.

Point on Fluid Surface (P) Area

The **Point on Fluid Surface (P)** area of this dialog box is used to specify the X, Y, and Z coordinates of the point that is on the top of the fluid surface for creating the hydrostatic pressure. You can specify these coordinates in their respective **X**, **Y**, and **Z** edit boxes. Note that only the elements below this point will be affected by the magnitude of the pressure. The magnitude of the pressure on the elements of surfaces below this point will vary and will depend upon the distance from the point on the fluid surface and the density of fluid. The distance from the point on the top of the fluid surface will be multiplied by the value of the density of the fluid to calculate the magnitude of pressure on an element.

By using the **Point Selector** button in this area of the dialog box, you can also select a point directly from the drawing area as the top point on the fluid surface.

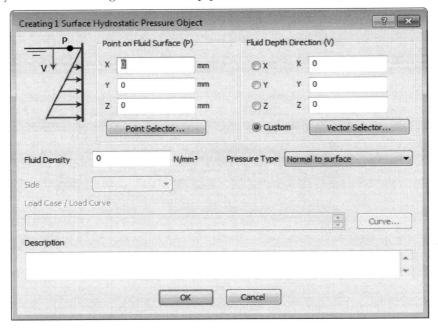

*Figure 8-51 The **Creating Surface Hydrostatic Pressure Object** dialog box*

Fluid Depth Direction (V) Area

The **Fluid Depth Direction (V)** area of the dialog box is used to specify the direction vector for the depth of fluid. You can specify the direction vector normal to the X, Y, or Z direction by selecting the respective radio button from this area. For example, on selecting the **X** radio button, the direction of fluid depth will be toward the X direction with respect to the global coordinate system.

You can also specify the direction of fluid depth other than the X, Y, and Z directions by using the **Custom** radio button. To do so, select the **Custom** radio button; the **X**, **Y**, and **Z** edit boxes of the dialog box will be enabled. In these edit boxes, you can specify the X, Y, and Z coordinates of the direction vector.

In addition to this, you can also specify the direction of fluid depth by selecting two points in the drawing area. To do so, choose the **Vector Selector** button available on the right of the **Custom** radio button in the dialog box; the dialog box will disappear. Now, you can specify direction vector by selecting two points. Specify the start point and then the end point toward which the direction will point. As soon as you specify the endpoint for the direction vector, the dialog box will be displayed again with the coordinate values of the direction vector entered in the **X**, **Y**, and **Z** edit boxes.

Fluid Density Edit Box
The **Fluid Density** edit box is used to specify the density of the fluid which will create the hydrostatic pressure.

Pressure Type Drop-down List
The **Pressure Type** drop-down list is used to select the way the hydrostatic pressure will be applied on the selected surface. By default, the **Normal to surface** option is selected in this drop-down list. As a result, the direction of the magnitude applied will be normal to the selected surfaces.

On selecting the **Full pressure in horizontal** option from this drop-down list, the magnitude of the hydrostatic pressure will be applied only in the horizontal direction. In other words, no pressure will be applied parallel to the direction vector of the fluid depth.

Figure 8-52 shows the representation of hydrostatic pressure applied on selecting the **Normal to surface** option from the **Pressure Type** drop-down list. Figure 8-53 shows the hydrostatic pressure applied horizontally by selecting the **Full pressure in horizontal** option.

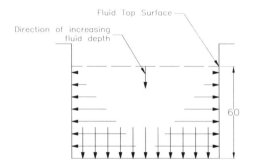

Figure 8-52 *Hydrostatic pressure applied on surfaces with the **Normal to surface** option selected*

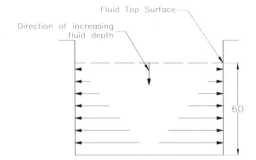

Figure 8-53 *Hydrostatic pressure applied on surfaces with the **Full pressure in horizontal** option selected*

If you select the **Horizontal component only** option from the **Pressure Type** drop-down list, the hydrostatic pressure will be applied horizontally only on the horizontal components of the selected surfaces.

Note
*If you select the **Horizontal component only** option, the magnitude of the hydrostatic pressure will be equal to the fluid density multiplied by the depth of fluid from the top surface and sin(θ), Where θ is the angle between the normal to the surface selected and the normal to the direction vector of the fluid depth.*

After specifying all the parameters for the hydrostatic pressure, choose the **OK** button from the dialog box.

Remote Load & Constraint

Ribbon: Setup > Loads > Remote Load & Constraint

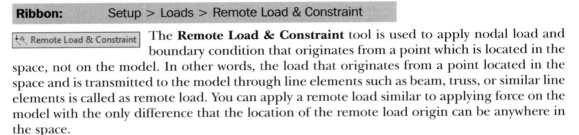 The **Remote Load & Constraint** tool is used to apply nodal load and boundary condition that originates from a point which is located in the space, not on the model. In other words, the load that originates from a point located in the space and is transmitted to the model through line elements such as beam, truss, or similar line elements is called as remote load. You can apply a remote load similar to applying force on the model with the only difference that the location of the remote load origin can be anywhere in the space.

You can select vertices, lines, edges, surfaces, and parts for applying remote load. Regardless of the element selected, the remote load will be applied to the vertices of the selection. To apply remote load, choose the **Remote Load & Constraint** tool from the expanded **Loads** panel of the **Setup** tab in the **Ribbon**; the **Create Remote Load** dialog box will be displayed, as shown in Figure 8-54. The options in this dialog box are discussed next.

*Figure 8-54 The **Create Remote Load** dialog box*

Attribute Area

The **Part**, **Surface**, and **Layer** spinners of the **Attribute** area are used to specify the part, surface, and layer attributes for the load elements. Make sure that the part number specified in the **Part** spinner is not already specified to other part of the model.

Load Location Area

The **Load Location** area is used to specify the location from where the remote load will originate. You can specify the X, Y, and Z coordinates of the point in their respective **X**, **Y**, and **Z** edit boxes of the **Load Location** area. Note that this point will act as the point of originating remote load. Alternatively, you can select a vertex or construction vertex from the graphic area and choose the **Use Selected Point** button from the **Load Location** area of the dialog box; the coordinates of the selected point will be entered in the **X**, **Y**, and **Z** edit boxes of the dialog box.

Load Destination Area

The **Load Destination** area is used to specify the destination for the remote load. To specify the destination for the remote load, select the destination for it from the graphic area. The destination can be vertices, lines, edges, surfaces, or parts. Next, choose the **Add** button from the **Load Destination** area; the selected geometry will be selected as the destination for the remote load. Also, the name of the selected geometry will be displayed in the selection area of the **Load Destination** area of the dialog box. Similarly, by using the **Add** button, you can also add multiple geometries as destinations for the remote load. Note that all the vertices of the selections will be calculated for distributing the remote load.

You can also remove the selected geometry from the list of geometries added in the **Load Destination** area. To do so, select the geometry to be removed from the selection area of the **Load Destination** area and then choose the **Remove** button; the selected geometry will be removed.

Note
*To select a vertex, line, edge, surface, or part, click on the **Selection** tab in the **Ribbon**. Next, choose the **Point or Rectangle** tool from the **Shape** panel and then the **Vertices**, **Lines**, **Edges**, **Surfaces**, or **Parts** tool from the **Select** panel of the **Selection** tab in the **Ribbon**.*

Generate Load Elements Button

The **Generate Load Elements** button of the dialog box is used to generate the load elements that transmit the load from its originating location to the nodes of the model. As soon as you choose this button, a new geometry and a node at the specified point in the space will be created. Also, the lines representing the load elements from the remote load location to the model will be displayed in the graphic area, refer to Figure 8-55.

Note
*The new geometry created on choosing the **Generate Load Elements** button will also be displayed in the **Tree View**. Therefore, you need to define its element type, element definition, and the material to continue with the analysis process.*

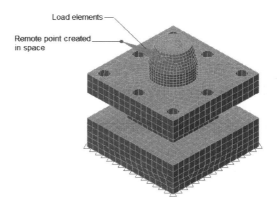

Figure 8-55 *The remote load applied on the model*

Add Load Button

The **Add Load** button of the dialog box is used to select the type of load or the boundary conditions to be applied at the remote load location. Note that this button will be enabled only after the load elements are generated.

To select the type of load or the boundary condition to be applied, click on the side arrow available on the **Add Load** button; a flyout will be displayed, as shown in Figure 8-56. You can select the required type of remote load or the boundary conditions from this flyout. Note that depending upon the type of remote load or boundary condition you select from the flyout, the dialog box for specifying the parameters will be displayed.

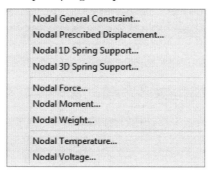

Figure 8-56 *The flyout displayed on clicking on the arrow on the **Add Load** button*

After defining all the parameters for creating the remote load, exit from the **Create Remote Load** dialog box by choosing the **Close** button; the new part with the specified part number will be displayed in the **Tree View** in red, refer to Figure 8-57. Also, the graphical representation of the line elements transforming the load from the point in space to the model will be displayed, refer to the Figure 8-58.

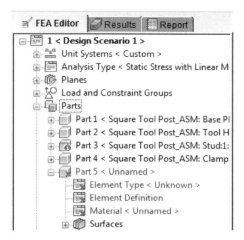
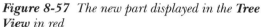

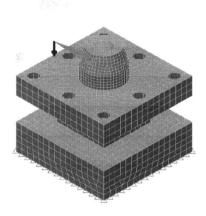

Figure 8-57 *The new part displayed in the Tree View in red*

Figure 8-58 *The graphical representation of the remote load applied on the model*

Now, you need to define the element type, element definition, and the material for the newly created part in order to continue with the analysis process.

To define the element type, select the **Element Type** sub-node under the newly created part node in the **Tree View** and right-click on it; a shortcut menu will be displayed with the options of different element types, refer to Figure 8-59. You can select the required element type from the shortcut menu displayed.

 Note

1. If you have applied moment as the remote load, then you need to select the beam elements because the moment can be transmitted only through the beam elements. The other elements such as truss, gap, and so on do not have rotational degrees of freedom and cannot transmit moments and torques.

Figure 8-59 *A shortcut menu displayed with different element type options*

2. The options displayed in the shortcut menu, refer to Figure 8-58 and 8-59, will be depends upon the type of analysis selected.

Similarly, you can define the element definition and the material for the line elements that define the remote load.

Remote Force

Ribbon:	Setup > Loads > Remote Force

 The **Remote Force** tool is used to apply force that originates from a point which is located in the space, not on the model. You can apply remote force similar to applying a force on to the model with the only difference that the location of the remote force origin can be anywhere in the space.

You can select surfaces of the model for applying remote force. Note that regardless of the surface selected, the remote force will be distributed to all its nodes. To apply remote force, select the surface on which you want to simulate the effects of remote force and then choose the **Remote Force Object** tool from the **Loads** panel of the **Setup** tab in the Ribbon; the **Creating Surface Remote Force** dialog box will be displayed, as shown in Figure 8-60. The options in this dialog box are discussed next.

Magnitude Edit Box
The **Magnitude** edit box of the dialog box is used to specify the magnitude of the remote force.

Remote Point Area
The **Remote Point** area of the dialog box is used to specify the location from where the remote force will be applied. You can specify a point for originating remote force by specifying its X, Y, and Z coordinates in their respective **X**, **Y**, and **Z** edit boxes of the dialog box.

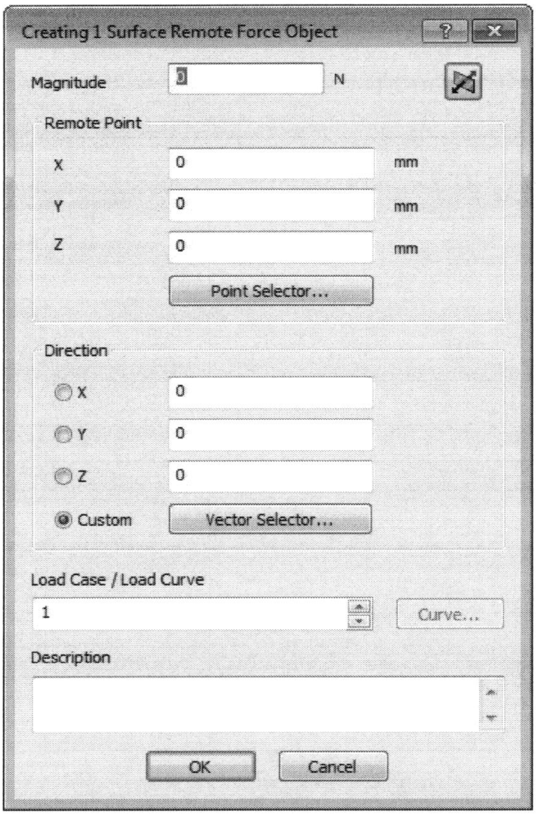

Figure 8-60 The **Create Surface Remote Force** dialog box

Alternatively, you can select a vertex or a construction vertex from the graphic area and choose the **Point Selector** button from this area of the dialog box; the coordinates of the point will be displayed in the **X**, **Y**, and **Z** edit boxes of the dialog box. Also, the point will be selected as the point of origination of the remote force.

Direction Area

The **Direction** area of the dialog box is used to specify the direction of force. You can specify the direction of force along the X, Y, or Z direction by selecting the respective radio button from this area. For example, on selecting the **X** radio button, the direction of force will be toward the X direction with respect to the Global Coordinate System.

You can also specify the direction of force other than the X, Y, or Z directions by using the **Custom** radio button. To do so, select the **Custom** radio button; the **X**, **Y**, and **Z** edit boxes of the dialog box will be enabled. In these exit boxes, you can specify the X, Y, and Z coordinates of the force direction.

In addition to this, you can also specify the direction of force by specifying two points in the graphic area. To do so, choose the **Vector Selector** button available in this area; the dialog box will disappear and you will be prompted to specify the direction vector for the force to be applied. Specify the start point and then the endpoint for defining the direction. As soon as you specify the endpoint for the direction vector, the **Create Surface Remote Force** dialog box will be displayed again. Note that the coordinate values of the specified direction vector are entered in the X, Y, and Z edit boxes.

Load Case/Load Curve Spinner

The **Load Case/Load Curve** spinner is used to select the required load case / load curve to be associated with the force being applied.

Note
*The **Curve** button that is available on the right of the **Load Case/Load Curve** spinner of the dialog box will be enabled only on performing MES with nonlinear material analysis and the static stress with nonlinear material analysis. This **Curve** button is used to define the load curve to be followed by the force.*

Description

The **Description** area of the dialog box is used to enter a comment or description about the force being applied.

After specifying all the parameters for applying the remote force, choose the **OK** button from the dialog box; the remote force will be applied on the selected surfaces and an arrow representing the direction of force will be displayed in the drawing area.

Figure 8-61 shows a model with remote force applied on the circular face of the pulley.

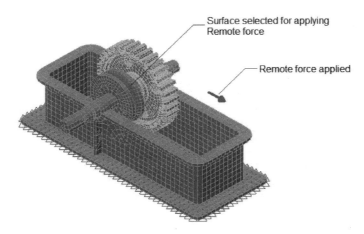

Figure 8-61 *Remote force applied on the model*

Moment

| **Ribbon:** | Setup > Loads > Moment |

 A moment is a load or force that causes rotation of the nodes to which it is applied in a model. In Autodesk Simulation Mechanical, you can select nodes or surfaces for applying moment along a direction specified by a vector. Note that the nodes connected to the nodes on which the moment is applied will also be affected by the applied moment and will tend to rotate about the specified vector. The procedure of applying nodal or surface moment is discussed next.

Applying Nodal Moment

To apply nodal moment, select the node of the model and then choose the **Moment** tool from the **Loads** panel of the **Setup** tab in the **Ribbon**; the **Creating Nodal Moment Object** dialog box will be displayed, refer to Figure 8-62. You can also select multiple nodes for applying nodal moment by pressing the CTRL key.

Alternatively, you can also invoke this dialog box by selecting a node of the model and right clicking in the drawing area to display a shortcut menu. Next, select **Add > Nodal Moments** from the shortcut menu displayed.

Note
*1. On selecting more than one node, the name of the dialog box invoked on choosing the **Moment** tool will be changed to **Creating** (number of nodes selected) **Nodal Moment Objects**.*

2. You can select nodes for applying moment either before or after invoking the dialog box.

The options available in the dialog box are same as discussed earlier. Specify the parameters of the nodal moment such as magnitude, axis of rotation, and so on in their respective edit boxes. Note that the direction of rotation of moment can be controlled by specifying the positive or negative magnitude value. On specifying the positive magnitude value, the rotation of moment will be in the clockwise direction and on specifying the negative magnitude value, the rotation of moment will be in the counterclockwise direction. After specifying all the parameters, choose the **OK** button. Figure 8-63 shows a meshed model in FEA Editor environment with moment applied on a corner node at the top right edge of the model. Figure 8-64 shows the same model in the Result environment with displacement representation that occurred due to the moment applied on the node. Note that the magnitude of the displacement is higher on the node where moment has been applied than the other nodes. In other words, the more the distance travelled by moment, the less will be its effect.

*Figure 8-62 The **Creating Nodal Moment Object** dialog box*

Note
You can apply nodal moment only to plate and beam elements. Also, if you select multiple nodes to apply moment in a model, the magnitude of the moment will be applied to each selected node. For example, if you select 5 nodes and specify 10 as the magnitude of moment in the model units about the X axis, then the applied magnitude to the model will be 50 (5*10) in model units.

Figure 8-63 The nodal moment applied on a node at the top right edge of the model

Figure 8-64 Model in the Result environment after applying moment on a node of the model

Applying Surface Moment

Similar to applying nodal moment on the selected nodes you can also apply moment on surfaces. To do so, select the surface on which you want to apply moment and then choose the **Moment** tool from the **Loads** panel of the **Setup** tab in the **Ribbon**; the **Creating Surface Moment Object** dialog box will be displayed, refer to Figure 8-65. Alternatively, you can invoke this dialog box by selecting the surface of the model and right-click in the drawing area to display a shortcut menu. Next, select **Add > Surface Moment** from the shortcut menu displayed.

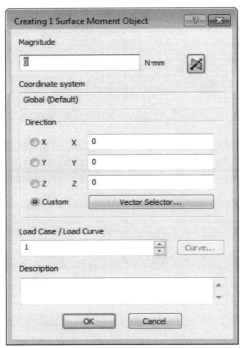

Figure 8-65   *The* **Creating Surface Moment Object** *dialog box*

The options available in the dialog box are same as discussed earlier. Specify the parameters of the surface moment such as magnitude, axis of rotation, and so on in their respective edit boxes. Next, choose the **OK** button. Figure 8-66 shows a meshed model in the FEA Editor environment with the moment applied on its front face and Figure 8-67 shows the model in the Result environment. You will learn more about the Result environment of Autodesk Simulation Mechanical in the later chapters.

Note
You can apply surface moment only on plate, brick, and tetrahedron elements.

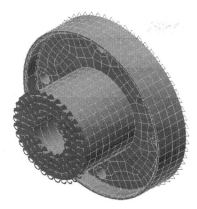

Figure 8-66 *Moment applied on the front surface of the model*

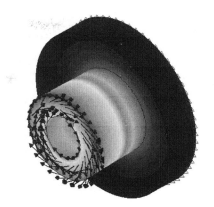

Figure 8-67 *The model in the Result environment*

Bearing Load

Ribbon: Setup > Loads > Bearing

 A bearing load is defined as a compressive load that causes radial, thrust, or combination of radial and thrust loading between the contact areas of shafts and bearing or bushings. To apply the bearing load: radial and thrust, select a cylindrical surface of the model where you want to apply bearing load from the drawing area and then choose the **Bearing** tool from the **Loads** panel of the **Setup** tab in the **Ribbon**; the **Creating Bearing Load Object** dialog box will be displayed, refer to Figure 8-68. In this dialog box, you can specify the parameters for applying thrust and radial loading. The left half of the dialog box displays the options for specifying the thrust load and right half of the dialog box displays the options for specifying the radial load.

Note
*You can also select multiple cylindrical surfaces for applying the bearing load by pressing the CTRL key. However, depending upon the number of surfaces selected, name of the dialog box invoked will be modified and will display the count number. For example, on selecting one surface, the name of the dialog box invoked will be displayed as **Creating 1 Bearing Load Object**, refer to Figure 8-68. On the other hand, on selecting two surfaces, the name of the dialog box invoked will be displayed as **Creating 2 Bearing Load Objects**.*

To apply the thrust/radial load on the selected surface, specify the magnitude of thrust/radial load in their respective **Magnitude** edit boxes. The **Magnitude** edit box available at the left in the dialog box is used to specify the magnitude of the thrust load whereas the **Magnitude** edit box at the right in the dialog box is used to specify the magnitude of the radial load. Note that in case of thrust load, the direction vector of the load being applied will automatically be selected in the axial direction of the surfaces selected. Therefore, the **X**, **Y**, and **Z** edit boxes, available on the left side in the dialog box, for specifying the direction vector for the thrust load will not be enabled. However, you can reverse the default direction of the thrust direction vector by choosing the **Toggle Vector Direction** button.

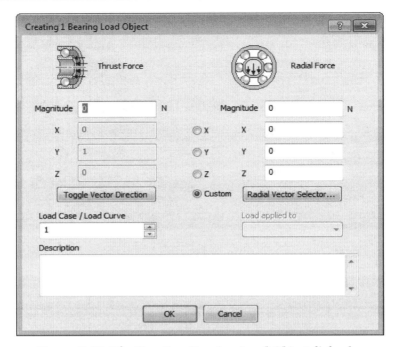

Figure 8-68 *The **Creating Bearing Load Object** dialog box*

To specify the direction vector for the radial load, select the **X**, **Y**, or **Z** radio button. You can also choose the **Custom** radio button to specify the radial direction vector other than the global X, Y, and Z directions by entering the value of x, y, and z vectors manually in their respective edit boxes. These edit boxes will be enabled on choosing the **Custom** radio button. Note that the radial direction vector should not be parallel to the thrust direction vector.

You can also specify the load case/load curve to be followed for the bearing load (thrust/radial load) being applied by using the **Load Case/Load Curve** spinner of the dialog box. After specifying all the parameters for applying the bearing load (thrust/radial load), choose the **OK** button; the bearing load will be applied to the selected cylindrical surface.

Figure 8-69 shows the thrust load applied on one half of the cylindrical surface in the FEA Editor environment and Figure 8-70 shows the same model in the Result environment after performing the analysis. You will learn more about Result environment in the later chapters.

Figure 8-71 shows the radial load applied on one half of the cylindrical surface in the FEA Editor environment and Figure 8-72 shows the same model in the Result environment after performing the analysis.

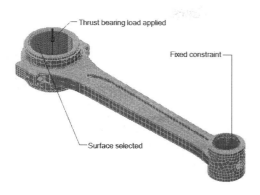

Figure 8-69 *Thrust load applied on a surface of the model in the FEA Editor environment*

Figure 8-70 *The model in the Result environment after applying the thrust load*

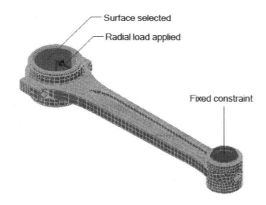

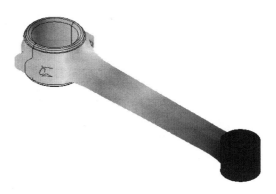

Figure 8-71 *Radial load applied on a surface of the model in FEA Editor environment*

Figure 8-72 *The model in the Result environment after applying the radial load*

Gravity

Ribbon:	Setup > Loads > Gravity

Gravity is defined as force per unit mass. It applies constant acceleration along the direction to a part that has mass. In Autodesk Simulation Mechanical, you can simulate the force of gravity by defining the acceleration occurred due to the body force. To define gravity or acceleration load for a model, choose the **Gravity** tool from the **Loads** panel of the **Setup** tab in the **Ribbon**; the **Analysis Parameters** dialog box will be displayed with the **Gravity/Acceleration** tab chosen, refer to Figure 8-73.

To specify the standard acceleration due to gravity on earth, choose the **Set for standard gravity** button from the **Gravity/Acceleration** tab. As soon as you choose this button, the standard acceleration due to the gravity on earth as per the units defined for the model will be entered automatically in the **Acceleration due to body force** edit box of the **Gravity/Acceleration** tab in the dialog box. You can also enter a different acceleration magnitude as per the requirement in the **Acceleration due to body force** edit box.

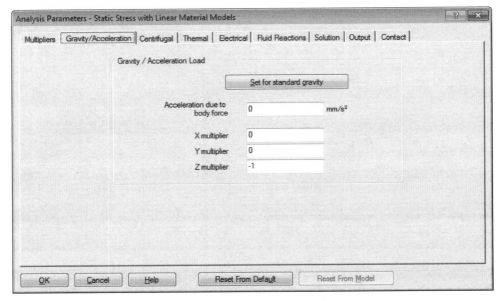

Figure 8-73 *The **Analysis Parameters** dialog box with the **Gravity/Acceleration** tab chosen*

After specifying the acceleration, you need to define the direction along which the acceleration will be applied. The **X multiplier**, **Y multiplier**, and **Z multiplier** edit boxes of the **Gravity/ Acceleration** tab in the dialog box are used to define the vector along which the acceleration will be applied. To specify the direction of acceleration toward positive direction of X axis, specify a positive value in the **X multiplier** edit box of the dialog box. To specify the direction of acceleration toward negative direction of X axis, specify a negative value in the **X multiplier** edit box. Similarly, you can specify the direction of acceleration toward positive or negative direction of Y and Z axes. Apart from specifying the acceleration direction along the X, Y, or Z axes, you can also specify the acceleration along an arbitrary direction by specifying the value in more than one edit box. For example, if you enter 1 in the X multiplier and 1 in the Y multiplier, the direction of acceleration will be bisecting the X and Y axes.

Note

*The value entered in the **Acceleration due to body force** edit box will be multiplied by the values entered in the **X multiplier**, **Y multiplier**, and **Z multiplier** edit boxes before it is applied to the model in the direction of acceleration of gravity.*

The **Reset From Default** button of the dialog box is used to reset the values to default.

After defining the gravity and its direction, choose the **OK** button from the dialog box; the gravity will be applied and the direction arrow will be displayed in the model.

Note

*You may need to change the display style of the model to view the direction arrow of the applied gravity to Mesh or Edges display style. To do so, click on the **View** tab in the **Ribbon** and then choose the **Mesh** or **Edges** option from the **Visual Style** drop-down list in the **Appearance** panel of the **View** tab.*

Figure 8-74 shows a model with an arrow pointing downward, representing the acceleration direction of the gravity applied.

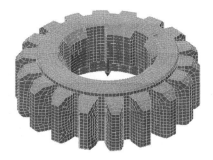

Figure 8-74 *A meshed model with direction arrow of the gravity applied*

TUTORIALS

Tutorial 1

In this tutorial, you will open the meshed model of Tutorial 1 of Chapter 5, refer to Figure 8-75. You will then assign Stainless Steel (AISI 405) material to the model. After assigning the material, you will apply the boundary conditions such as force, pin constraint, and frictionless constraint to the model, refer to Figure 8-76. While applying the Pin constraint, you need to fix the radial and axial movement. **(Expected time: 30 min)**

Figure 8-75 *The meshed model for Tutorial 1*

The following steps are required to complete this tutorial:

a. Start Autodesk Simulation Mechanical and Open Tutorial 1 of Chapter 5
b. Save Tutorial 1 of Chapter 5 model in the *c08* folder with the name *c08_tut01*.
c. Assign material to the model.
d. Apply the pin constraint.
e. Apply the frictionless constraint.
f. Apply force.
g. Save and close the model.

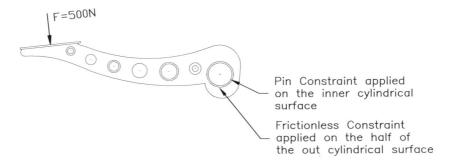

Figure 8-76 Required boundary conditions of the model

Starting Autodesk Simulation Mechanical and Opening the Tutorial File

As the required input file is saved in the *c05* folder, you need to browse to this folder and then open the *c05_tut01.fem* file in Autodesk Simulation Mechanical.

1. Start Autodesk Simulation Mechanical and then choose the **Open** button from the **Quick Access Toolbar**; the **Open** dialog box is displayed.

2. Browse to *C:\Autodesk Simulation Mechanical\c05\Tut01* folder.

3. Select the *c05_tut01.fem* file and then choose the **Open** button from the dialog box; the selected *fem* file is opened in the Autodesk Simulation Mechanical.

Saving the .fem File in the c08 Folder

When you open a file created in some other chapter, it is recommended to first save the file with a different name in the folder of the current chapter (document) before modifying it. On saving the file in the current chapter folder, the original file of the other chapter will not get modified.

1. Choose the **Save As** button from the **Application Menu**; the **Save As** dialog box is displayed.

2. Browse to the *\Autodesk Simulation Mechanical* folder. Next, create a new folder with the name *c08* by using the **New Folder** button. Create one more folder with the name *Tut01* inside the *c08* folder. Make the newly created *Tut01* folder the current folder by double-clicking on it.

3. Enter **c08_tut01** as the new name of the file in the **File name** edit box and then choose the **Save** button to save the file.

The file is saved with the new name and is now opened in the graphics area of Autodesk Simulation Mechanical.

Assigning Material

Now, you need to assign material to the model.

1. Expand the **Part** node by clicking on the **+** sign available on its left in the **Tree View**. Next, right-click on the **Material <Unnamed>** option available under the **Part** node in the **Tree**

View and then choose the **Edit Material** option from the shortcut menu displayed; the **Element Material Selection** dialog box is displayed, refer to Figure 8-77.

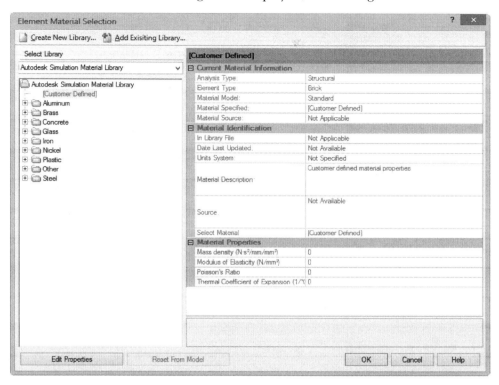

*Figure 8-77 The **Element Material Selection** dialog box*

2. Make sure **Autodesk Simulation Material Library** is selected in the **Select Library** drop-down list of the dialog box. Next, expand the **Steel** material family node by clicking on the + sign available on the left of the **Steel** node.

3. Expand the **Stainless** sub-node from the **Steel** material family and then select the **Stainless Steel (AISI 405)** material; all default properties of the Stainless Steel (AISI 405) material are displayed on the right in the dialog box, as shown in Figure 8-78.

4. Choose the **OK** button from the dialog box; all the properties of the selected material are assigned to the model.

Applying Pin Constraint

Now, you need to apply the pin constraint to the inner cylindrical surface of the model, refer to Figure 8-76.

1. Change the current orientation of the model similar to the one shown in Figure 8-79 by using the ViewCube.

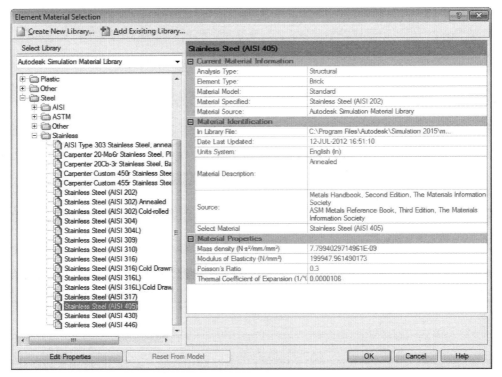

Figure 8-78 *The* ***Element Material Selection*** *dialog box with the* ***Stainless Steel (AISI 405)*** *material selected*

Figure 8-79 *Changed orientation of the model*

2. Click on the **Selection** tab of the **Ribbon** and then choose the **Circle** tool from the **Shape** panel of the **Selection** tab. Next, choose the **Surfaces** tool from the **Select** panel of the **Selection** tab in the **Ribbon**.

3. Draw a circle around the large hole of the model, refer to Figure 8-80. All the surfaces that are completely enclosed inside the circle are selected, refer to Figure 8-81.

4. Choose the **Setup** tab of the **Ribbon** and then choose the **Pin Constraint** tool from the **Constraints** panel of the **Setup** tab in the **Ribbon**; the **Creating Pin Constraint Objects** dialog box is displayed, as shown in Figure 8-82.

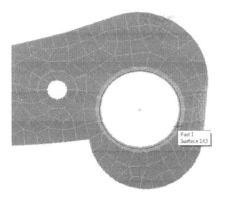

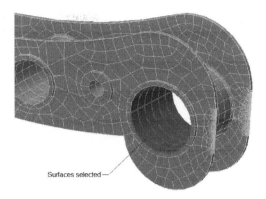

Figure 8-80 *The circle being drawn around the large hole of the model*

Figure 8-81 *Surfaces enclosed completely inside the circle selected*

5.   Select the **Fix Radial** and **Fix Axial** check boxes from the **Constraints** area of the dialog box. Next, choose the **OK** button; the pin constraint is applied. Figure 8-83 shows the rotated view of the model with the pin constraint applied.

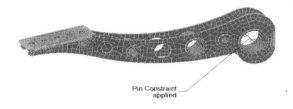

Figure 8-82 *The **Creating Pin Constraint Objects** dialog box*

Figure 8-83 *The rotated view of the model with pin constraint applied*

Applying Frictionless Constraint

Now you need to apply frictionless constraint to one half of the larger outer cylindrical surface of the model, refer to Figure 8-76.

1.   Change the current orientation of the model similar to the one shown in Figure 8-84 by using the ViewCube.

2.   Click on the **Selection** tab of the **Ribbon** and then choose the **Point or rectangle** tool from the **Shape** panel of the **Selection** tab. Next, choose the **Surfaces** tool from the **Select** panel of the **Selection** tab in the **Ribbon**.

3.   Select the bottom surface of the outer cylinder by clicking the left mouse button, refer to Figure 8-85.

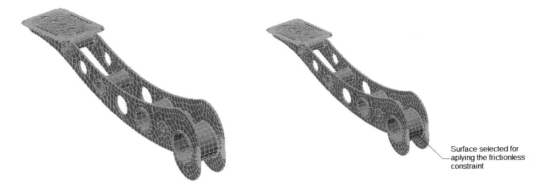

Figure 8-84 *Rotated view of the model*   **Figure 8-85** *Surface selected for applying the frictionless constraint*

4.   Choose the **Setup** tab of the **Ribbon**. Next, expand the **Constraints** panel of the **Setup** tab in the **Ribbon** by clicking on the down arrow available on the **Constraints** panel bar.

5.   Choose the **Frictionless** tool from the expanded **Constraints** panel, refer to Figure 8-86; the **Creating Frictionless Constraint Object** dialog box is displayed, as shown in Figure 8-87.

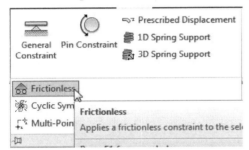

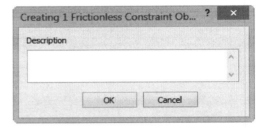

Figure 8-86 *The expanded **Constraints** panel of the **Setup** tab*   **Figure 8-87** *The **Creating Frictionless Constraint Object** dialog box*

6.   Choose the **OK** button from the **Creating Frictionless Constraint Object** dialog box; the frictionless constraint is applied to the selected surface of the model, refer to Figure 8-88.

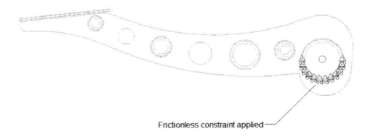

Frictionless constraint applied

Figure 8-88 *The frictionless constraint applied on the surface*

Note

*In Figure 8-88, the visual style of the model has been changed to edges visual style for clarity. You can do so by choosing the **Edges** tool from the **Visual Style** drop-down in the **Appearance** panel of the **View** tab in the **Ribbon**.*

Applying Force

Now you need to apply force of 500 N on the top planar surface of the model, refer to Figure 8-76.

1. Select the top planar surface of the model for applying the force of 500 N by clicking the left mouse button, refer to Figure 8-89.

2. Choose the **Setup** tab of the **Ribbon** and then choose the **Force** tool from the **Loads** panel; the **Creating Surface Force Object** dialog box is displayed.

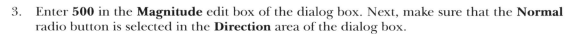

3. Enter **500** in the **Magnitude** edit box of the dialog box. Next, make sure that the **Normal** radio button is selected in the **Direction** area of the dialog box.

4. Choose the **OK** button from the dialog box; the force of magnitude 500 N is applied to the selected surface and the arrows representing the force are displayed in the graphics area, as shown in Figure 8-90.

Figure 8-89 *Surface selected for applying force*

Figure 8-90 *The arrows representation of the force applied*

Saving the Model

1. Choose the **Save** button from the **Quick Access Toolbar** to save the model.

2. Choose **Close** from the **Application Menu** to close the file.

Tutorial 2

In this tutorial, you will open the model that was meshed in Tutorial 1 of Chapter 7, refer to Figure 8-91. You will then assign Aluminum Alloy 6061-O material to all the parts of the model. After assigning the material, you will define the cross-section area for the truss element used in pin joint part of the model to 2 square millimeter. Next, apply the boundary conditions such as force of magnitude 500 N and 1D Spring constraint with stiffness value 1000 N/mm, refer to Figure 8-92. **(Expected time: 30 min)**

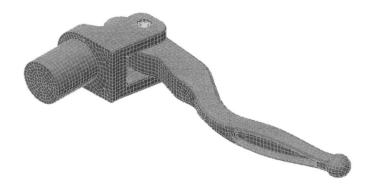

Figure 8-91 The model for Tutorial 2

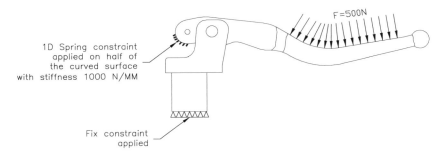

Figure 8-92 Required boundary conditions of the model

The following steps are required to complete this tutorial:

a. Start Autodesk Simulation Mechanical and Open Tutorial 1 of Chapter 7.
b. Save the opened model in the *c08* folder with the name *c08_tut02*.
c. Assign material to the model.
d. Define cross-section for truss element in pin joint.
e. Apply 1D spring support constraint.
f. Apply fixed constraint.
g. Apply force.
h. Save and close the model.

Starting Autodesk Simulation Mechanical and Opening the Tutorial File

As the required input file is saved in the *c07* folder, you need to browse to this folder and then open the *c07_tut01.fem* file in Autodesk Simulation Mechanical.

1. Start Autodesk Simulation Mechanical, if not started already. Next, choose the **Open** button from the **Quick Access Toolbar**; the **Open** dialog box is displayed.

2. Browse to *C:\Autodesk Simulation Mechanical\c07\Tut01* folder.

3. Select the *c07_tut01.fem* file and then choose the **Open** button from the dialog box; the selected *fem* file is opened in the Autodesk Simulation Mechanical.

Saving the .fem File in the c08 Folder

When you open a file created in some other chapter, it is recommended that before modifying the file, you must first save the file with a different name in the folder of the current chapter (document). The reason to do so is that if you save the file in the folder of the current chapter, the original file will not get modified.

1. Choose the **Save As** button from the **Application Menu**; the **Save As** dialog box is displayed.

2. Browse to the *Autodesk Simulation Mechanical* folder. Next, create a new folder with the name *c08* by using the **New Folder** button, if not created in Tutorial 1 of this chapter. Next, create a folder with the name *Tut02* inside the *c08* folder. Make the newly created *Tut02* folder the current folder by double-clicking on it.

3. Enter **c08_tut02** as the new name of the file in the **File name** edit box and then choose the **Save** button to save the file.

    The file is saved with the new name and is now opened in the graphics area of Autodesk Simulation Mechanical.

Assigning Material

Now, you need to assign material to all parts of the model.

1. Expand the **Part 1** and **Part 2** nodes in the **Tree View** and then select the **Material <Unnamed>** options of both the parts (**Part 1** and **Part 2**) by pressing the CTRL key.

2. Right-click to display a shortcut menu and select the **Edit Material** option from the shortcut menu displayed; the **Element Material Selection** dialog box is displayed.

3. Expand the **Aluminum** material family node by clicking on the **+** sign available on the left of the **Aluminum** node; the materials available in this family are displayed.

4. Select the **Aluminum Alloy 6061-O** material from the list of **Aluminum** material family; all the properties of the selected material are displayed on the right side of the dialog box.

5. Choose the **OK** button from the dialog box; all properties of the **Aluminum Alloy 6061-O** material are assigned to the selected parts of the model.

6. Similarly, assign the **Aluminum Alloy 6061-O** material to the **Part 3** of the model.

Defining Cross-section for Truss Element in Pin Joint

Now, you need to define cross-section for the truss elements used for creating the pin joint in the model.

1. Expand the **Part 3** node in the **Tree View**, if not expanded already. Next, select the **Element Definition** option and right-click to display a shortcut menu.

2. Select the **Edit Element Definition** option from the shortcut menu; the **Element Definition - Truss** dialog box is displayed, as shown in Figure 8-93.

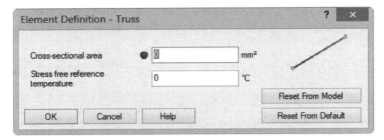

Figure 8-93 *The **Element Definition - Truss** dialog box*

3.   Enter **2** in the **Cross-sectional area** edit box of the dialog box. Next, choose **OK**; the cross-section of 2 square millimeter is defined for the truss elements.

Applying 1D Spring Support Constraint

Now, you will apply 1D Spring support constraint.

1.   Change the current orientation of the model similar to the one shown in Figure 8-94 by using the ViewCube.

2.   Click on the **Selection** tab of the **Ribbon** and then choose the **Point or Rectangle** tool from the **Shape** panel of the **Selection** tab. Next, choose the **Surfaces** tool from the **Select** panel of the **Selection** tab in the **Ribbon**.

3.   Select the curved surface by using the left mouse button, refer to Figure 8-95.

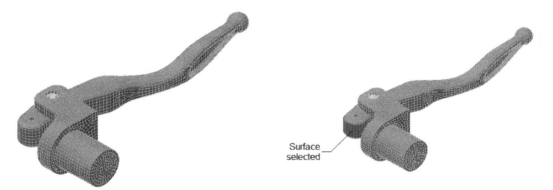

Figure 8-94 *Changed orientation of the model*   *Figure 8-95* *Surface selected for applying the 1D spring support constraint*

4.   Choose the **1D Spring Support** tool from the **Constraints** panel of the **Setup** tab in the **Ribbon**; the **Creating Surface 1D Spring Support Element** dialog box is displayed, as shown in Figure 8-96.

5.   Make sure that the **Translation** radio button is selected in the **Type** area and the **Custom** radio button is selected in the **Direction** area of the dialog box, refer to Figure 8-96.

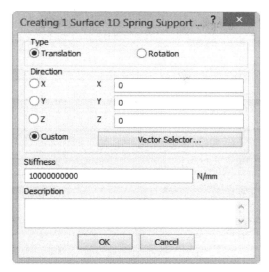

*Figure 8-96 The **Creating Surface 1D Spring Support Element** dialog box*

6.   Enter **1** in both the **X** and **Y** edit boxes of the **Direction** area in the dialog box.

7.   Enter **1000** in the **Stiffness** edit box of the dialog box. Next, choose the **OK** button; the 1D spring support is applied to the selected surface of the model, refer to Figure 8-97.

Applying Fixed Constraint

Now, you need to apply fixed constraint to the surface of the model, refer to Figure 8-92.

1.   Select the surface of the model for applying the fixed constraint, refer to Figure 8-98.

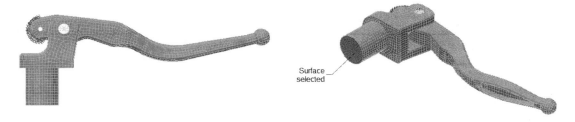

Figure 8-97 The 1D spring support applied to the selected surface of the model

Figure 8-98 Surface selected for applying the fixed constraint

2.   Select the **General Constraint** tool from the **Constraints** panel of the **Setup** tab in the **Ribbon**; the **Creating Surface General Constraint Object** dialog box is displayed, as shown in Figure 8-99.

3.   Choose the **Fixed** button from the **Predefined** area of the dialog box; all the check boxes available in the **Constrained DOFs** area of the dialog box are selected.

4.   Choose the **OK** button; the fixed constraint is applied to the selected surface.

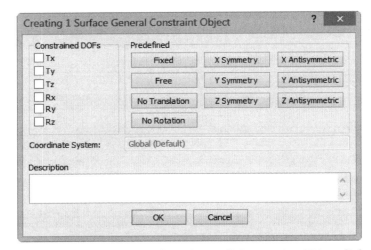

*Figure 8-99 The **Creating Surface General Constraint Object** dialog box*

Applying Force

Now, you need to apply force of 500 N on the top planar surface of the model, refer to Figure 8-92.

1. Select the top planar surfaces of the model for applying the force of 500 N by pressing the CTRL key, refer to Figure 8-100.

2. Choose the **Force** tool from the **Loads** panel of the **Setup** tab in the **Ribbon**; the **Creating Surface Force Objects** dialog box is displayed.

3. Enter **500** in the **Magnitude** edit box of the dialog box. Next, make sure that the **Normal** radio button is selected in the **Direction** area of the dialog box.

4. Choose the **OK** button from the dialog box; the force of magnitude 500 N is applied to the selected surfaces and the arrows representing the force are displayed in the graphics area, as shown in Figure 8-101.

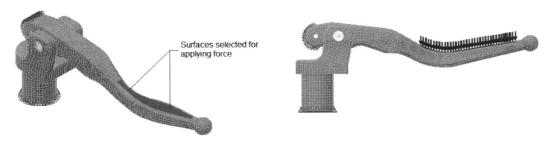

Figure 8-100 *Surfaces selected for applying force*   **Figure 8-101** *The arrows representation of force applied*

Saving the Model

1. Choose the **Save** button from the **Quick Access Toolbar** to save the model.

2. Choose **Close** from the **Application Menu** to close the file.

Self-Evaluation Test

Answer the following questions and then compare them to those given at the end of this chapter:

1. The general constraint applied on a node is called _____.

2. The _____ tool is used to constrain the radial, tangential, and axial degrees of freedom of the selected cylindrical surface.

3. On choosing the **Remote Loads and Constraint** tool, the _____ dialog box will be displayed.

4. In Autodesk Simulation Mechanical, you can apply _____ or _____ pressure to the selected surface by using the **Pressure** dialog box.

5. On choosing the _____ button from the **Predefined** area of the **Creating Nodal General Constraint Object** dialog box, the translation degree of freedom along the X, Y, and Z axes of the selected node will be constrained.

6. In Autodesk Simulation Mechanical, the default properties of a material cannot be modified. (T/F)

7. In Autodesk Simulation Mechanical, the material library file is saved in the *.mlb* file format. (T/F)

8. You can create a copy of an existing material library to create a new material library. (T/F)

9. The **General Constraint** tool is used to constrain the degree of freedom of nodes. (T/F)

10. In Autodesk Simulation Mechanical, by default, you are provided only with Autodesk Simulation Material Library. (T/F)

Review Questions

Answer the following questions:

1. Which of the following dialog boxes is displayed on choosing the **Variable Pressure** tool?

(a) **Variable Pressure**   (b) **Creating Variable Pressure**
(c) **Creating Surface Variable Pressure**   (d) None of these

2. Which of the following dialog boxes is displayed on choosing the **Gravity** tool?

    (a) **Analysis Gravity**   (b) **Analysis Parameters**
    (c) **Gravity**   (d) None of these

3. Which of the following dialog boxes is displayed on choosing the **Edit Properties** button from the **Element Material Selection** dialog box?

    (a) **Element Material Specification**   (b) **Edit Specification**
    (c) **Edit Material Specification**   (d) None of these

4. Which of the following radio buttons of the **Creating Surface Pressure/Traction Object** dialog box is used to apply a pressure that is pointed along a specific directions?

    (a) **Pressure**   (b) **Traction**
    (c) **Pressure/Traction**   (d) None of these

5. On choosing the _____ button from the **Creating Surface Pressure/Traction Object** dialog box, the **Multiplier Table Editor** dialog box will be displayed.

6. On choosing the **Import CSV** button from the **Multiplier Table Editor** dialog box, you can import the data of an existing load curve saved in the _____ file format.

7. In Autodesk Simulation Mechanical, in addition to the default libraries, you can create a new material library. (T/F)

8. In Autodesk Simulation Mechanical, you cannot delete a material from the user-defined material libraries. (T/F)

9. You can apply pin constraint to the cylindrical and planar surfaces of the model. (T/F)

EXERCISES

Exercise 1

In this exercise, you will open the model created in Tutorial 3 of Chapter 7, refer to Figure 8-102. You will then assign Stainless Steel (AISI 202) material to all parts of the model. After assigning the material, apply the boundary conditions such as pressure of magnitude 1000 N inside the model and the fixed constraint at the base end of the model, refer to Figure 8-103.

(Expected time: 30 min)

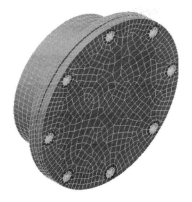

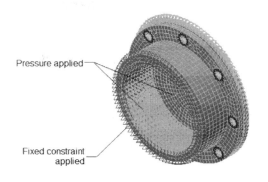

Figure 8-102 *The model for Exercise 1*

Figure 8-103 *The model after applying boundary conditions*

Exercise 2

In this exercise, you will open the model created in Tutorial 2 of Chapter 7, refer to Figure 8-104. You will then assign Stainless Steel (AISI 446) material to all parts of the model. After assigning the material, define the cross-section area for the truss elements used in pin joint parts of the model to 1 square millimeter. Next, apply the boundary conditions such as force of magnitude 650 N at the top planar face of the model in the downwards direction and fixed constraint at the base of the model, refer to Figure 8-105. **(Expected time: 30 min)**

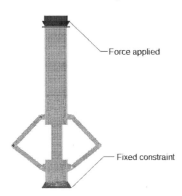

Figure 8-104 *The model for Exercise 2*

Figure 8-105 *The model after applying boundary conditions*

Answers to Self-Evaluation Test

1. nodal general constraint, **2. Pin Constraint**, **3. Create Remote Load**, **4.** pressure traction, **5. No Translation, 6.** F, **7.** T, **8.** T, **9.** T, **10.** F

Chapter 9

Performing Analysis and Viewing Results

After completing this chapter, you will be able to:

- *Specify multiple load cases*
- *Specify solver for analysis*
- *Specify result output files to be saved*
- *Change the analysis type*
- *Review the displacement results*
- *Review different types of stress and strain results*
- *Review the reaction force results*
- *Review the current result for nodes, faces, and parts*
- *Add probes to display current results*
- *Review results in graphical path format*
- *Animate current displayed result*
- *Modify the legend properties*
- *Control the display of loads and constraints in the graphics area*
- *Create presentations*
- *Capture image of the displayed result*
- *Generate, configure, and save report*

INTRODUCTION

Performing analysis and viewing results are the Solution and Postprocessor phases of any finite element analysis. In the solution phase, the FEA software generates matrices for each element, computes nodal values and derivatives, and stores the results. Also, the solution phase is completely automatic and the resultant data stored in this phase is used in the postprocessor phase for reviewing different analysis results and making the design decisions.

In Autodesk Simulation Mechanical, after preparing the model for analysis or solution phase by generating mesh, defining material properties, applying boundary conditions, defining contacts, and so on, you can run the analysis to solve the analysis problem for the defined boundary conditions by using the **Run Simulation** tool. Once the analysis problem has been performed, the model will be automatically displayed in the Result environment of Autodesk Simulation Mechanical, where you can view different analysis results and make the design decision. However, before you run the analysis process, it is important to specify the analysis parameters such as load cases (if any) solver to be used, data to be included in the output files, and so on. The procedure to specify some of the analysis parameters is discussed next.

Specifying Multiple Load Cases

Ribbon: Setup > Model Setup > Parameters

In Autodesk Simulation Mechanical, you can specify multiple load cases for an applied load. This will help you to get the results for multiple load combinations in a single run of analysis. For example, if you want the results for different force magnitudes on the same model, then instead of running the analysis separately for all magnitudes, you can specify multiple load cases defining different magnitudes of the force. After defining different load cases, when the analysis is performed, a separate set of results will be displayed for each specified load case in the Result environment of Autodesk Simulation. You will learn more about the Result environment later in this chapter.

To specify multiple load cases, choose the **Parameters** tool from the **Model Setup** panel of the **Setup** tab in the **Ribbon**; the **Analysis Parameters** dialog box will be displayed, refer to Figure 9-1. In the **Analysis Parameters** dialog box, by default, the **Multipliers** tab is chosen. In this tab, the **Load Case Multipliers** table is used to add multiple load cases for the applied loads. Note that the rows of this table represent load cases. You can add multiple rows for representing different load cases by using the **Add Row** button of the **Multipliers** tab. The options available in the **Load Case Multipliers** table are discussed next.

Note
*The options available in the **Analysis Parameters** dialog box will be dependent on the type of analysis selected. Figure 9-1 shows the **Analysis Parameters** dialog box when the Static Stress with Linear Material Models analysis is being performed.*

Index

The **Index** column of the **Load Case Multipliers** table represents the load case number. The load case numbers are generated automatically in an order.

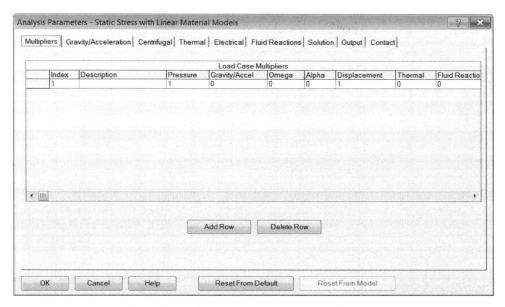

*Figure 9-1 The **Analysis Parameters** dialog box*

Description

The **Description** column of the **Load Case Multipliers** table is used to specify the description of the load case.

Pressure

The **Pressure** column of the **Load Case Multipliers** table is used to specify the pressure multiplier for a particular load case. The value entered in the field of the **Pressure** column will be multiplied with the magnitude of pressure/traction, force, hydrostatic pressure, variable pressure, and distributed loads, if applied to surfaces of the model. Note that the resulting magnitude will be applied to the model while performing the results for the respective load case.

For example, if you have applied a force of 1000 N to a model and specified a load case with multiplier value 2 in the **Pressure** column then the actual force applied to the model will be 2000 N (1000 x 2).

Note

*While applying nodal/edge forces and nodal moments to a model, you can specify the load cases by using the **Load Case / Load Curve** spinner in their respective dialog boxes.*

Gravity/Accel

The **Gravity/Accel** column of the **Load Case Multipliers** table is used to specify the gravity/acceleration multiplier for a particular load case. The value entered in this column will be multiplied with the magnitude of acceleration load defined in the **Acceleration due to body force** edit box of the **Gravity/Acceleration** tab in the **Analysis Parameters** dialog box. Note that the resulting magnitude will be applied to the model while performing the results for the respective load case.

Note

*The value entered in the **Gravity/Accel** column will not be multiplied with the magnitude of the centrifugal loading.*

Omega

The **Omega** column is used to specify the angular velocity (Omega) multiplier for the load case. The value entered in this column will be multiplied with the magnitude of the angular velocity (Omega) that is specified in the **Magnitude** edit box of the **Angular Velocity (Omega)** area in the **Centrifugal** tab of the **Analysis Parameters** dialog box. The resulting magnitude will be applied to the model while performing the results for the respective load case.

Alpha

The **Alpha** column is used to specify the angular acceleration (Alpha) multiplier for the load case. The value entered in this column will be multiplied with the magnitude of the angular acceleration (Alpha) that is specified in the **Magnitude** edit box of the **Angular Acceleration (Alpha)** area in the **Centrifugal** tab of the **Analysis Parameters** dialog box.

Displacement

The **Displacement** column is used to specify the prescribed displacement multiplier for the load case. The value entered in this column will be multiplied with the magnitude of the prescribed displacement, if applied to the model.

Thermal

The **Thermal** column is used to specify the thermal multiplier for the load case. The value entered in this column will be multiplied with the magnitude of the thermal loads applied to the model.

Electrical

The **Electrical** column is used to specify the electrical multiplier for the load case. The value entered in this column will be multiplied with the magnitude of voltages specified to the model.

Figure 9-2 shows the **Analysis Parameters** dialog box in which multiple load cases have been specified.

Note

After specifying multiple load cases for linear material model, when you perform the analysis, a separate set of results will be displayed for individual load case in the Results environment.

While applying loads for non-linear material models, you can specify different load cases defined by their respective load curves.

Specifying Solver for Analysis

Ribbon:	Setup > Model Setup > Parameters

As discussed earlier, a matrix is created for each meshed element. The solution of such matrices in FEA is done by a solver. In Autodesk Simulation Mechanical, there are different type of solvers that can be used to solve the equations of matrices depending upon the type of analysis selected. To select a solver type, choose the **Parameters** tool from the **Model Setup** panel of the **Setup** tab in the **Ribbon**; the **Analysis Parameters** dialog box will be displayed with the **Multipliers** tab chosen. Choose the **Solution** tab of this dialog box; the options available in the **Solution** tab will be displayed, as shown in Figure 9-3. In the **Solution** tab, the **Type of solver** drop-down list is used to select the type of solver to solve the equations. By default, the **Automatic** option is selected in this drop-down list. As a result, Autodesk Simulation Mechanical automatically chooses the best suitable type of solver depending upon the type of analysis, number of equations to be solved, size of the model, and so on. Some of the solvers of Autodesk Simulation Mechanical are discussed next.

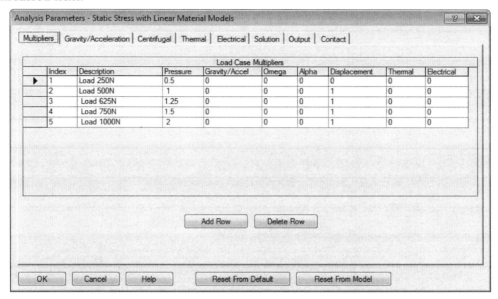

*Figure 9-2 The **Analysis Parameters** dialog box with multiple load cases*

Note
*The options available in the **Analysis Parameters** dialog box depend on the type of analysis selected. Figure 9-3 shows the **Analysis Parameters** dialog box displayed while the Static Stress with Linear Material Models analysis is being performed.*

Sparse
The Sparse solver provides direct solution to the matrix. As a result, you need not to define the maximum number of iteration to be performed for solving the equations. The Sparse solver iterates for the solutions until all the equations have not been solved. As there is no limit for the number of iteration to be performed, you will definitely get the solution, provided the model is set up properly for analysis. The Sparse solver is used for providing fast solution to midsize models.

Iterative
The Iterative solver provides solution to the matrix depending upon the number of iteration specified. As a result, you need to define the maximum number of iteration to be performed in order to solve the equations in the **Iterative Solver** area in the **Solution** tab of the dialog box. The Iterative solver only performs the number of specified iterations to solve the equations. As this solver does not exceed the specified iterations for solving the equation, there is no surety of solution. The Iterative solver is used to provide fast solution to large models and requires less memory than the Sparse solver.

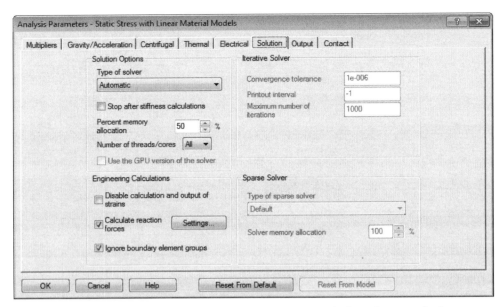

*Figure 9-3 The **Analysis Parameters** dialog box with the **Solution** tab chosen*

Specifying Result Output Files for Saving

Ribbon: Setup > Model Setup > Parameters

As discussed in earlier chapters, when you open a CAD model in Autodesk Simulation Mechanical, the system automatically saves the model with the same name as that of the original file with the **.fem* file extension at the same location. In addition to saving **.fem* file, a folder with **.ds_data* file extension is also created in the same directory automatically. This folder contains or stores all the files required for the input of processor, results of the analysis, and so on. In addition to the files created by default in this folder, you can also specify additional output files to be created by using the options available in the **Output** tab of the **Analysis Parameters** dialog box. To do so, invoke the **Analysis Parameters** dialog box and then choose the **Output** tab, refer to Figure 9-4. On selecting the required check boxes from the **Results in Output File** area of the **Output** tab, the respective output file will be generated as a text file. Similarly, you can generate files of input data specified by selecting the respective check boxes from the **Input Data in Output File** area. After selecting the required check boxes for generating output or input data files other than the default generated files, choose the **OK** button.

Changing the Analysis Type

Ribbon: Analysis > Change > Type drop-down

When you import a CAD model for analysis, the **Choose Analysis Type** dialog box is displayed, as shown in Figure 9-5. By default, the **Static Stress with Linear Material Models** analysis type is selected in this dialog box and you can select the required analysis type from this dialog box. Further, in Autodesk Simulation Mechanical, you can change the analysis type at any point of the analysis. To do so, choose the **Analysis** tab in the **Ribbon** to display the tools available in this tab and then choose the required type of analysis from the **Type** drop-down of the **Change** panel, refer to Figure 9-6. As soon as you choose the analysis type, the **Autodesk Simulation Mechanical** message window is displayed, refer to Figure 9-7. On choosing the **Yes** button from this message box, a new design scenario will be created with the selected type of analysis and its default name will be displayed in the **Tree View**. Note that the newly created design scenario will be activated by default.

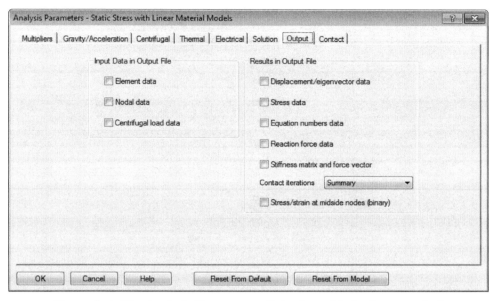

*Figure 9-4 The **Analysis Parameters** dialog box with the **Output** tab chosen*

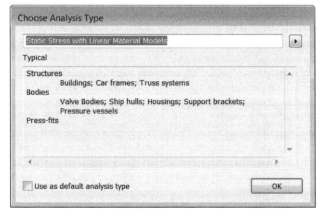

*Figure 9-5 The **Choose Analysis Type** dialog box*

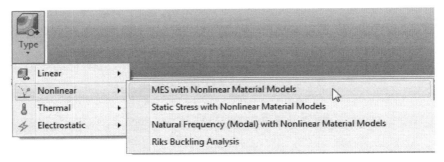

Figure 9-6 The **Type** *drop-down*

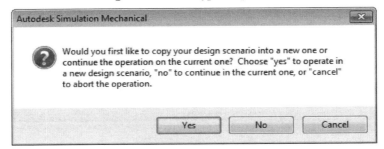

Figure 9-7 The **Autodesk Simulation Mechanical** *message box*

 Note
*You can create different design scenarios with different analysis types and boundary conditions and run analysis results for each design scenario. To activate a design scenario, double-click on its name displayed in the **Browser Tree View**.*

On choosing the **No** button from the **Autodesk Simulation Mechanical** message box, the change in the analysis type will be made in the currently activated design scenario (analysis study) instead of creating the new design scenario (analysis study). Note that if a warning message is displayed, choose the **Yes** button to continue with the process of changing the analysis type.

Alternatively, you can change the analysis type by using the **Browser Tree View**. To do so, select the **Analysis Type** node from the **Browser Tree View** in the **FEA Editor** environment and then right-click; a shortcut menu will be displayed. You can move the cursor over the **Set Current Analysis Type** option in the shortcut menu displayed; a cascading menu will be displayed, refer to Figure 9-8. From this menu, you can select the required type of analysis.

PERFORMING ANALYSIS (SOLUTION PHASE)

After preparing the model for analysis or solution phase by meshing, defining material properties, boundary conditions, contacts, and so on, you can run the analysis to solve the analysis problem for the defined boundary conditions by using the **Run Simulation** tool that is available in the **Analysis** panel of the **Analysis** tab in the **Ribbon**. This tool is discussed next.

 The **Run Simulation** tool is used to start the procedure of calculating the results. On choosing the **Run Simulation** tool from the **Analysis** panel of the **Analysis** tab in the **Ribbon**, the process of computing results will be started and the status will be displayed as the **Running** in the **Tree View**, refer to Figure 9-9.

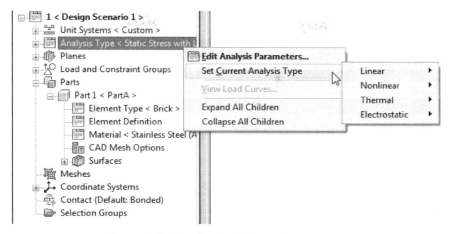

Figure 9-8 The **Analysis Type** shortcut menu

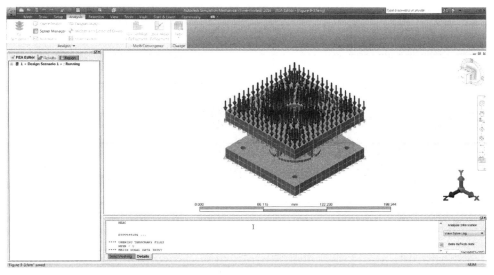

Figure 9-9 The status of analysis displayed as **Running** in the **Tree View**

In the process of computing results or solution phase, Autodesk Simulation Mechanical generates matrices for each element describing their behavior based on the input data specified including the type of analysis, properties of the materials, types and behavior of finite elements of the model, boundary conditions, and so on. After generating the matrices for each element, Autodesk Simulation Mechanical combines these matrices into a large matrix equation that represents the finite element structure and solves this equation to determine the displacement occurred due to the defined boundary conditions. Next, based on the displacement calculated, Autodesk Simulation Mechanical calculates results for strains, stresses, reaction forces, and so on. Note that all the resulting data calculated will be stored in the files which will be used for reviewing results and making the design decisions later in the postprocessor phase of analysis.

Once the process of computing results has been completed, the Result environment will be invoked automatically and the analyzed model will be displayed in it with the displacement results. The Result environment of Autodesk Simulation Mechanical is provided with various tools for viewing different analysis results, refer to Figure 9-10.

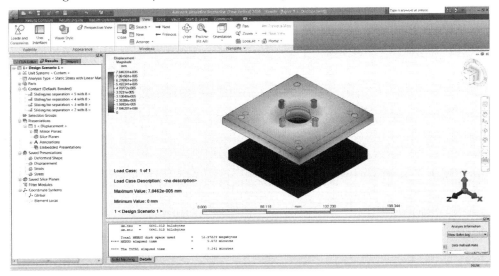

Figure 9-10   *The model in the Result environment*

VIEWING RESULTS (POSTPROCESSOR PHASE)

Viewing results of an analyzed model is the postprocessor phase of the analysis. In this phase, you can review and analyze the results through graphics display, tabular listings, and so on in the Result environment. This is one of the most important phase in the analysis process because you need to make the decisions based on the results. As discussed earlier, the Result environment of Autodesk Simulation Mechanical has various tools that are used to review results and help you make right design decisions easily.

For linear static analysis problems, you can review the results such as displacements, strains, stresses, and reaction forces. However, for modal, transient dynamic, and nonlinear analysis problems, you can review the results, such as natural frequencies, displacements, relative stresses, strains, reaction forces, and so on.

The methods for reviewing the results such as displacements, strains, stresses, and reaction forces are discussed next.

REVIEWING THE DISPLACEMENT RESULTS

As discussed earlier, once the process of computing results has been completed, the model will be displayed in the Result environment with the displacement results. This is because a displacement tool is activated by default in the **Displacement** panel of the **Results Contours** tab, refer to Figure 9-10. The tools used to review displacement results are grouped under different drop-downs. These drop-downs and the tools are discussed next.

Displacement Drop-Down

The tools available in the **Displacement** drop-down are used to review the translational displacement results of the nodes along the X, Y, and Z axes. The translational displacement results will be calculated for nodes on those elements that have translational degree of freedom free. The tools available in this drop-down are discussed next.

Magnitude

The **Magnitude** tool is used to review the total displacement results of the nodes along the X, Y, and Z axes by using the magnitude contours in the graphics area of the Result environment, refer to Figure 9-11.

Vector Plot

The **Vector Plot** tool is used to review the displacement results of the nodes along the X, Y, and Z axes by using vectors in the graphics area of the Result environment, refer to Figure 9-12.

Figure 9-11 Displacement results displayed using magnitude contours

Figure 9-12 Displacement results displayed using vectors

X

The **X** tool is used to review the total displacement result of the nodes in the X direction.

Y

The **Y** tool is used to review the total displacement result of the nodes in the Y direction.

Z

The **X** tool is used to review the total displacement result of the nodes in the Z direction.
Figures 9-13 through 9-15 show the displacement results along the X, Y, and Z directions, respectively, by using the magnitude contours.

Figure 9-13 Displacement results in X direction

Figure 9-14 Displacement results in Y direction

Figure 9-15 *Displacement results in Z direction*

Rotation Drop-Down

The tools available in the **Rotation** drop-down are used to review the rotational displacement results of the nodes. The rotational displacement results will be calculated for nodes on elements having free rotational degree of freedom such as beam, shell, and so on. This drop-down is available in the expanded **Displacement** panel, refer to Figure 9-16. The tools available in this drop-down are discussed next.

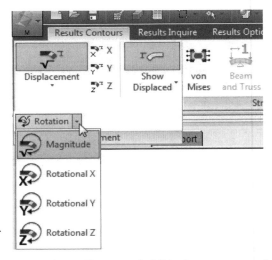

Magnitude

 The **Magnitude** tool of the **Rotation** drop-down in the **Displacement** panel is used to review the total rotational displacement results of the nodes about X, Y, and Z axes, refer to Figure 9-17.

Figure 9-16 *The expanded **Displacement** panel with the **Rotation** drop-down*

Rotational X

 The **Rotational X** tool is used to review the rotational displacement results of the nodes about X axis.

Rotational Y

 The **Rotational Y** tool is used to review the rotational displacement results of the nodes about Y axis, refer to Figure 9-18.

Rotational Z

 The **Rotational Z** tool is used to review the rotational displacement results of the nodes about Z axis.

Show Displaced Drop-Down

The tools available in the **Show Displaced** drop-down are used to show and control the deflection in the 3D model and are discussed next.

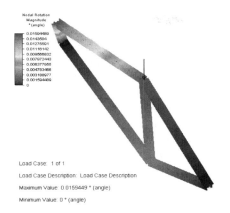

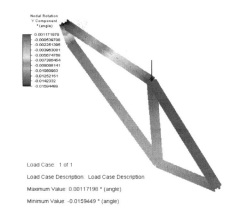

Figure 9-17 *Total rotational displacement results about the X, Y, and Z axes*

Figure 9-18 *Rotational displacement results about Y axis*

Show Displaced

The **Show Displaced** tool is used to show the displacement in the 3D model as per the default scale factor. By default, this tool is activated. As a result, the model under analysis shows the deformed shape in the graphics area. Note that the displaced scale factor can be set by using the **Displaced Options** tool in the **Show Displaced** drop-down.

Displaced Options

The **Displaced Options** tool is used to set the displaced scale factor and other related settings for the model to be deformed in the graphics area. On invoking the **Displaced Options** tool from the **Show Displaced** drop-down, the **Displaced Model Options** dialog box will be displayed, as shown in Figure 9-19. The options available in the dialog box are discussed next.

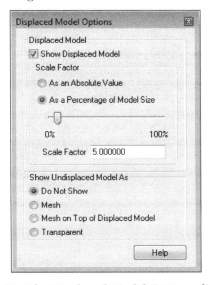

Figure 9-19 *The **Displaced Model Options** dialog box*

Show Displaced Model

By default, the **Show Displaced Model** check box is selected in the **Displaced Model** area. As a result, the **Show Displaced** tool is activated by default in the **Displacement** panel of the **Results Contours** tab in the **Ribbon** and the deformed shape of the model is displayed in the graphics area. Also, as this check box is selected, the options available in the **Displaced Model Options** dialog box are enabled. These options are discussed next.

As an Absolute Value: Select this radio button to enter the absolute value of the scale factor in the **Scale Factor** edit box. In other words, on selecting this radio button, you can specify the scale factor of the deflections in the **Scale Factor** edit box.

As a Percentage of Model Size: Select this radio button to enter the percentage value as per the model size for specifying the scale factor in the **Scale Factor** edit box. In other words, on selecting this radio button, you can specify the scale factor of the deflections in terms of the percentage of the model size in the **Scale Factor** edit box. You can also specify the percentage value of the scale factor by using the Slider available below the **As a Percentage of Model Size** radio button.

Do Not Show: By default, the **Do Not Show** radio button is selected in the **Show Undisplaced Model As** area. As a result, the original shape of the model before deformation is not displayed in the drawing area.

Mesh: On selecting this radio button, the wireframe display of the original shape of the model before deformation will be displayed in the drawing area, refer to Figure 9-20.

Mesh on Top of Displaced Model: On selecting this radio button, the edges of the original shape of the model that are hidden due to the deformed shaded shape will also be displayed in the graphics area, refer to Figure 9-21.

Figure 9-20 *The original shape model along with deformed shape on selecting the* **Mesh** *radio button*

Figure 9-21 *The model displayed after selecting the* **Mesh on Top of Displaced Model** *radio button*

Transparent: On selecting the **Transparent** radio button, the shaded transparent display of the original shape of the model before deformation will be displayed in the drawing area, refer to Figure 9-22.

REVIEWING THE STRESS RESULTS

Once the displacement results have been reviewed, you can review the stress results by using the tools available in the **Stress** panel of the **Results Contours** tab. The tools used to review stress results are discussed next.

von Mises

The **von Mises** tool is used to display the von Mises stress results. The von Mises stress results can be displayed for 2D, beam, shell, brick, and tetrahedral elements. Choose the **von Mises** tool from the **Stress** panel of the **Results Contours** tab; the von Mises stress results will be displayed in the drawing area, refer to Figure 9-23.

Figure 9-22 *The shaded transparent display of original shape*

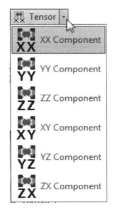

Figure 9-23 *The von Mises stress results displayed*

Tensor

Tensor tools are used to display stress results of the components in a particular direction. The stress tensor results can be displayed for 2D, beam, shell, and brick elements. The tools for displaying the stress tensor results in any particular direction are grouped together into the **Tensor** drop-down available in the **Stress** panel of the **Ribbon**, refer to Figure 9-24. All these tools are discussed next.

Figure 9-24 *The **Tensor** drop-down*

XX Component

The **XX Component** tool is used to display the resultant normal stress in the global X direction. Positive stress value indicates tension where as the negative stress value indicates compression. Figure 9-25 shows the stress results of the component in the global X direction.

YY Component

The **YY Component** tool is used to display the resultant normal stress in the global Y direction. Positive stress value indicates tension whereas the negative stress value indicates compression. Figure 9-26 shows the stress results of the component in the global Y direction.

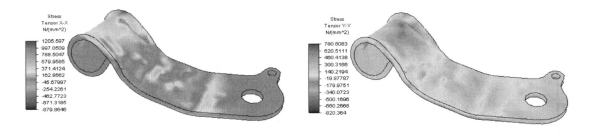

Figure 9-25 *The stress results in the global X direction*

Figure 9-26 *The stress results in the global Y direction*

ZZ Component

The **ZZ Component** tool is used to display the resultant normal stress in the global Z direction. Positive stress value indicates tension whereas the negative stress value indicates compression.

XY Component

The **XY Component** tool is used to display the resultant shear stress in the global XY direction, where X represents the direction normal to the face and Y represents the direction of the shear stress.

YZ Component

The **YZ Component** tool is used to display the resultant shear stress in the global YZ direction, where Y represents the direction normal to the face and Z axis represents the direction of the shear stress.

ZX Component

The **ZX Component** tool is used to display the resultant shear stress in the global ZX direction, where Z represents the direction normal to the face and X represents the direction of the shear stress.

Tresca*2

The **Tresca*2** tool is used to calculate the shear stress and the resultant value should be twice the maximum shear stress or tresca stress. The equation for calculating the tresca stress is given below.

$$0.5 \times MAX[\,|(S1-S2)|,\ |(S2-S3)|,\ |(S3-S1)|\,]$$

Where S1 = Maximum principal stress
S2 = Intermediate principal stress
S3 = Minimum principal stress

As the tresca*2 value calculated is twice the maximum shear stress or tresca stress, the yielding occurs in a component when this value is reached. To calculate the actual tresca stress value, you need to multiply the tresca*2 stress values by 0.5.

To calculate the tresca*2 stress values for the analyzed component in the Result environment, choose the **Tresca*2** tool from the expanded **Stress** panel, refer to Figure 9-27. Figure 9-28 shows the model with tresca*2 result displayed.

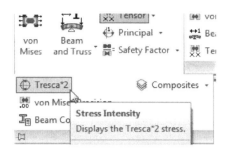

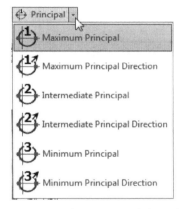

Figure 9-27 *Expanded* **Stress** *panel*   *Figure 9-28* *The tresca*2 stress results*

Principal Stress

Principal stress tools are used to calculate or display the maximum, intermediate, and minimum principal stresses on the component. The principal stress results can be displayed for 2D elements, plates elements, brick elements, and so on. The tools for calculating the principal stresses results are grouped together into **Principal** drop-down available in the **Stress** panel of the **Ribbon**, refer to Figure 9-29. All these tools are discussed next.

Figure 9-29 *The* **Principal** *drop-down*

Maximum Principal

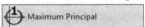

The **Maximum Principal** tool of the **Principal** drop-down in the **Stress** panel is used to display the resultant maximum principal stress normal to the maximum principal plane. Positive stress values indicate tension whereas the negative stress values indicate compression. Figure 9-30 shows the maximum principal stress results for a component.

Maximum Principal Direction

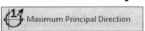

The **Maximum Principal Direction** tool of the **Principal** drop-down in the **Stress** panel is used to display the vector plot for the direction of the maximum principal stress in each element, refer to Figure 9-31.

Intermediate Principal

The **Intermediate Principal** tool of the **Principal** drop-down in the **Stress** panel is used to display the resultant intermediate principal stress. The stress in the direction normal to the minimum and maximum principal stresses is known as intermediate principal stress. Figure 9-32 shows the intermediate principal stress results for the component.

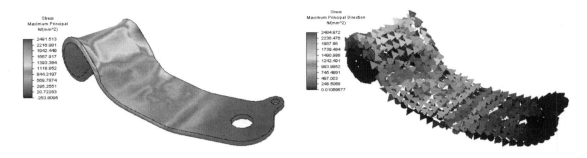

Figure 9-30 *The maximum principal stress results*

Figure 9-31 *Model displaying the direction of maximum principal stress in each element*

Intermediate Principal Direction

The **Intermediate Principal Direction** tool of the **Principal** drop-down in the **Stress** panel is used to display a vector plot for the directions of the intermediate principal stress in each element, refer to Figure 9-33.

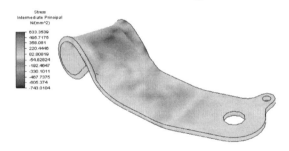

Figure 9-32 *The intermediate principal stress results*

Figure 9-33 *Direction of the intermediate principal stress in each element*

Minimum Principal

The **Minimum Principal** tool of the **Principal** drop-down in the **Stress** panel is used to display the resultant minimum principal stress normal to the minimum principal plane. Positive stress value indicates tension whereas the negative stress value indicates compression. Figure 9-34 shows the minimum principal stress results for the component.

Minimum Principal Direction

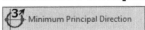

The **Minimum Principal Direction** tool of the **Principal** drop-down is used to display a vector plot for the direction of the minimum principal stress in each element, refer to Figure 9-35.

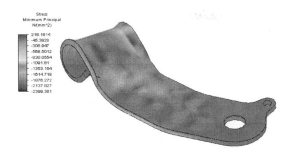

Figure 9-34 *The minimum principal stress results*

Figure 9-35 *The direction of minimum principal stress in each element*

Safety Factor

The **Safety Factor** tool is used to display the factor of safety values for the stress result. The factor of safety is defined as the ratio of allowable stress value to the actual working stress. By default, the yield stress of the material is used as the allowable stress for calculating the safety factor.

To display the safety factor for the current displayed von Mises, tresca, maximum and minimum principal stresses, or so on, choose the **Safety Factor** tool available in the **Stress** panel. Figure 9-36 shows the safety factor for von Mises stress result.

Figure 9-36 *The safety factor for von Mises stress result*

As mentioned above by default, the yield stress value of the material is defined as the allowable stress for calculating the safety factor. However, you can modify the values of allowable stress as per your requirement by using the **Set Allowable Stress Values** tool, which is discussed next.

Modifying Allowable Stress Values

To modify the default defined allowable stress value for calculating the safety factor, click on the down arrow available next to the **Safety Factor** tool in the **Stress** panel; a drop-down will be displayed. Choose the **Set Allowable Stress Values** tool from the drop-down; the **Allowable Stress Values** dialog box will be displayed, as shown in Figure 9-37.

By default, in this dialog box, the yield stress value of the material is defined as the allowable stress for calculating the safety factor. You can choose the **Load Ultimate Stress** button of the dialog box to set the ultimate stress value of the material as the allowable stress for calculating the safety factor. You can also manually enter the allowable stress value in the field corresponding to the **Allowable Stress** column and the name of the component row in the dialog box. After setting the value for allowable stress, choose the **OK** button.

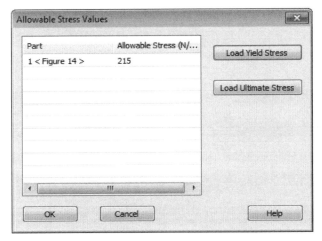

Figure 9-37 The ***Allowable Stress Values*** *dialog box*

Beam and Truss

The tools available in the **Beam and Truss** drop-down are used to review the axial, bending, and worst stresses in truss, beam, and pipe elements. Note that the truss and beam elements can be linear and nonlinear elements. Figure 9-38 shows the **Beam and Truss** drop-down. The tools available in this drop-down are discussed next.

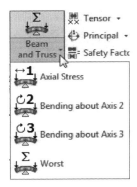

Axial Stress

The **Axial Stress** tool of the **Beam and Truss** drop-down is used to display the axial stress in beam or truss elements. The positive stress value indicates tension whereas the negative stress value indicates compression. Figure 9-39 shows the axial stress results for the beam elements.

Figure 9-38 The ***Beam and Truss*** *drop-down*

Figure 9-39 The axial stress result for beam

Bending about Axis 2

The **Bending about Axis 2** tool of the **Beam and Truss** drop-down is used to display the bending stress result. This stress is caused due to bending moment about the local axis 2 in the beam elements. Note that the local axis 2 passes through the k-node and is perpendicular to the beam, refer to Figure 9-40. Figure 9-41 shows the bending stress results for the beam element.

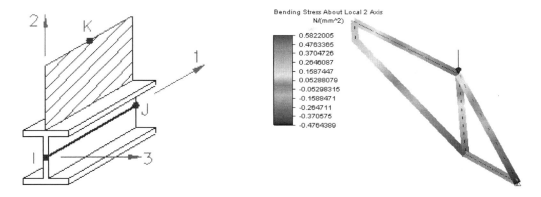

Figure 9-40 *A beam element*

Figure 9-41 *The bending stress results for the beam element*

 Tip. *A beam element is a three node element where two nodes (nodes I and J) define the element geometry and the third node (node K) is used to allow bending moment in 3D space.*

Note
The truss elements do not allow bending moment, therefore, the tools for displaying bending stress will not be enabled in case of truss elements.

Bending about Axis 3
The **Bending about Axis 3** tool of the **Beam and Truss** drop-down is used to display the bending stress caused due to the bending moment about the local axis 3 in the beam elements. Note that the local axis 3 is perpendicular to local axis 1 and 2, where axis 1 is from nodes I to J, refer to Figure 9-40. Figure 9-42 shows the bending stress results for the beam element about axis 3.

Worst
The **Worst** tool of the **Beam and Truss** drop-down is used to display the worst stress result due to the combined effect of axial stress, bending stress about axis 2, and bending stress about axis 3. Figure 9-43 shows the worst stress results.

REVIEWING THE STRAIN RESULTS
Similar to reviewing stress results, you can review the strain results such as von Mises strain, tensor strain, tresca strain, and principal strain by using the tools available in the **Strain** panel. The tools used to review various type of strain results are discussed next.

von Mises
The **von Mises** tool of the **Strain** panel in the **Results Contours** tab of the **Ribbon** is used to display the von Mises strain results. The von Mises strain results can be displayed for 2D, beam, shell, brick, and tetrahedral elements. On choosing the **von Mises** tool

from the **Strain** panel of the **Results Contours** tab, the von Mises strain results will be displayed in the graphics area, refer to Figure 9-44.

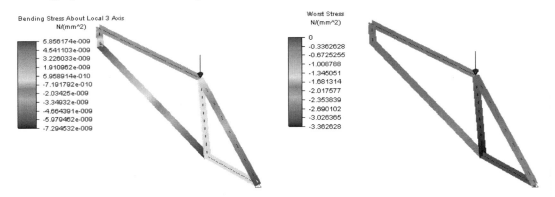

Figure 9-42 *The bending stress results for the beam element about axis 3*

Figure 9-43 *The worst stress result*

Figure 9-44 *Results of von Mises strain*

Tensor

 Tensor tools in the **Tensor** drop-down in the **Strain** panel, refer to Figure 9-45, are used to display strain results for the components in a particular direction. The strain tensor results can be displayed for 2D, beam, shell, and brick elements. All these tools are discussed next.

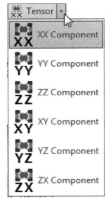

XX Component

 The **XX Component** tool of the **Tensor** drop-down in the **Strain** panel is used to display the resultant normal strain in the global X direction. Positive strain value indicates tension whereas the negative strain value indicates compression.

Figure 9-45 *The **Tensor** drop-down*

YY Component

The **YY Component** tool of the **Tensor** drop-down in the **Strain** panel is used to display the resultant normal strain in the global Y direction. Positive strain value indicates tension whereas the negative strain value indicates compression.

Similarly, you can display the normal strain results in other global directions by using their respective tools available in the **Tensor** drop-down of the **Strain** panel.

Tresca*2

 The **Tresca*2** tool of the expanded **Strain** panel is used to calculate the shear strain. Note that the resultant strain is twice the maximum shear strain or tresca strain.

To calculate the tresca*2 strain values for the analyzed component in the Result environment, choose the **Tresca*2** tool from the expanded **Strain** panel.

Principal Strain

Principal strain tools are used to calculate or display the maximum, intermediate, and minimum principal strain results for a component. The principal strain results can be displayed for various elements such as 2D elements, plates elements, and brick elements. The tools to calculate the principal strain results are grouped together in **Principal** drop-down available in the **Strain** panel of the **Result Contours** tab. All these tools are discussed next.

Maximum Principal

 The **Maximum Principal** tool in the **Principal** drop-down of the **Strain** panel is used to display the maximum principal strain which occurs normal to the maximum principal plane.

Maximum Principal Direction

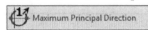 The **Maximum Principal Direction** tool in the **Principal** drop-down of the **Strain** panel is used to display the vector plot for the direction of the maximum principal strain in each element.

Similarly, you can display the results for the intermediate principal strain, vector plot for the direction of intermediate principal strain in each element, minimum principal strain, and vector plot for the direction of minimum principal strain in each element by using their respective tools available in the **Principal** drop-down of the **Strain** panel.

Beam and Truss

The tools available in the **Beam and Truss** drop-down of the **Strain** panel are used to review the axial, bending, and worst strain results in truss, beam, and pipe elements. The truss and beam elements can be linear and nonlinear elements. These tools are discussed next.

Axial Strain

The **Axial Strain** tool is used to display the axial strain in beam or truss elements.

Bending about Axis 2

The **Bending about Axis 2** tool is used to display the bending strain resulted due to bending moment about the local axis 2 in the beam elements.

Bending about Axis 3

The **Bending about Axis 3** tool is used to display the bending strain resulted due to the bending moment about the local axis 3 in the beam elements.

Worst

The **Worst** tool is used to display the worst strain resulted due to the combined effect of axial strain, bending strain about axis 2, and bending strain about axis 3. Note that the combined effect of axial strain, bending strain about axis 2, and bending strain about axis 3 causes greater amount of strain at one of the regions and is known as worst strain.

REVIEWING THE REACTION FORCE RESULTS

After reviewing displacement, stress, and strain results, it is important to review the results for reactions developed due to the applied load such as internal force, applied force, internal moment, and so on. The tools to review the results of the reactions due to the applied load are grouped together into the **Reactions** drop-down of the **Other Results** panel, refer to Figure 9-46. These tools are discussed next.

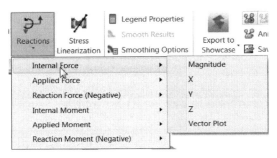

*Figure 9-46 The **Reactions** drop-down with the **Internal Force** cascading menu*

Note
*For nonlinear analysis, the **Reactions** drop-down displays only the **Reaction Force** and **Reaction Moment** menus, refer to Figure 9-47.*

*Figure 9-47 The **Reactions** drop-down for nonlinear material models*

Internal Force

The internal forces are the forces developed by the body against the applied load such as tension, compression, and so on. You can review the magnitude of the internal force developed on the entire body or the individual components along any particular global axes. To review the internal force values due to the applied load, click on the down arrow available at the lower side of the **Reactions** tool in the **Other Results** panel; the **Reactions** drop-down will be displayed, refer to Figure 9-46. Next, move the cursor over the **Internal Force** option; a cascading menu will be displayed, refer to Figure 9-46. Next, choose the required tool from this cascading menu to display the respective internal force. On choosing the **Magnitude** tool, the magnitude of the internal force will be displayed, refer to Figure 9-48. However, if you choose the X, Y, or Z tool, the internal force along the chosen global axis will be displayed. Figure 9-49 shows the internal force results in the global X axis.

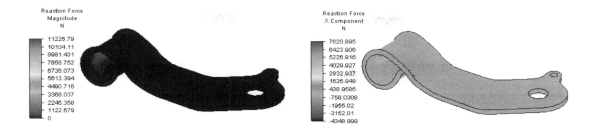

Figure 9-48 *The internal force magnitude*   **Figure 9-49** *The internal force result in global X axis*

On choosing the **Vector Plot** tool from the **Reactions** drop-down, the internal force result at each node will be displayed or represented as arrows. The length and color of the arrow represents the magnitude of the result and the direction of arrows represents the direction of result, refer to Figure 9-50.

Figure 9-50 *The arrow showing the internal force result at each node*

Applied Force

You can review the magnitude of the applied force on the model or the individual components along any particular global axes. To review the applied force values, click on the down arrow available at the lower side of the **Reactions** tool in the **Other Results** panel; the **Reactions** drop-down will be displayed, refer to Figure 9-51. Next, move the cursor over the **Applied Force** option; a cascading menu will be displayed, refer to Figure 9-51. Next, choose the required tool from this cascading menu to display the corresponding applied force results.

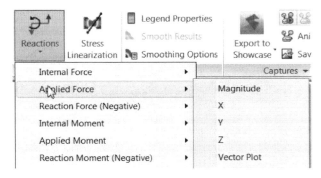

Figure 9-51 *The **Reactions** drop-down with the **Applied Force** cascading menu*

On choosing the **Magnitude** tool, the magnitude of the applied force on the model is displayed, refer to Figure 9-52. However, if you choose the X, Y, or Z tool, the applied force result for the respective components along the respective global X, Y, or Z axis will be displayed. Figure 9-53 shows the applied force results along the global Y axis.

Figure 9-52 *The applied force magnitude*   *Figure 9-53* *The applied force result in global Y axis direction*

Reaction Force (Negative)

You can review the magnitude of the residual force results on the model or on individual components of a model along the particular global axes by using the reaction force tools. To review the reaction/residual force values, click on the down arrow available at the lower side of the **Reactions** tool in the **Other Results** panel; the **Reactions** drop-down will be displayed. Next, move the cursor over the **Reaction Force (Negative)** option, a cascading menu will be displayed. Next, choose the required tool from this cascading menu to display the reaction/residual force results.

On choosing the **Magnitude** tool, the magnitude of the reaction/residual force on a model will be displayed, refer to Figure 9-54. However, if you choose the X, Y, or Z tool, the reaction/residual force result for the respective components along the respective global X, Y, and Z axes will be displayed. Figure 9-55 shows the reaction/residual force results along the global X axis.

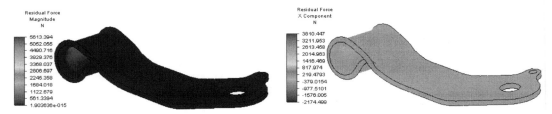

Figure 9-54 *The reaction/residual force magnitude*   *Figure 9-55* *The reaction/residual force result along global X axis*

Internal Moment

You can review the magnitude of the internal moment reaction results on to a model or on the individual components of a model along a particular global axes by using the internal moment tools. To review the internal moment reaction values, click on the down arrow available at the lower side of the **Reactions** tool in the **Other Results** panel; the **Reactions** drop-down will be displayed. Next, move the cursor over the **Internal Moment** option; a cascading menu will be displayed. Next, choose the required tool from this cascading menu to display the internal moment reaction results.

On choosing the **Magnitude** tool, the magnitude of the internal moment reaction on to a model is displayed, refer to Figure 9-56. However, if you choose the X, Y, or Z tool, the internal moment reaction result for the respective components along the respective global X, Y, and Z axes will be displayed. Figure 9-57 shows the internal moment reaction results along the global Y axis.

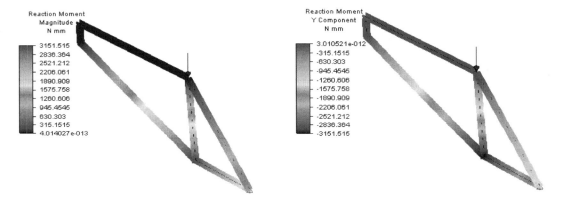

Figure 9-56 *The internal moment reaction magnitude*

Figure 9-57 *The internal moment reaction result along the global Y axis*

Note
For linear analysis, the internal moment tools are used to display the internal moment reaction at each node and the results displayed will not be considered as support reactions. However, for nonlinear analysis, the internal moment tools display the reaction moment at each node. The reaction moment is the moment that the model exerts on the surroundings and it can be considered as support reaction for nonlinear analysis.

Applied Moment
You can review the magnitude of the applied moment on a model or on individual components of a model along a particular global axes by using the applied moment tools. To review the applied moment values, click on the down arrow available at the lower side of the **Reactions** tool in the **Other Results** panel; the **Reactions** drop-down will be displayed. Next, move the cursor over the **Applied Moment** option; a cascading menu will be displayed. Next, choose the required tool from this cascading menu to display the applied moment reaction results.

Reaction Moment (Negative)
You can review the magnitude of the residual moment results on a model or on the individual components of a model along a particular global axis by using the reaction moment tools. To review the reaction/residual moment values, invoke the **Reactions** drop-down. Next, move the cursor over the **Reaction Moment (Negative)** option; a cascading menu will be displayed. Next, choose the required tool from this cascading menu to display respective reaction/residual moment results.

On choosing the **Magnitude** tool, the magnitude of the reaction/residual moment on a model is displayed, refer to Figure 9-58. However, if you choose the X, Y, or Z tool, the reaction/ residual moment result for the respective components along the respective global X, Y, and Z axes will be displayed. Figure 9-59 shows the reaction/residual moment results along the global Y axis.

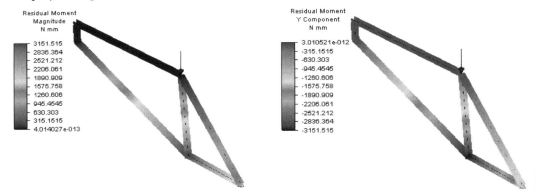

Figure 9-58 The reaction/residual moment magnitude

Figure 9-59 The reaction/residual moment result in the global Y axis direction

REVIEWING THE CURRENT RESULT FOR NODES AND FACES

In Autodesk Simulation Mechanical, you can review the current result displayed in graphics area for particular node and element of the model by using the **Current Results** tool. Sometimes reviewing results for individual node or element becomes important in order to make right decisions for analysis. To do so, first review or display the required result for the model in the graphics area. Once the required result for the model has been displayed in the graphics area, choose the **Results Inquire** tab in the **Ribbon**; the tools available in the **Results Inquire** tab are displayed. Next, choose the **Current Results** tool; the **Inquire Results** dialog box will be displayed, as shown in Figure 9-60.

Select the node or element whose results are to be inquired; the result of the selected node or element will be displayed in the dialog box. You can also select multiple nodes or elements by pressing the CTRL key to review their result. For a node, depending upon the current result displayed, the **Inquire Results** dialog box provides information about its coordinates, name of the element of the selected node, and the current result value. For an element, depending upon the current result displayed, the **Inquire Results** dialog box provides the information about the element number, and coordinates and current result values for all its nodes. Figure 9-61 shows the nodal results for a selected node and Figure 9-62 shows the element results for an element selected when the current displayed result for displacement.

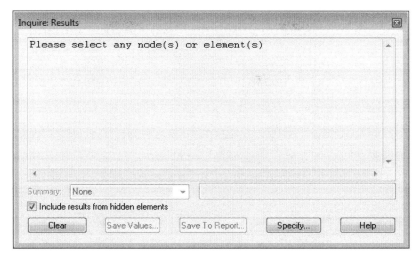

Figure 9-60 The **Inquire Results** dialog box

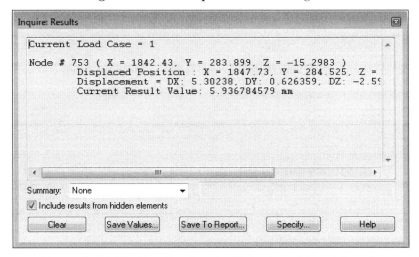

Figure 9-61 The **Inquire Results** dialog box with nodal results

Note

*To select a node/element of a model, choose the **Selection** tab in the **Ribbon**. Next, choose the **Point or Rectangle** tool from the **Shape** panel and the **Nodes/Elements** tool from the **Select** panel of the **Selection** tab in the **Ribbon**.*

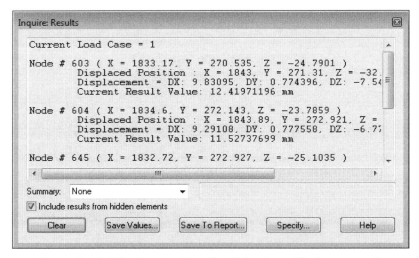

Figure 9-62 *The **Inquire Results** dialog box with element results*

The options available in the **Inquire Results** dialog box are discussed next.

Display Area

The display area of the dialog box is used to display results for selected nodes or elements depending upon the type of current results displayed.

Summary

The **Summary** drop-down list is used to display different results such as maximum magnitude, minimum magnitude, sum, and equilibrium temperature for selected nodes/elements. By default, the **None** option is selected in the **Summary** drop-down list, refer to Figure 9-62. As a result, no summary result will be displayed for the selected nodes/elements in the area available in front of the **Summary** drop-down list. If you select the **Maximum** option from this drop-down list, the maximum magnitude will be displayed from the selection in the field available next to the drop-down list, refer to Figure 9-63. On selecting the **Maximum Magnitude** option, the value of the maximum magnitude will be displayed for the selections node. The **Mean** option displays the average result of the selection (sum of results of all the selected nodes divided by number of nodes). The **Minimum** option displays the minimum result value for the selections. The **Range** option of the **Summary** drop-down list displays the range calculated by subtracting the minimum result value to the maximum result value. The **Sum** option displays the sum of the results. Similarly, you can display other respective results for the selected nodes/elements by selecting their respective option from the **Summary** drop-down list.

Include results from hidden elements

By default, the **Include results from hidden elements** check box is selected. As a result, the hidden elements will also be included in the results displayed in the **Inquire Results** dialog box. If you clear the **Include results from hidden elements** check box, then the results will only be displayed for the visible elements.

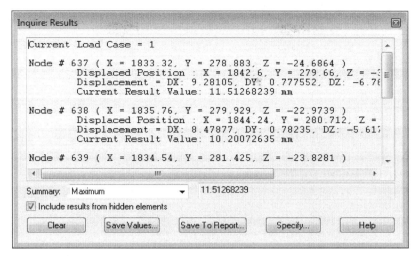

*Figure 9-63 The **Inquire Results** dialog box with the **Maximum** option selected*

Clear

The **Clear** button of the **Inquire Results** dialog box is used to clear the current displayed results from the **Display** area of the dialog box.

Save Values

The **Save Values** button of the **Inquire Results** dialog box is used to save the results in a text file. When you choose the **Save Values** button, the **Export Results** dialog box will be displayed. Browse to the location where you want to save the file and then specify the name of the file in the **File name** edit box of the dialog box. By default, the **File Out (*.out)** option is selected in the **Save as type** edit box. As a result, the result file will be saved with the *.out* file extension. You can also save the result file with *.csv* file format.

Save To Report

The **Save To Report** button of the dialog box is used to save the result displayed in the **Inquire Results** dialog box in the final report file. You will learn more about report file later in this chapter. When you choose the **Save To Report** button, the **Export Results** dialog box will be displayed. Note that this dialog box does not allow you to browse to any location to save the file and the file can only be saved with the *.out* file extension. Specify the name of the file to be saved in the **File name** edit box of the dialog box and then choose the **Save** button.

Specify

The **Specify** button is used to specify the node number or the location whose results are to be displayed. When you choose the **Specify** button, the **Specify Node(s)** dialog box will be displayed, as shown in Figure 9-64. In this dialog box, you can specify the node number in the **Node Number(s)** edit field whose result is to be displayed in the **Inquire Results** dialog box. Also, you can specify the location of the node whose result is to be displayed by selecting the **Location** radio button of the dialog box. As soon as you select the **Location** radio button, the **X**, **Y**, **Z**, and **Radius** edit boxes will be enabled. In these edit boxes, specify the X, Y, and Z coordinates of the location. The value entered in the **Radius** edit box will define the region around the location that is specified by coordinate values entered in the **X**, **Y**, and **Z** edit boxes. Note that the result for all the nodes that are enclosed inside the defined region will be displayed. Once the node number or location of the node has been specified, choose the **OK** button; the respective result is displayed in the **Inquire Results** dialog box.

REVIEWING THE CURRENT RESULT FOR A PART

You can review the results for individual parts by using the **Results by Part** tool. To review the result for individual parts, choose the **Results Inquire** tab in the **Ribbon**; the tools available in the **Results Inquire** tab are displayed. Next, choose the **Results by Part** tool; the

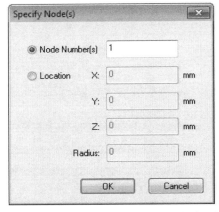

*Figure 9-64 The **Specify Node(s)** dialog box*

Inquire Results By Part dialog box will be displayed, refer to Figure 9-65. The list of all the parts of the model analyzed will be displayed in the top panel of the dialog box. Note that, by default, all the parts displayed in the top panel of the dialog box are selected. As a result, on choosing the **Inquire** button in the dialog box, the minimum and maximum current results for all the selected parts with their node numbers and location coordinates will be displayed in the lower panel of the dialog box.

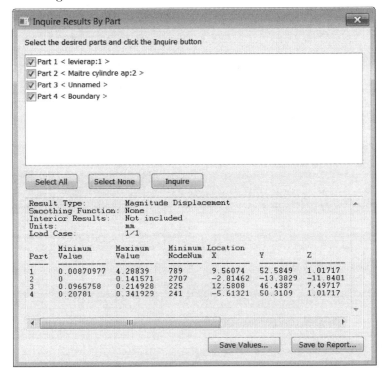

*Figure 9-65 The **Inquire Results By Part** dialog box*

You can clear the check boxes corresponding to those parts, whose results you do not want to review. On choosing the **Select None** button in the dialog box, all the check boxes of all parts displayed will be cleared. On choosing the **Select All** button, all the parts will be selected for reviewing the results. You can also select or clear the check boxes manually as per the requirement by clicking the left mouse button. The usage/functions of the **Save Values** and **Save to Report** buttons in this dialog box are the same as discussed earlier in this chapter.

ADDING PROBES TO DISPLAY CURRENT RESULTS

A probe is used to display the current result value of a node. To display the current result of the node in the graphics area in the probe, choose the **Probe** tool from the **Probes** panel of the **Results Inquire** tab in the **Ribbon**; the probe mode will be activated. Now, move the cursor over the node; a probe will be displayed with the current result of that node. Note that the probe displays the result of the node on which cursor is positioned, refer to Figure 9-66. To tag or add the probe on a node, right-click when a probe is displayed in the graphics area. Next, choose the **Add Probe** option from the shortcut menu displayed on right-clicking; the displayed probe will be fixed in the graphics area. Figure 9-67 shows the probe added on three nodes. Note that to delete the added probes, right-click and then choose the **Delete All Probes** option from the shortcut menu displayed.

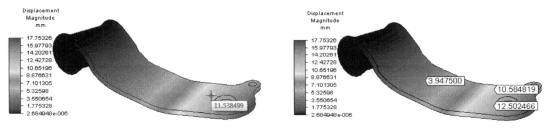

Figure 9-66 Probe displaying the current result of a node

Figure 9-67 Probe with current result value added on three nodes

You can also add probe results in the graphics area for those nodes that have maximum and minimum result values by using the **Maximum** and **Minimum** tools in the **Probes** panel of the **Results Inquire** tab in the **Ribbon**. To add probe to the nodes having maximum and minimum result values, use the **Maximum** and **Minimum** tools in the **Probes** panel of the **Results Inquire** tab in the **Ribbon**, respectively. Figures 9-68 shows the probe added on the node having maximum result value and Figure 9-69 shows the probe added on the node having minimum result value.

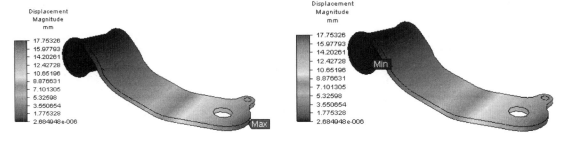

Figure 9-68 Probe added on the node having maximum result value

Figure 9-69 Probe added on the node having minimum result value

REVIEWING RESULTS IN GRAPHICAL PATH FORMAT

In Autodesk Simulation Mechanical, you can review the result on nodes or elements in graphical representation. The graphical representation can have a path form or a bar form.

Graphical Representation of Results in Path Form

In the Result environment of Autodesk Simulation Mechanical, you can review the result of nodes in graphical format along the path. Reviewing results in path form helps in comparing the results of multiple nodes. Also, you can compare the result of a single node for multiple load cases. To display results in graphical format along the nodal path, choose the **Create Path Plot** tool from the **Graphs** panel of the **Results Inquire** tab; the **Path Plot Definition** dialog box will be displayed, as shown in Figure 9-70. Select the nodes from the graphics area whose results are to be displayed in graphical format. Next, choose the **Add Selected** button from the dialog box; the node number of all the nodes selected will be displayed in the **Nodes** display area of the dialog box. Once the node numbers have been added, you can modify the order of the nodes in the **Nodes** display area to control the path of the plot, as required. This is because the plotted path uses the distance between the nodes on the horizontal axis of the graph. To modify the order of the nodes, select the node whose order is to be changed and then right- click; a shortcut menu will be displayed, as shown in Figure 9-71.

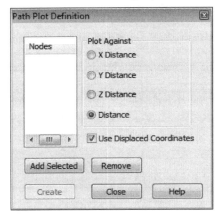

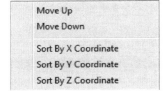

Figure 9-70 The **Path Plot Definition** *dialog box*   *Figure 9-71* *The shortcut menu*

On choosing the **Move Up** option from the shortcut menu displayed, the selected node will move up in the order. If you select the **Move Down** option, the selected node will be moved down in the order. On choosing the **Sort By X Coordinate**, **Sort By Y Coordinate**, or **Sort By Z Coordinate** option, the list of the nodes will be sorted or filtered accordingly. Note that if the shortcut menu is displayed by selecting a node in the **Nodes** display area, then on choosing the **Sort By X Coordinate**, **Sort By Y Coordinate**, or **Sort By Z Coordinate** option from the shortcut menu, the order of all the nodes listed in the **Nodes** display area of the dialog box will be sorted as per the X, Y, or Z coordinate, respectively. However, if you invoke the shortcut menu by selecting two or more than two nodes, then on selecting the **Sort By X Coordinate**, **Sort By Y Coordinate**, or **Sort By Z Coordinate** option, only the order of selected nodes will be sorted depending upon the option selected from the shortcut menu.

After selecting the nodes to be plotted and arranging the order of nodes, select the **X Distance**, **Y Distance**, **Z Distance**, or **Distance** radio button from the dialog box depending upon the abscissa to be used.

Note

*On selecting the **X Distance**, **Y Distance**, or **Z Distance** radio button, the abscissa along the horizontal axis in the graph to be plotted will be represented as the distance magnitude for X, Y, or Z direction components along the nodal path, respectively. However, on selecting the **Distance** radio button, the abscissa along the horizontal axis in the graph to be plotted will be represented as the distance magnitude along the path of the selected nodes.*

After specifying all the options, choose the **Create** button of the dialog box; the **Results Graph** window will be displayed with the plotted graph in it, refer to Figure 9-72. Figure 9-72 shows the graph plotted such that all the nodes are sorted in X coordinate and the abscissa along the horizontal axis is defined as the distance magnitude for X direction components by selecting the **X Distance** radio button.

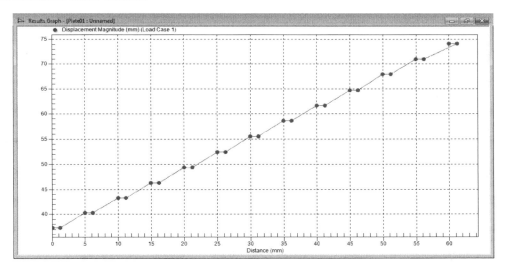

*Figure 9-72 The **Results Graph** window with graph plotted*

Graphical Representation of Results in Bar Form

In the Result environment of Autodesk Simulation Mechanical, you can also review the results of nodes in bar form. To do so, select the nodes from the graphics area whose results are to be compared through bar form graphical representation and then right-click; a shortcut menu will be displayed, as shown in Figure 9-73. Choose the **Graph Value(s)** option from the shortcut menu displayed; the **Results Graph** window will be displayed with graphical representation for each node result in bar form, refer to Figure 9-74.

Note

For the non-linear analysis, a line graph will be created against the time. It tracks the result at a node throughout the duration of the analysis. Figure 9-75 shows the line graph for non-linear material analysis for a single node and Figure 9-76 shows the line graph results for three nodes.

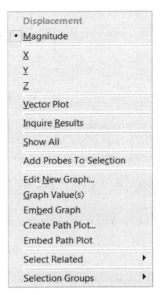

Figure 9-73 A shortcut menu displayed

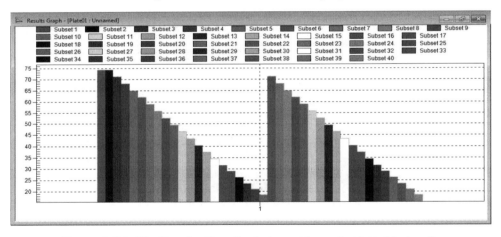

*Figure 9-74 The **Results Graph** window with results plotted in bar form*

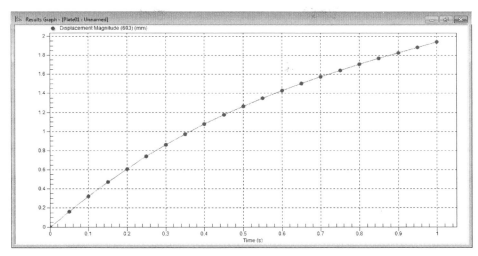

Figure 9-75 *The line graphical result for a single node in non-linear material analysis*

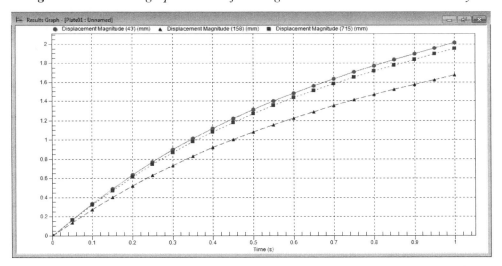

Figure 9-76 *The line graphical result for three nodes in non-linear material analysis*

Graphical Representation of Combined Nodal Results

You can also display the graphical representation of the combined results of multiple nodes. To do so, select the nodes whose results are to be reviewed in graphical form. Next, right-click in the graphics area; a shortcut menu will be displayed, refer to Figure 9-73. Choose the **Edit New Graph** option from the shortcut menu displayed; the **Edit Curve** dialog box will be displayed along with the **Results Graph** window. This window displays the results for the selected nodes in bar form, refer to Figure 9-77.

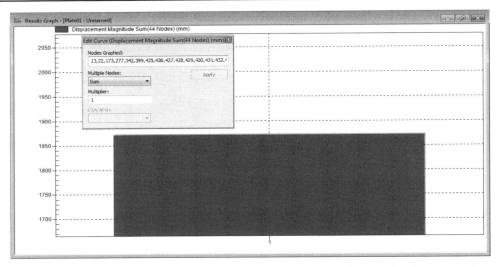

Figure 9-77 The Edit Curve dialog box displayed along with the Results Graph window

In the **Edit Curve** dialog box, the **Nodes Graphed** display area is used to display the node number of all the selected nodes separated by a comma. By default, the **Sum** option is selected in the **Multiple Nodes** drop-down list. As a result, the graph plotted in the **Results Graph** window is the combined result of all the selected nodes. On selecting the **Maximum** option, the graph will be plotted for the maximum result. If you select the **Maximum Magnitude** option, the graph will be plotted for the absolute value of all the nodal results and represents the maximum value. If you select the **Mean** option, the graph plotted displays average result value for all nodal results. On selecting the **Minimum** option, the graph plotted displays the minimum result values among all the selected nodes. If you select the **Range** option, the graph plotted displays the difference between the maximum and minimum result values.

You can also define the multiplier for the combined nodal result in the **Multiplier** edit box of the **Edit Curve** dialog box. By default, value **1** is entered in the **Multiplier** edit box. You can enter the multiplier value as required in this edit box and then choose the **Apply** button; the result value will be multiplied with the multiplier value and a graph will be plotted accordingly.

Note
For non-linear material analysis, the graphical representation for the combined result of multiple nodes is displayed against the time and in path form. Figure 9-78 shows a graph plotted for combined results of three nodes.

ANIMATING RESULTS

In Autodesk Simulation Mechanical, you can animate the effect of load applied on a model by using the **Start** tool of the **Captures** panel in the **Results Contours** tab of the **Ribbon**. To animate the result to get the displaced shapes due to the load applied, choose the **Start** button; the FEA model will start animating in the graphics area with the default settings.

You can modify the default settings for animating the FEA model by using the **Setup** tool available in the **Animate** drop-down in the **Captures** panel of the **Results Contours** tab. To modify the default animation settings, choose the down arrow displayed next to the **Animate** tool of the

Captures panel; the **Animate** drop-down will be displayed, as shown in Figure 9-79. Choose the **Setup** tool from the drop-down; the **Animation Settings** dialog box will be displayed, as shown in Figure 9-80.

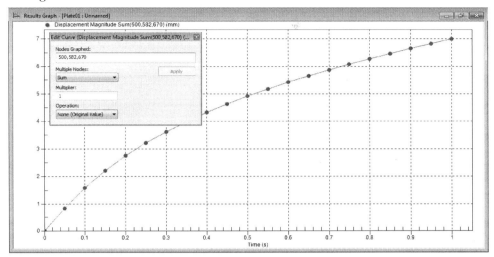

Figure 9-78 A graph plotted for combined results of three nodes at each time interval

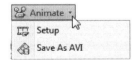

*Figure 9-79 The **Animate** drop-down*

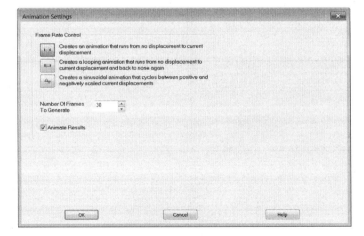

*Figure 9-80 The **Animation Settings** dialog box*

By default, the **Creates an animation that runs from no displacement to current displacement** button is activated in the **Animation Settings** dialog box. As a result, the FEA component starts animating from its original position to the maximum deformation position in the given number of time steps. You can specify the number of time steps to be taken for completing the animation from the original position to the maximum deformation shape in the **Number Of Frames To Generate** edit box of the dialog box. Note that after completing the first animation cycle, the animation starts again from the original position to the maximum deformation position.

On choosing the **Creates a looping animation that runs from no displacement to current displacement and back to none again** button, the process of animation, after completing the first animation cycle, will continue from the maximum deformation position back to the original position. On choosing the **Creates a sinusoidal animation that cycles between positive and negatively scaled current displacements** button, the animation starts from the original position to the highest displacement in the progress of every even time step and then after completing the first cycle it animates back to the original position in the same manner. After choosing the required button to animate the FEA model, choose the **OK** button from the **Animation Settings** dialog box. Next, choose the **Start** button from the **Captures** panel in the **Results Contours** tab to start the animation. To stop the animation, choose the **Stop** button from the **Captures** panel. You can also pause the animation at any time by choosing the **Pause** tool.

Note
As the non-linear material analysis is a time dependent analysis, the results change in every time step. While animating the result in the current time step is displayed in the legend that is available at the top left corner of the graphics area.

MODIFYING LEGEND PROPERTIES

A legend is used to display the maximum and minimum values of the result in the graphics area. Also, it displays results for different regions of the model with different color schemes, refer to Figure 9-81. By default, it is available at the top left corner of the graphics area with the default color scheme. You can modify the default properties of the legend such as position, color scheme, and so on, as required. To modify the legend properties, choose the **Legend Properties** tool available in the **Settings** panel of the **Results Contours** tab in the **Ribbon**; the **Plot Settings** dialog box will be displayed, as shown in Figure 9-82. Some of the options of the **Plot Settings** dialog box are discussed next.

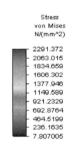

Figure 9-81 A legend for stress result

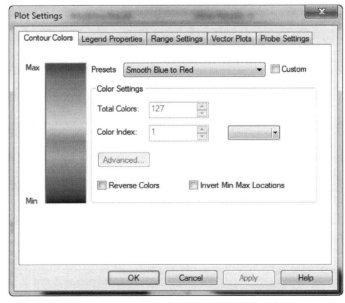

*Figure 9-82 The **Plot Settings** dialog box*

Contour Colors Tab

This tab is chosen by default. The options available in the **Contour Colors** tab are used to control the color scheme of the legend and are discussed next.

Presets

You can select the required predefined color scheme from the presets drop-down list. On doing so, the preview image of the legend displayed at the left of the dialog box will be modified as per the scheme selected.

Custom

The **Custom** check box available next to the **Presets** drop-down list is used to customize the predefined color scheme settings. When you select the **Custom** check box, the options used to specify the color settings such as **Total Colors**, **Color Index**, and **Color Switch** will be enabled in the **Color Settings** area of the dialog box. By using these options, you can define the desired color settings for a legend.

Reverse Colors

On selecting the **Reverse Colors** check box of the dialog box, you can reverse the order of colors defined in the legend.

Invert Min Max Locations

By default, the maximum result value is displayed at the top side in the legend and the minimum result value is displayed on the bottom of the legend. However, on selecting the **Invert Min Max Locations** check box of the dialog box, you can invert the location of the maximum and minimum result values in the legend.

Legend Properties Tab

The options available in the **Legend Properties** tab of the dialog box are used to control the position and appearance of the legend and are discussed next.

Show Legend Box

The **Show Legend Box** check box of the **Legend Properties** tab in the dialog box is used to control the visibility of a legend in the graphics area. By default, this check box is selected. As a result, the legend with respective result value will be displayed in the graphics area. However, if you clear this check box, the display of legend in the graphics area will be turned off.

Position

The radio buttons available in the **Position** area of the **Legend Properties** tab in the dialog box are used to control the position of the legend in the graphics area. By default, the **Upper Left** radio button is selected in the **Position** area of the **Legend Properties** tab. As a result, the legend is displayed at the upper left side of the graphics area. You can change the position of the legend as required by using the radio buttons available in the **Position** area.

Appearance

The options available in the **Appearance** area of the **Legend Properties** tab are used to control the appearance of the legend in the graphics area. You can specify the number of divisions or result values in the legend by entering the required value in the **Tick Marks** edit box of the **Appearance** area. To define the precision of the result values, enter the required precision value in the **Precision** edit box of this area. On selecting the **Border** check box of this area, a border will appear around the legend in the graphics area. To define the color of the border, you can use **Border Color Switcher** that is available next to the **Border** check box in the **Appearance** area of the dialog box.

You can also define the background color for the legend to be displayed in the graphics area by using the **Background** check box and **Background Color Switcher** in the **Appearance** area.

Range Settings Tab

The options available in the **Range Settings** tab of the dialog box are used to specify the lowest and highest range of the legend to display the result. By default, the **Automatically calculate value range** check box is selected in this tab. As a result, the lowest and the highest range of the legend will be considered as the minimum and maximum result values. To define the required range value for the legend, clear the **Automatically calculate value range** check box. On clearing this check box, the **Low** and **High** edit boxes below the check box will be enabled. In these edit boxes, you can specify the lowest and highest ranges for the minimum and maximum results to be displayed in the legend.

Vector Plots Tab

As discussed earlier, the display of FEA model for various results such as displacement and reaction forces can be represented by vector arrows in the graphics area. The options to control the settings for these vector arrows are available in the **Vector Plots** tab of the dialog box. You can control the display of vectors size by using the **Size of vectors** slider and to control the arrow head size, you can use the **Size of arrow heads** slider of the **Vector Plots** tab in the dialog box.

Probe Settings Tab

The options available in the **Probe Settings** tab are used to control the settings of probes such as font, style, and decimal places.

DISPLAYING LOADS AND CONSTRAINTS

In Autodesk Simulation Mechanical, you can turn on or off the visibility of the applied loads and constraints in the FEA model displayed in the Result environment. To do so, click on the **View** tab and then choose the **Loads and Constraints** tool from the **Visibility** panel in the **View** tab; the load and constraints will be displayed in the graphics area, refer to Figure 9-83.

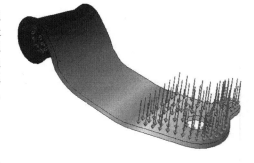

CREATING PRESENTATIONS

In Autodesk Simulation Mechanical, you can create presentations for different results such as presentation for stress, strain, displacement, and reaction forces. By default, the presentation for

Figure 9-83 *The model with the visibility of applied loads and constraints turned on*

deformed shape, displacement, strain, and stress is saved under the **Saved Presentations** node of the **Tree View** in the Result environment. However, out of all the presentations saved by default, the displacement presentation is created and loaded in the **Presentations** node of the **Tree View**, refer to Figure 9-84. As a result, by default, the displacement result is displayed in the graphics area of the Result environment. To activate or create the other presentation, select the saved presentation to be activated under **Saved Presentations** node of the **Tree View** and then right-click; a shortcut menu will be displayed. Choose the **Activate** option from the shortcut menu displayed; the selected presentation is activated and loaded in the **Presentations** node of the **Tree View**, refer to Figure 9-85. Also, the result of the activated presentation will be displayed in a separate window. To display the result of any loaded presentation in the graphics area, click on the name of the presentation in the **Presentations** node of the **Tree View** to be displayed in the graphics area. Note that each presentation loaded will display the result in a separate window.

Once the required saved presentation is activated and loaded in the **Presentations** node of the **Tree View**, you need to save it. To do so, select the presentation to be saved from the **Presentations** node and then right-click; a shortcut menu will be displayed, refer to Figure 9-86. In the shortcut menu, on selecting the **Save with Model** option, the respective presentation will be saved and listed in the **Saved Presentations** node of the **Tree View** for the model only. Note that the presentations that are saved as **Save with Model** will only be listed in the **Saved Presentations** node of the **Tree View** when you open the same model next time. However, the presentations that are saved by using the **Save with System** option from the shortcut menu will not only be listed in the **Saved Presentations** node for the same model but will also be listed for other models with similar type of analysis.

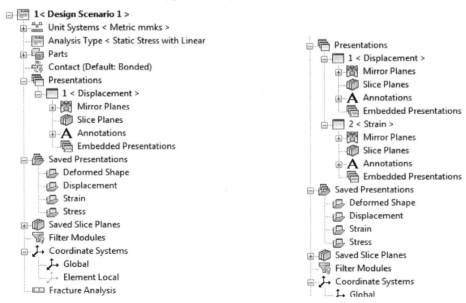

Figure 9-84 *The **Tree View** with the displacement presentation loaded by default*

Figure 9-85 *The **Tree View** with the displacement and strain presentation loaded*

Note
*The presentations created in the Result environment and displayed under the **Presentations** node of the **Tree View** will also be added in the report.*

To rename the presentation, you can choose the **Rename** option from the shortcut menu displayed and enter the new name.

Note that under every presentation in the **Tree View**, various sub-nodes and options are available such as **Mirror Planes**, **Slice Plane**, and **Annotations**, refer to Figure 9-87. Figure 9-87 shows the partial view of the **Tree View** showing the **Displacement** presentation. The options available under the **Presentation** node of the **Tree View** are discussed next.

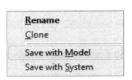

Figure 9-86 *A shortcut menu*

Figure 9-87 *The **Displacement** presentation in the **Tree View***

Mirror Planes

The **Mirror Planes** sub-node of a presentation node in the **Tree View** is used to create the mirror image of a model about a mirroring plane. Expand the **Mirror Planes** sub-node by clicking on the plus sign available on its left, refer to Figure 9-88. Note that in the expanded **Mirror Planes** sub-node, the names of the three default mirroring planes are struck through. It means that these mirroring planes are not activated. To create a mirror image of the model about a mirroring plane, double-click on the required mirroring plane available in the **Mirror Planes** sub-node; the mirroring plane will be activated and a mirror image of the model about the activated mirroring plane is displayed

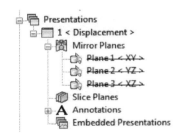

Figure 9-88 *The expanded **Mirror Planes** sub-node in the **Tree View***

in the graphics area. Figure 9-89 shows a model to be mirrored and Figure 9-90 shows the resultant model after mirroring. To deactivate the mirror image of the model, double-click again on the same mirroring plane in the **Tree View**.

Figure 9-89 *A model before mirroring*   *Figure 9-90* *The model after mirroring*

Slice Planes

The **Slice Planes** option of the presentation node in the **Tree View** is used to cut the model about a cutting plane. To do so, select the **Slice Planes** option of the **Presentations** node and then right-click; a shortcut menu will be displayed, refer to Figure 9-91. Next, move the cursor over the **Add Slice Plane** option; a cascading menu will be displayed, as shown in Figure 9-91.

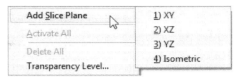

Figure 9-91 The shortcut menu displayed

Next, in the cascading menu, select the required plane as the cutting plane; the sliced image of the model is displayed in the graphics area. Also, the cutting plane is added under the **Slice Planes** sub-node in the **Presentations** node of the **Tree View**. You can delete this cutting plane by selecting the **Delete** option from the shortcut menu displayed by right-clicking on it. On deleting the cutting plane, the full view of the model is displayed in the graphics area. Figure 9-92 shows the model before cutting it and Figure 9-93 shows the same model after cutting it.

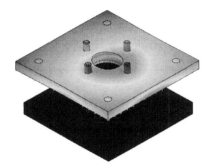

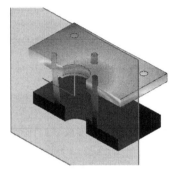

Figure 9-92 The full view of the model   *Figure 9-93 The sliced view of the model*

Annotations

The **Annotations** sub-node of the presentation node in the **Tree View** is used to add text or annotations to the graphics area. By default, in the graphics area of the Result environment, three type of annotations are available in the lower left corner. In addition to the default annotations added, you can also add annotations as per your requirement. To do so, right-click on the **Annotations** option of the **Tree View** and then choose the **Add** option from the shortcut menu displayed; the **Annotation** dialog box will be displayed, as shown in Figure 9-94.

In the **Annotation** dialog box, you can write text, specify font, write description, and specify justification in their respective edit boxes. After writing the text, choose the **OK** button from the dialog box; the preview of the text will be attached to the cursor. Next, move the cursor to the location in the graphics area and then click to specify its placement. Also, the default name of the text added is placed under the **Annotation** node in the **Tree View**. You can rename the default text name by choosing the **Rename** option from the shortcut menu displayed on right clicking on the default name in the **Tree View**.

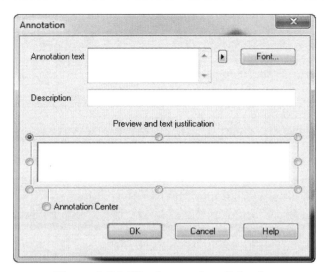

Figure 9-94 *The **Annotation** dialog box*

You can also edit the existing text written in the graphics area. To do so, expand the **Annotation** node in the **Tree View** and then right-click on the name of the text to be edited. Next, the **Edit** option from the shortcut menu displayed; the **Annotation** dialog box will be displayed. By using the options available in this dialog box, you can edit the existing text.

CREATING AN IMAGE FILE

In Autodesk Simulation Mechanical, you can save the image file of the results in standard file format. These saved image files of the result can be added in the analysis report. You will learn more about generating report in the next topic of this chapter. To create the image file of the result displayed in the graphics area, choose the **Save Image** tool in the **Captures** panel of the **Results Contours** tab in the **Ribbon**; the cursor is changed into a camera cursor ($+_\text{\tiny{圙}}$). Now you need to define the area by specifying two diagonally opposite corners. As soon as you define the area to be captured, the **Save image as** dialog box will be displayed. Browse to the location where you want to save the image file and then specify the name of the image file in the **File name** edit box of the dialog box. Next, choose the **Save** button.

GENERATING REPORT

In Autodesk Simulation Mechanical, you can generate the report of the results in the Report environment. To invoke the Report environment, choose the **Report** tab available at the top of the **Tree View**; the Report environment will be invoked and the report of different results and input parameters will be generated, refer to Figure 9-95.

The Report environment of Autodesk Simulation Mechanical is used to create the report of different analysis results including all the input parameters. You can save the report in different file formats such as PDF (*.pdf), WORD (*.doc), and HTML (*.htm) by selecting the respective tool from the **Save As** panel of the **Report** tab in the **Ribbon**.

Note that in the left side of the display area of the Report environment, the default contents that are added in the report are displayed in the **Tree View**. You can select the content whose result you want to display in the display area. You can control the display of contents in the report by using the **Configure** tool of the **Setup** panel in the **Report** tab of the **Ribbon**. To do so, choose the **Configure** tool; the **Configure Report** dialog box will be displayed, refer to Figure 9-96.

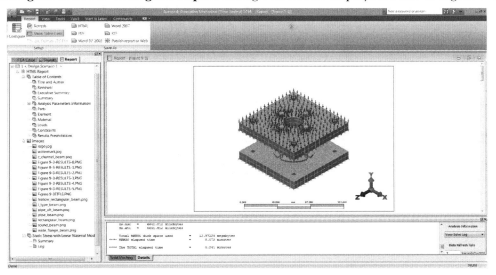

Figure 9-95 *The Report environment*

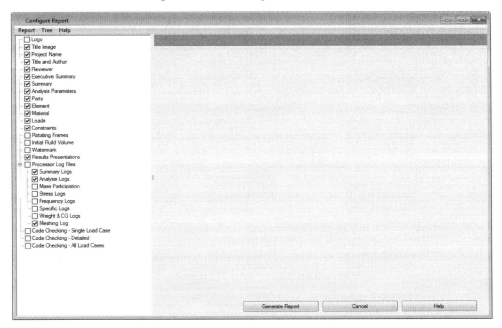

Figure 9-96 *The Configure Report dialog box*

Note that in the report only those parameters/headings will be available that are selected in the left panel of the **Configure Report** dialog box. You can select or clear the check boxes of the headings as per your requirement and then choose the **Generate Report** button of the dialog box. The report will be generated depending upon the heading selected in the **Configure Report** dialog box.

In the **Configure Report** dialog box, you can change the order of the headings. To do so, select the heading to be reordered and then drag and drop the selected heading in the required position; the selected heading will be reordered. Note that the report generated displays the headings in the same order as they are given in the **Configure Report** dialog box.

You can also add new text headings, images, and so on in the list of headings available in the **Configure Report** dialog box by using the options available in the **Tree** menu of the **Configure Report** dialog box.

TUTORIALS

Tutorial 1

In this tutorial, you will open the structure created in Tutorial 3 of Chapter 4, refer to Figure 9-97. You will then assign Stainless Steel (AISI 202) material to the structural model. After assigning the material, you will apply the boundary conditions such as force and fix constraint, refer to Figure 9-97. Also, the structure contains the beam elements of two different cross-section angles: 8 X 8 X 1 and 6 X 6 X 1.   (**Expected time: 30 min**)

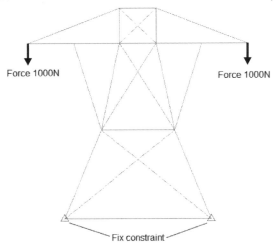

Figure 9-97 The model

The following steps are required to complete this tutorial:

a. Start Autodesk Simulation Mechanical.
b. Open Tutorial 3 of Chapter 4 in Autodesk Simulation Mechanical.
c. Define element type.
d. Assign stainless steel (AISI 202) material.
e. Define cross-section for beam elements.
f. Apply fixed constraint.
g. Apply force.
h. Run analysis.
i. Review axial stress.
j. Review axial strain.
k. Save and close the model.

Opening Tutorial 3 of Chapter 4

As the required input file is saved in the *c04* folder, you need to browse to this folder and then open the *c04_tut03.fem* file in Autodesk Simulation Mechanical.

1. Start Autodesk Simulation Mechanical and then choose the **Open** button from the **Quick Access Toolbar**; the **Open** dialog box is displayed, if it is not displayed by default.

2. Browse to the *C:\Autodesk Simulation Mechanical\c04\Tut03* folder.

3. Select the *c04_tut03.fem* file and then choose the **Open** button from the dialog box; the selected *fem* file is opened in Autodesk Simulation Mechanical.

Saving the .fem File in the c09 Folder

When you open a file of some other chapter, it is recommended that you first save the file with some other name in the folder of the current chapter (document) before modifying the file. This is because on saving the file in the folder of the current chapter, the original file will not get modified.

1. Choose the **Save As** button from the **Application Menu**; the **Save As** dialog box is displayed.

2. Browse to the *C:\Autodesk Simulation Mechanical* folder. Next, create a new folder with the name *c09* by using the **New Folder** button. Create one more folder with the name *Tut01* inside the *c09* folder. Make the newly created *Tut01* folder the current folder by double-clicking on it.

3. Enter **c09_tut01** as the new name of the file in the **File name** edit box and then choose the **Save** button to save the file.

   The file is saved with the new name and is now opened in the graphics area of Autodesk Simulation Mechanical.

Defining Element Type

Now, you need to define the beam elements for the structure.

1. Expand the **Part 1** node in the **Tree View**, if it is not expanded by default.

2. Select the **Element Type <Unknown>** option from the expanded **Part 1** node and then right-click to display a shortcut menu. Next, choose the **Beam** option from the shortcut menu displayed; the beam elements for the structure are defined.

Assigning Material

Now, you need to assign material to the entire structure model.

1. Select the **Material <Unnamed>** option from the expanded **Part 1** node of the **Tree View** and then right-click to display a shortcut menu.

2. Choose the **Edit Material** option from the shortcut menu displayed; the **Element Material Selection** dialog box is displayed.

3. Expand the **Steel** material family node by clicking on the **+** sign available on the left of the **Steel** node; the categories available in the Steel material family are displayed.

4. Expand the **Stainless** category; the materials available in the expanded category are displayed.

5. Select the **Stainless Steel (AISI 202)** material from the list of **Steel** material family; all the properties of the selected material are displayed on the right side of the dialog box.

6. Choose the **OK** button from the dialog box; all the properties of the selected material are assigned to the structure.

Defining Cross-Section for Beam Elements

Now, you need to define cross-section for the beam elements. Note that in this structure, the beam elements are made up of two cross-sections 8 X 8 X 1 and 6 X 6 X 1.

1. Select the **Element Definition** option from the expanded **Part 1** node of the **Tree View** and then right-click to display a shortcut menu.

2. Choose the **Edit Element Definition** option from the shortcut menu; the **Element Definition - Beam** dialog box is displayed, refer to Figure 9-98.

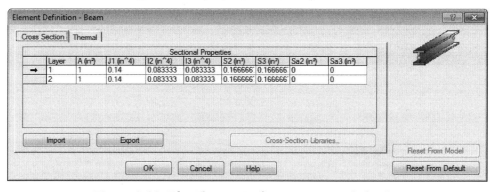

Figure 9-98 The Element Definition - Beam dialog box

3. Move the cursor to the field available in front of the Layer 1 row in the dialog box and then click to select it when the cursor changes to an arrow cursor, refer to Figure 9-98. On selecting the Layer 1 row of the dialog box, the **Cross-Section Libraries** button is enabled.

4. Choose the **Cross-Section Libraries** button; the **Cross-Section Libraries** dialog box is displayed.

5. Select the **aisc 2001** option from the **Section database** drop-down list of the **Cross-Section Libraries** dialog box.

6. Select the **L** option from the **Section type** drop-down list and then select the **L8X8X1** option from the **Section list** area of the dialog box, refer to Figure 9-99.

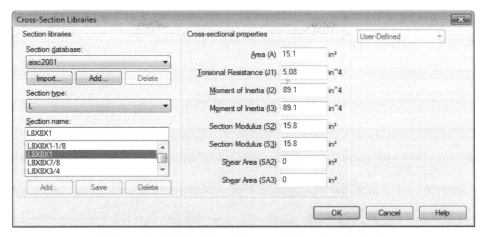

*Figure 9-99 The **Cross-Section Libraries** dialog box*

7. Choose **OK**; the cross-section 8X8X1 is defined for the beam elements of Layer 1.

8. Similarly, define the cross-section 6X6X1 for the beam elements of Layer 2. After defining the cross-sections for the beam element, exit the **Element Definition** dialog box by choosing the **OK** button.

Applying Fixed Constraint

Now, you will apply fixed constraint.

1. Changethe orientation of the model to the one hown in Figure 9-100, by using the ViewCube.

2. Choose the **Selection** tab in the **Ribbon** and then choose the **Point or Rectangle** tool from the **Shape** panel of the **Selection** tab. Next, choose the **Vertices** tool from the **Select** panel of the **Selection** tab.

3. Select the lower right and lower left corners of the structure using the left mouse button. Note that for multiple selections, you need to press the CTRL key.

4. Choose the **General Constraint** tool of the **Constraints** panel of the **Setup** tab in the **Ribbon**; the **Creating 2 Nodal General Constraint Objects** dialog box is displayed.

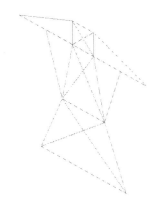

Figure 9-100 Changed orientation of the model

5. Choose the **Fixed** button in the **Creating 2 Nodal General Constraint Objects** dialog box; the fixed constraint is applied to the selected vertices of the structure. Next, choose the **OK** button and press the ESC key. Figure 9-101 shows the structure after applying the fixed constraint.

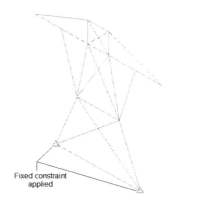

Figure 9-101 *Structure model after applying the fixed constraint*

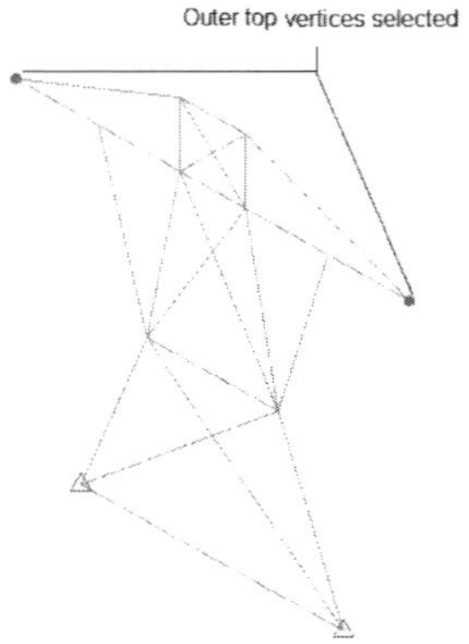

Figure 9-102 *Outer top vertices selected for applying force*

Applying Force

Now, you need to apply a force of 500 N on the outer top vertices of the structure model.

1. Select the top outer vertices of the structure by using the CTRL key, refer to Figure 9-102.

2. Choose the **Force** tool from the **Loads** panel of the **Setup** tab in the **Ribbon**; the **Creating Nodal Force Objects** dialog box is displayed.

3. Enter **-1000** in the **Magnitude** edit box of the dialog box. Next, select the **Y** radio button in the **Direction** area of the dialog box. Note that the negative and positive magnitude values are used to define the direction of the force applied.

4. Choose the **OK** button from the dialog box; the force of magnitude 1000 N is applied and the arrows representing the force are displayed in the graphics area, as shown in Figure 9-103.

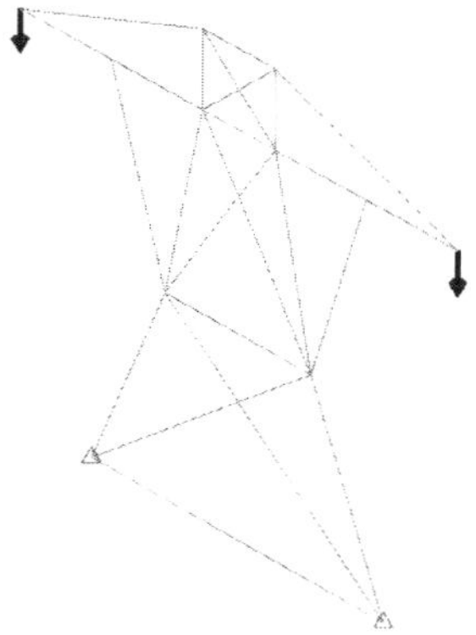

Running Analysis

After specifying all the required input parameters such as loads, constraints and material, you can run the analysis and review the results.

Figure 9-103 *Structure model after applying the force of magnitude 1000 N*

1. Choose the **Analysis** tab in the **Ribbon** and then choose the **Run Simulation** tool in the **Analysis** panel; the process of structural analysis is started.

2. Once the process of analysis is completed, the **Result** environment of Autodesk Simulation Mechanical is invoked with displacement result displayed in the graphics area, refer to Figure 9-104.

Reviewing Axial Stress

After reviewing the displacement result, you need to review the axial stress of the structure.

1. Click on the down arrow available on the right of the **Beam and Truss** tool in the **Stress** panel of the **Results Contours** tab in the **Ribbon**; a flyout is displayed.

2. Choose the **Axial Stress** tool from the flyout displayed; the axial stress result is displayed in the graphics area, refer to Figure 9-105.

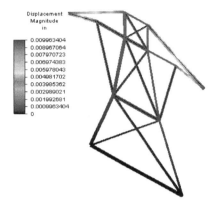

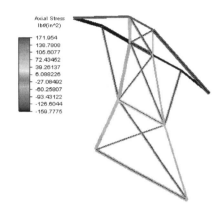

Figure 9-104 *Displacement result displayed in the graphics area*

Figure 9-105 *Axial stress result displayed in the graphics area*

Reviewing Axial Strain

After reviewing the displacement and axial stress results, you need to review the axial strain result for the structure.

1. Click on the down arrow available on the right of the **Beam and Truss** tool in the **Strain** panel of the **Results Contours** tab in the **Ribbon**; a flyout is displayed.

2. Choose the **Axial Strain** tool from the flyout displayed; the axial strain result is displayed in the graphics area, refer to Figure 9-106.

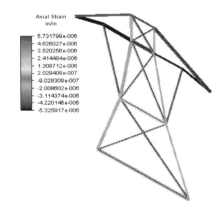

Saving the Model

1. Choose the **Save** button from the **Quick Access Toolbar** to save the model.

Figure 9-106 *Axial strain result displayed in the graphics area*

2. Choose **Close** from the **Application Menu** to close the file.

Tutorial 2

In this tutorial, you will open the model of Tutorial 1 of Chapter 8, refer to Figure 9-107 and then run the analysis. After completion of the analysis process, you need to review the results such as displacement, stress, and strain. Also, you will create different presentations of the displacement, stress, and strain results and generate a report.   (**Expected time: 30 min**)

Figure 9-107 *The model for Tutorial 2*

Note
In Figure 9-107, the visibility of loads and constraints has been turned off for clarity. To do so, choose the **View** *tab of the* **Ribbon** *and then choose the* **Loads and Constraints** *tool in the* **Visibility** *panel.*

The following steps are required to complete this tutorial:

a. Open Tutorial 1 of Chapter 8 in Autodesk Simulation Mechanical.
b. Save the Tutorial 1 of Chapter 8 model in the *c09* folder with the name *c09_tut02*.
c. Run the analysis.
d. Review the displacement result in the X, Y, and Z direction.
e. Review the stress and strain results.
f. Create presentation.
g. Generate report.
h. Save and close the model.

Opening Tutorial 1 of Chapter 8

As the required input file is saved in the *c08* folder, you need to browse to this folder and then open the *c08_tut01.fem* file in Autodesk Simulation Mechanical.

1. Start Autodesk Simulation Mechanical and then choose the **Open** button from the **Quick Access Toolbar**; the **Open** dialog box is displayed, if it is not displayed by default.

2. Browse to the *C:\Autodesk Simulation Mechanical\c08\Tut01* folder.

3. Select the *c08_tut01.fem* file and then choose the **Open** button from the dialog box; the selected *fem* file is opened in Autodesk Simulation Mechanical.

Saving the .Fem File in the c09 Folder

Now, you need to save the model in the c09 folder.

1. Choose the **Save As** button from the **Application Menu**; the **Save As** dialog box is displayed.

2. Browse to the *C:\Autodesk Simulation Mechanical* folder. Next, create a new folder with the name *c09* using the **New Folder** button, if not created in Tutorial 1 of this chapter. Next, create one more folder with the name *Tut02* inside the *c09* folder. Make the newly created *Tut02* folder the current folder by double-clicking on it.

3. Enter **c09_tut02** as the new name of the file in the **File name** edit box and then choose the **Save** button to save the file.

   The file is saved with the new name and is now opened in the graphics area of Autodesk Simulation Mechanical.

Running Analysis

As all the required input parameters such as loads, constraints, and material have been defined in the Tutorial 1 of Chapter 8, you can directly run the analysis and review the results.

1. Choose the **Analysis** tab in the **Ribbon** and then choose the **Run Simulation** tool in the **Analysis** panel; the process of structural analysis is started.

Note

*In the analysis, if the **FEA Data Warning** message window is displayed, choose the **Yes** button to view the log file. On choosing the **Yes** button, a notepad file with detailed information about the warning is displayed. After reviewing the log file, exit it.*

2. Once the analysis process is over, the Result environment of Autodesk Simulation Mechanical is invoked with displacement result displayed in the graphics area, refer to Figure 9-108.

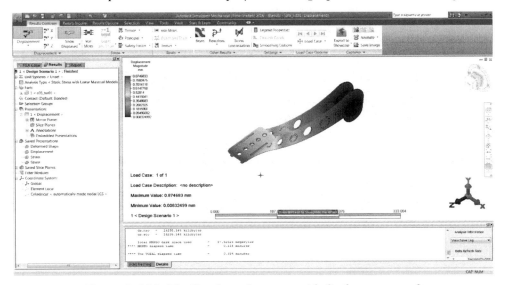

Figure 9-108 *The Result environment with displacement result*

Reviewing the Displacement Result in the X, Y, and Z Directions

It is evident from the Figure 9-108 that when the Result environment is invoked, the displacement magnitude result is displayed by default. Now, you need to review the displacement results in X, Y, and Z directions.

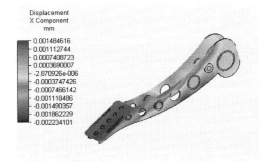

Figure 9-109 *Displacement result in the X direction*

1. Choose the **X** tool from the **Displacement** panel of the **Results Contours** tab; the displacement result in the X direction is displayed in the graphics area, refer to Figure 9-109.

2. Similarly, choose the **Y** and **Z** tools from the **Displacement** panel of the **Results Contours** tab; the displacement results in the Y and Z directions are displayed in the graphics area, refer to Figures 9-110 and 9-111.

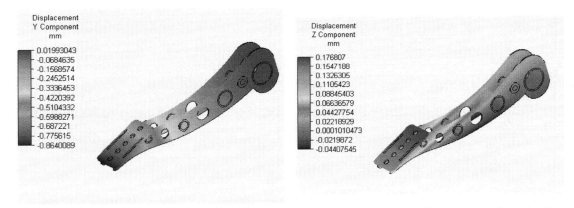

Figure 9-110 *Displacement result in the Y direction*

Figure 9-111 *Displacement result in the Z direction*

Reviewing the Stress and Strain Results

After reviewing the displacement result, you need to review the stress and strain results.

1. Choose the **von Mises** tool available in the **Stress** panel of the **Results Contours** tab in the **Ribbon**; the von Mises stress result is displayed in the graphics area, refer to Figure 9-112.

After reviewing the stress result, you need to review the strain result.

2. Choose the **von Mises** tool available in the **Strain** panel of the **Results Contours** tab; the von Mises strain result is displayed in the graphics area, refer to Figure 9-113.

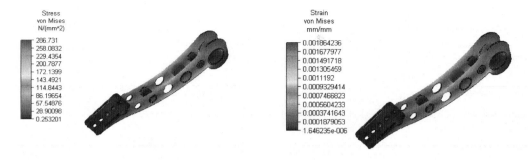

Figure 9-112 *Von Mises stress result*   **Figure 9-113** *Von Mises strain result*

Creating Presentation

After reviewing different results, you need to create a presentation for the results.

1. Expand the **Saved Presentations** node of the **Tree View** by clicking on the +sign available on its left, if not expanded by default.

2. Select **Stress** from the expanded **Saved Presentations** node of the **Tree View** and then right-click; a shortcut menu is displayed.

3. Choose the **Activate** option; the stress presentation is activated and added under the **Presentations** node of the **Tree View**. Also, the stress result is displayed in the graphics window.

4. Similarly, create the strain presentation.

Note

*The presentations created in the Result environment and displayed under the **Presentations** node of the **Tree View** will also be added in the report.*

Generating Report

Once the required results have been reviewed and different presentations have been created in the Result environment, you can generate the report.

1. Choose the **Report** tab available at the top of the **Tree View**; the **Report** environment is invoked and a report of different results and input parameters is generated, refer to Figure 9-114.

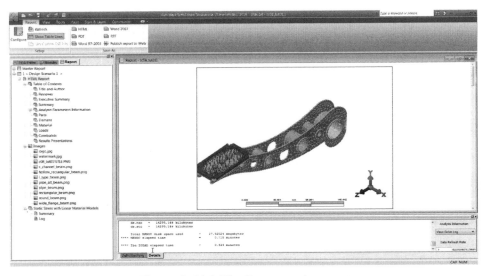

Figure 9-114 *The Report environment*

As discussed earlier, in the left of the Report environment, the default contents added in the report are displayed in the **Tree View**. You can select the content whose result you want to display in the display area. You can control the display of contents in the report by using the **Configure** tool in the **Setup** panel of the **Report** tab.

2.   Choose the **PDF** tool of the **Save As** panel of the **Report** tab in the **Ribbon**; the **Save As** dialog box is displayed.

3.   Browse to the location where you want to save the report file and then choose the **Save** button; the report is save in the pdf file format.

Saving the Model

1.   Choose the **Save** button from the **Quick Access Toolbar** to save the model.

2.   Choose **Close** from the **Application Menu** to close the file.

Self-Evaluation Test

Answer the following questions and then compare them to those given at the end of this chapter:

1.   On selecting the _____ solver, the maximum number of iterations to be performed in order to solve the equations are not required to be defined in Autodesk Simulation Mechanical.

2.   The _____ tool is used to show the displacement in the 3D model as per the default scale factor.

3. The _____ tool is used to display the axial stress in beam or truss elements.

4. You can animate the effect of load applied on to a model by using the _____ tool of the _____ panel in the **Results Contours** tab of the **Ribbon**.

5. You can turn on or off the visibility of the applied loads and constraints in the Result environment by using the _____ tool.

6. The _____ option of the presentation node in the **Tree View** of the Result environment is used to add text or annotations in the graphics area.

7. In Autodesk Simulation Mechanical, you can specify multiple load cases. (T/F)

8. In Autodesk Simulation Mechanical, you can change the analysis type at any point of analysis process. (T/F)

9. In Autodesk Simulation Mechanical, you cannot review the current result for particular nodes of the model. (T/F)

10. In Autodesk Simulation Mechanical, you cannot control or customize the contents of the report generated by default. (T/F)

Review Questions

Answer the following questions:

1. Which of the following dialog boxes is displayed on choosing the **Legend Properties** tool?

   (a) **Plot Settings**   (b) **Legend**
   (c) **Legend Properties**   (d) None of these

2. Which of the following dialog boxes is displayed on choosing the **Configure** tool in the Report environment?

   (a) **Generate Report**   (b) **Report**
   (c) **Configure Report**   (d) None of these

3. Which of the following tools is used to modify the default settings for animating the FEA model result?

   (a) **Animate**   (b) **Setup**
   (c) **Animate Setup**   (d) None of these

4. The tools available in the **Beam and Truss** drop-down of the **Strain** panel are used to review the _____ , _____ , and _____ strain results for truss, beam, and pipe elements, respectively.

5. The _____ tool is used to set the displaced scale factor and other related settings for the model to be deformed in the graphics area.

6. Viewing results of an analyzed model is the postprocessing phase of the analysis process. (T/F)

7. In Autodesk Simulation Mechanical, you cannot review the current result for individual parts of an assembly model. (T/F)

8. A probe is used to display the current result value of the entire model. (T/F)

9. You can add probe for the nodes that have maximum and minimum result values in the graphics area. (T/F)

10. In Autodesk Simulation Mechanical, the presentation added in the Result environment is displayed in the report generated. (T/F)

EXERCISES

Exercise 1

In this exercise, you will open the model of Exercise 1 of Chapter 8, refer to Figure 9-115 and then run the analysis. Once the process of analysis is completed, you need to review the results such as displacement, stress, strain, and reaction forces. Also, you will create different presentations of the displacement, stress, and strain results and generate a report. The final model after performinganalysis is shown in Figure 9-116.   **(Expected time: 30 min)**

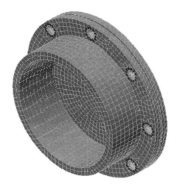   

Figure 9-115 *The model for Exercise 1*   *Figure 9-116* *The deformed model after performing analysis*

Exercise 2

In this exercise, you will open the model of Tutorial 2 of Chapter 8, refer to Figure 9-117 and then run the analysis. Once the process of analysis is completed, you need to review the results such as displacement, stress, strain, and reaction forces. Also, you will create different presentations of the displacement, stress, and strain results and generate a report. The final model after analysis is shown in Figure 9-118.   **(Expected time: 30 min)**

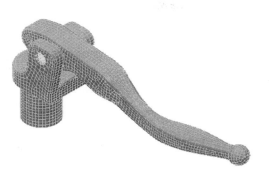

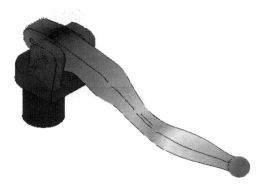

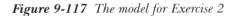

Figure 9-117 *The model for Exercise 2*

Figure 9-118 *The model after performing analysis*

Answers to Self-Evaluation Test

1. Sparse, 2. Show Displaced, 3. Axial Stress, 4. Start, Captures, 5. Loads and Constraints, 6. Annotations 7. T, 8. T, 9. F, 10. F

Chapter 10

Advanced Structural Analysis

Learning Objectives

After completing this chapter, you will be able to:

- *Understand dynamic analysis*
- *Understand different categories of dynamic analysis*
- *Perform modal analysis*
- *Perform harmonic/frequency response analysis*
- *Perform transient stress analysis*
- *Perform nonlinear analysis*

INTRODUCTION

In the earlier chapters, the static structural analysis problems were discussed in the tutorials. In this type of analysis, the load or the field condition did not vary with time, and were assumed to be applied gradually. However, in this chapter, dynamic analysis and nonlinear analysis problems will be covered where the load or the field condition varies with time.

DYNAMIC ANALYSIS

As discussed earlier, in the dynamic analysis, the load or field condition varies with time. Here, the assumption is that on a linear or nonlinear material system the load or the field condition is applied suddenly. The dynamic analysis is classified into three main categories discussed next.

Modal Analysis

The modal analysis is used to calculate the vibration characteristics such as natural frequencies and modal shapes (deformed shapes) of a structure or a machine component. The output of the modal analysis can be further used as input for the harmonic and transient analyses.

Harmonic/Frequency Response Analysis

The harmonic analysis is used to calculate the response of a structure to harmonically time varying loads.

Transient Analysis

The transient analysis is used to calculate the response of a structure to the arbitrary time varying loads (Impulse load).

In dynamic analysis, the following matrices are solved:

For system without any external loads.

$$[M] \times \text{Double Derivative of } [X] + [K] \times [X] = 0$$
where,
K = Stiffness Matrix
X = Displacement Matrix

For system with external loads.

$$[M] \times \text{Double Derivative of } [X] + [K] \times [X] = [F]$$
where,
K = Stiffness Matrix
X = Displacement Matrix
F = Load Matrix

These are called the force balance equations for a dynamic system. By solving the above set of equations, you can extract the natural frequencies of a system. The kinds of loads that can be applied in a dynamic analysis are same as those in the static analysis.

The outputs that can be expected from the software for advanced structural analysis are:

1. Natural frequencies
2. Mode shapes (Displacements)
3. Strains
4. Stresses
5. Reaction forces

Note that all the above outputs are obtained with respect to time.

Steps to Perform Dynamic Analysis
The steps to perform different types of dynamic analysis are discussed next.

Modal Analysis
When the natural frequency of a system is very close to the operating conditions or the excitation frequency, the component can attain resonance and can fail. So, to avoid resonance, you need to strengthen the component. But sometimes, strengthening the component may not be possible due to the design considerations. Also, in actual practice, the displacement at resonance may not be infinite due to damping. Therefore, you need to calculate the response of a system under time/frequency based loads. If the stress/strain/displacement response is less than the permitted limit, the component is not required to be strengthened or redesigned to avoid resonance.

The following steps are involved in the Modal analysis:

1. Create or import a solid model as discussed in the previous chapters.
2. Specify the type of analysis.
3. Define element attributes (element types, real constants, and material properties).
4. Define mesh attributes and mesh the solid model.
5. Specify the analysis options and apply loads.
6. Obtain the solution.
7. Review the results.

Note
In the modal analysis, neither displacement constraints nor loads are required. However, you can apply forces, pressures, temperatures, and other loads to calculate the load vector that is used for other analyses such as Harmonic and Transient.

Harmonic / Frequency Response Analysis
The Harmonic analysis is used to calculate the response of a structure to harmonically time varying loads (cyclic loads), thereby, helping you check your designs for resonance, fatigue, and other harmful effects of forced vibrations. This analysis is used by companies producing rotating machinery such as Turbines (gas, steam, wind, and water), Turbo pumps, Internal Combustion engines, Electric motors, Generators, Gas and Fluid pumps, Disc drives, and so on.

The following steps are involved in Harmonic analysis:

1. Create or import a solid model, as discussed in the previous chapters.
2. Specify the type of analysis.
3. Define element attributes (element types, real constants, and material properties).

4. Define mesh attributes and mesh the solid model.
5. Specify the analysis options and apply loads.
6. Obtain the solution.
7. Review results.

Transient Analysis

The transient analysis is used to calculate the response of a structure to arbitrary time varying loads (Impulse load). You can calculate time varying displacements, stresses, strains, and forces on a structure or on a component using the transient analysis. In transient analysis, inertia and damping effects are important.

The transient analysis takes more time to perform. Therefore, it is necessary to understand the problem for reducing the time involved in it. For example, if the problems contain nonlinearities, you need to understand how they affect the structure response by doing the static analysis first. For some problems, you can ignore nonlinearities. You can also perform the modal analysis and use natural frequencies for calculating the correct integration time steps.

The following steps are involved in transient analysis:

1. Create or import the solid model as discussed in the previous chapters.
2. Specify the type of analysis.
3. Define element attributes (element types, real constants, and material properties).
4. Define mesh attributes and mesh the solid model.
5. Specify the analysis options and apply loads.
6. Obtain the solution.
7. Review results.

NONLINEAR ANALYSIS

As discussed earlier, the linear structural analysis problems follows the Hooke's law. This law states that within the elastic limit stress is directly proportional to strain.

Stress / Strain = Constant (within the elastic limit)

Note that if the intensity of the stress is within the elastic limit, the member will regain its original shape after the removal of load, refer to Figure 10-1. However, if loading is so large that the intensity of stress the elastic limit is exceeded, the member will lose its elastic property. In such cases, when load is removed after exceeding the elastic limit, the member does not retain its original shape, refer to Figure 10-1. This means that once the material crosses the elastic limit, it will behave nonlinearly. There are many examples of non-linearity that can be observed in our daily life such as, crushing a plastic bottle, stamping a pin, and breaking a plastic ruler. Most of the plastic materials have nonlinearity.

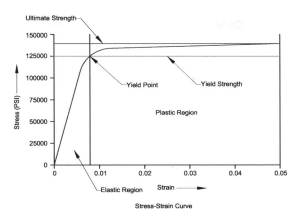

Figure 10-1 The Stress and Strain Curve

There are three types of nonlinearity: Geometric, Material, and Boundary (changing status). These are discussed next.

Geometric Nonlinearity

If a structure undergoes large deformations, its geometric configuration will change and it will start behaving nonlinearly. The following examples will clear the concept of geometric nonlinearity.

1. Consider a long beam with supports on both its ends. After a while, you will observe that the beam itself bends in the middle to attain the state of equilibrium, refer to Figure 10-2.

2. Consider a long cantilever beam with force applied at its free end. The beam under the action of force will undergo a large deformation and will behave nonlinearly, refer to Figure 10-3.

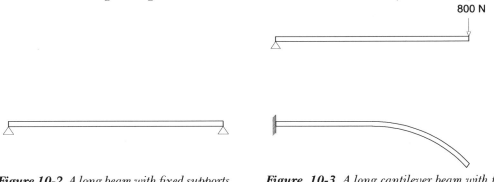

Figure 10-2 A long beam with fixed supports at both ends

Figure 10-3 A long cantilever beam with force applied at its free end

Material Nonlinearity

When the intensity of stress exceeds the elastic limit, the member will loose its elastic property. If loading is removed after exceeding the elastic limit, the member will not regain its original shape. This plastic behavior is called material nonlinearity.

The following material behaviors come under material nonlinearity:

1. Plasticity
2. Nonlinear elasticity
3. Creep
4. Viscoplasticity
5. Viscoelasticity

Boundary Nonlinearity (Changing Status)

Some structures may change their status, as a result, the stiffness of their material changes abruptly. This is called boundary nonlinearity. The following examples explain this concept:

1. When a cable gets slack, its stiffness becomes zero.
2. When two bodies suddenly come in contact, their overall stiffness changes drastically.

Steps to Perform Nonlinear Analysis

The following steps are involved in Nonlinear analysis:

1. Create or import a solid model, as discussed in the previous chapters.
2. Specify the type of analysis.
3. Define element attributes (element types, real constants, and material properties).
4. Define mesh attributes and mesh the solid model.
5. Apply loads.
6. Obtain the solution.
7. Review results.

TUTORIALS

Tutorial 1   Modal Analysis

In this tutorial, you will perform a simple Modal analysis on a cantilever beam shown in Figure 10-4, and then find the first five natural frequencies and mode shapes.

(Expected time: 30 min)

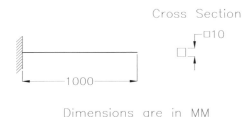

Figure 10-4 The cantilever beam for Tutorial 1

The following steps are required to complete this tutorial:

a. Start new FEA model in Autodesk Simulation Mechanical.
b. Set units for the analysis.

c.   Create cantilever beam as a 1D element.
d.   Define element types.
e.   Assign the material property.
f.   Define cross-section for the beam element.
g.   Apply fixed constraint.
h.   Set analysis parameters.
i.   Run analysis.
j.   Animate mode shapes
k.   Save and close the model.

Starting Autodesk Simulation Mechanical and a New FEA Model

Start Autodesk Simulation Mechanical and then open a new FEA model file of Autodesk Simulation Mechanical.

Note

Before you start with this tutorial, you need to create a new folder named c10 at the location C:\Autodesk Simulation Mechanical\. Next, create Tut01 folder inside the c10 folder.

1.   Start Autodesk Simulation Mechanical if not started already.

2.   Invoke the **New** dialog box, refer to Figure 10-5, if not invoked by default.

3.   Select the **Natural Frequency (Modal)** analysis type by clicking on the arrow pointing toward the right in the **Choose analysis type** area of the **New** dialog box, refer to Figure 10-5.

4.   Select the **FEA Model** option from the **New** dialog box and then choose the **New** button; the **Save As** dialog box is displayed.

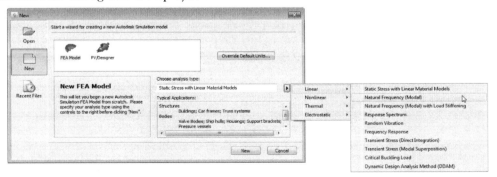

*Figure 10-5 The **New** dialog box*

5.   Browse to the location *C:\Autodesk Simulation Mechanical\c10\Tut01* and then enter **c10_tut01** in the **File name** edit box of the dialog box. Next, choose the **Save** button; a new FEA file with the name *c10_tut01* is saved at the specified location and a new FEA model environment of Autodesk Simulation Mechanical is invoked.

Setting Units for Analysis

Now you need to change the default Model unit system to Metric (mm).

1. Expand the **Unit Systems** node available in the **Tree View**. Next, set the units of the current session to Metric (mm) by double-clicking on the **Display Units < Metric mmks >** option.

Creating Cantilever Beam as a 1D Element

You need to invoke the sketching environment by selecting the XZ(-Y) plane as the sketching plane for creating the structure members.

1. Expand the **Planes** node available in the **Tree View** by clicking on its **+** sign. Next, double-click on the **Plane 3 XZ(-Y)** plane available in the expanded **Planes** node of the **Tree View**; the sketching environment is invoked and the selected XZ plane becomes parallel to the screen.

2. Click on the **Draw** tab of the **Ribbon**; all the tools of the **Draw** tab are displayed.

3. Choose the **Line** tool from the **Draw** panel of the **Draw** tab; the **Define Geometry** dialog box is displayed, as shown in Figure 10-6.

4. Make sure that value 1 is entered in the **Part**, **Surface**, and **Layer** edit boxes of the dialog box. Next, clear the **Use as Construction** check box in the **Attributes** area of the dialog box.

5. Make sure that **0, 0, 0** is entered in the **X**, **Y**, and **Z** edit boxes, respectively. Next, press ENTER; the first point of the line is specified and a rubber band line is attached to the cursor. Also, you are prompted to specify the second point of the line.

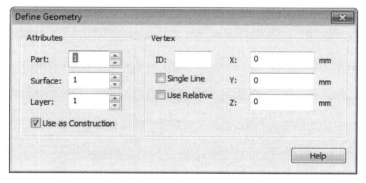

Figure 10-6 The Define Geometry dialog box

6. Enter **1000, 0, 0** in the **X**, **Y**, and **Z** edit boxes, respectively and then press ENTER; a line is created between the two specified points. Next, press the ESC key twice to terminate the creation of geometry and to exit the dialog box.

7. Choose the **Zoom (Fit all)** tool from the Navigation Bar in the graphic area.

8. Select the **Plane 3 <XZ(-Y)>** option from the expanded **Planes** node of the **Tree View**. Next, right-click and then choose the **Sketch** option from the shortcut menu displayed to exit the sketching environment.

Defining Element Type

Now, you need to define the beam elements for the line element.

1. Expand the **Part 1** node in the **Tree View** if not expanded by default.

2. Select the **Element Type <Unknown>** option from the expanded **Part 1** node. Next, right-click and choose the **Beam** option from the shortcut menu displayed; the beam elements for the structure are defined.

Assigning the Material

Now, you need to assign material to the structure model.

1. Select the **Material <Unnamed>** option from the expanded **Part 1** node of the **Tree View** and right-click to display a shortcut menu.

2. Choose the **Edit Material** option from the shortcut menu displayed; the **Element Material Selection** dialog box will be displayed.

3. Expand the **Steel** material family node by clicking on the + sign available on the left of the **Steel** node; the materials available in the **Steel** material family of Autodesk Simulation Mechanical will be displayed.

4. Select the **Stainless Steel (AISI 446)** material from the list in the **Stainless Steel** material node; all the properties of the selected material are displayed on the right side of the dialog box.

5. Choose the **OK** button from the dialog box; all the properties of the selected material are assigned to the beam member.

Defining Cross-Section for Beam Elements

Now, you need to define cross section for the beam elements 10 X 10.

1. Select the **Element Definition** option from the expanded **Part 1** node of the **Tree View** and right-click to display a shortcut menu.

2. Choose the **Edit Element Definition** option from the shortcut menu; the **Element Definition - Beam** dialog box is displayed, refer to Figure 10-7.

3. Move the cursor to the field available in front of the Layer 1 row in the dialog box and when the cursor changes to an arrow cursor, click to select the row, refer to Figure 10-7; the **Cross-Section Libraries** button is enabled in the dialog box.

4. Choose the **Cross-Section Libraries** button of the dialog box; the **Cross-Section Libraries** dialog box is displayed.

5. Select the **User Defined** option from the **Section database** drop-down list in the dialog box.

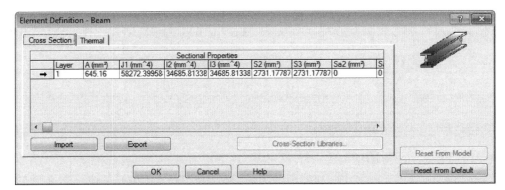

*Figure 10-7 The **Element Definition - Beam** dialog box*

6. Select the **Rectangular** option from the **Type** drop-down list available at the upper right corner of the dialog box, refer to Figure 10-8.

7. Enter **10** in the **b** and **h** edit boxes available below the **Type** drop-down list, refer to Figure 10-8.

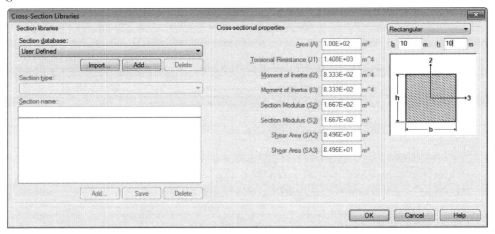

*Figure 10-8 The **Cross-Section Libraries** dialog box*

8. Choose **OK**; the user defined cross-section 10 X 10 is defined for the beam elements of Layer 1. Next, choose the **OK** button to exit the dialog box.

Applying the Fixed Constraint

Now, you will apply the fixed constraint at one end of the beam element.

1. Click on the **Selection** tab of the **Ribbon**. Next, choose the **Point or Rectangle** tool from the **Shape** panel if not already chosen. Next, choose the **Vertices** tool from the **Select** panel of the **Selection** tab.

2. Select the left end point of the cantilever beam by clicking the left mouse button.

3. Choose the **General Constraint** tool from the **Constraints** panel of the **Setup** tab in the **Ribbon**; the **Creating 1 Nodal General Constraint Objects** dialog box is displayed.

4. Choose the **Fixed** button in the **Creating 1 Nodal General Constraint Objects** dialog box; the fixed constraint is applied to the selected vertex. Next, choose the **OK** button. Figure 10-9 shows the cantilever beam after applying fixed constraint.

Figure 10-9 *The cantilever beam after applying fixed constraint*

Setting Analysis Parameters

Now, you need to specify the analysis parameters.

1. Click on the **Setup** tab of the **Ribbon**. Next, choose the **Parameters** tool in the **Model Setup** panel; the **Analysis Parameters - Natural Frequency (Modal)** dialog box is displayed, as shown in Figure 10-10.

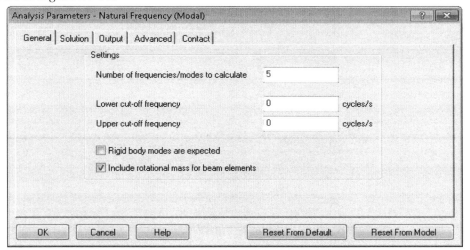

Figure 10-10 *The **Analysis Parameters - Natural Frequency (Modal)** dialog box*

2. Make sure that the number of frequencies to be calculated is set to 5 in the **Number of frequencies/modes to calculate** edit box of the dialog box.

3. Make sure that the lower cut off and upper cut off frequencies are set to 0 in the respective edit boxes of the dialog box.

   Note
   *The **Lower cut-off frequency** and **Upper cut-off frequency** edit boxes of the dialog box are used to specify the lower and upper cut off frequency range, respectively. Any frequency lower or higher than the specified frequency range is filtered out from the result. By default, value 0 is set in these edit boxes. As a result, no lower or higher frequency is filtered out from the result.*

4. After accepting the default parameters, choose the **OK** button.

Running Analysis

After specifying all the required input parameters such as loads, constraints, and material, you can run the analysis and review the results.

1. Choose the **Analysis** tab in the **Ribbon**. Next, choose the **Run Simulation** tool in the **Analysis** panel; the process of structural analysis is started.

2. Once the process of analysis is done, the **Result** environment is invoked with the frequency result of a mode, refer to Figure 10-11.

3. Click on the **Next** button in the **Load Case Options** panel of the **Results Contours** tab; the frequency result for the next mode is displayed, refer to Figure 10-12.

Note that for better understanding of displacement of the model, you can show the undisplaced model in the graphics area. To do so, invoke the **Displaced model options** dialog box by choosing the **Displaced Options** tool from the **Normalized Displacement** panel of the **Result Contours** tab. Next, choose the **Mesh** radio button from the **Show Undisplaced Model As** area in the **Displaced Model Options** dialog box; the undisplaced model will be displayed.

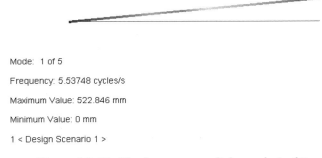

Mode: 1 of 5

Frequency: 5.53748 cycles/s

Maximum Value: 522.846 mm

Minimum Value: 0 mm

1 < Design Scenario 1 >

Figure 10-11 *The frequency result for mode 1 of 5*

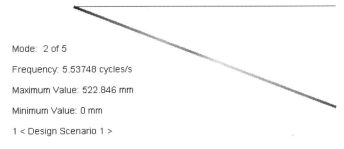

Mode: 2 of 5

Frequency: 5.53748 cycles/s

Maximum Value: 522.846 mm

Minimum Value: 0 mm

1 < Design Scenario 1 >

Figure 10-12 *The frequency result for mode 2 of 5*

4. Similarly, you can review the frequency for different modes by clicking the **Next** button ▶ in the **Load Case Options** panel.

Note
*To review the frequency of the previous mode, choose the **Previous** button ◄ in the **Load Case Options** panel.*

Animating Mode Shapes

Now, you need to animate the frequency mode shapes.

1. Choose the **Start** button 📸 from the **Captures** panel of the **Results Contours** tab in the **Ribbon**; the animation gets started in the graphics area.

2. Once the animated results have been reviewed, stop the animation by choosing the **Stop** button 📸 from the **Captures** panel of the **Results Contours** tab.

Saving the Model

1. Choose the **Save** button from the **Quick Access Toolbar** to save the model.

2. Choose **Close** from the **Application Menu** to close the file.

Tutorial 2   Modal Analysis

In this tutorial, you will perform a simple Modal analysis on a cantilever beam shown in Figure 10-13, and then find the first five natural frequencies and mode shapes.

Download the input file of this chapter from *www.cadcim.com*. The complete path for downloading the file is as follows:

Textbooks > CAE Simulation > Autodesk Simulation Mechanical > Autodesk Simulation Mechanical 2016 for Designers > Input Files > c10_simulation_2016_input.zip   (**Expected time: 30 min**)

Figure 10-13 The cantilever beam for Tutorial 2

The following steps are required to complete this tutorial:

a. Download the zipped file containing tutorial input files of Chapter 10.
b. Start Autodesk Simulation Mechanical.
c. Import the input file of this tutorial into Autodesk Simulation Mechanical.
d. Set the units of the analysis.
e. Generate mesh on the model.
f. Assign material.
g. Apply fixed constraint.
h. Set analysis parameters.

   i.   Run analysis.
   j.   Animate mode shapes.
   k.   Save and close the model.

Downloading the File

Before you start the tutorial, you need to download the part files of chapter 10.

1.   Create a folder with the name *Tut02* at the location *C:\Autodesk Simulation Mechanical\c10*.

2.   Download the zipped file *c10_simulation_2016_input* from *www.cadcim.com*. The complete path for downloading the file is as follows:

Textbooks > CAE Simulation > Autodesk Simulation Mechanical > Autodesk Simulation Mechanical 2016 for Designers > Input Files

3.   Extract the downloaded *c10_simulation_2016_input* file. Next, copy the *c10_tut02.step* file from the extracted folder and paste it at the location *C:\Autodesk Simulation Mechanical\c10\ Tut02*.

Importing the STEP File into Autodesk Simulation Mechanical

Now, you need to open Autodesk Simulation Mechanical and import the c10_tut02.step file into it.

1.   Start Autodesk Simulation Mechanical and then choose the **Open** button from the **Launch panel** of the **Start & Learn** tab; the **Open** dialog box is displayed.

2.   Select the **STEP Files (*.stp; *.ste; *.step)** file extension from the **Files of type** drop-down list of the **Open** dialog box.

3.   Browse to the location *C:\Autodesk Simulation Mechanical\c10\Tut02*; the *c10_tut02.step* file is displayed in the **Open** dialog box.

4.   Select the **c10_tut02.step** file and then choose the **Open** button from the **Open** dialog box; the **Choose Analysis Type** dialog box is displayed.

5.   Select the **Natural Frequency (Modal)** analysis type, refer to Figure 10-14.

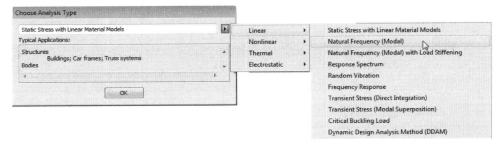

Figure 10-14 *Selecting the **Natural Frequency (Modal)** analysis type*

Note
*When you open a file; message window(s) may be displayed. Choose the **Yes** button from these window(s).*

6. Choose the **OK** button from the dialog box; the selected file is opened into the Autodesk Simulation Mechanical. Next, by using the ViewCube, change the orientation of the model similar to the one shown in Figure 10-15.

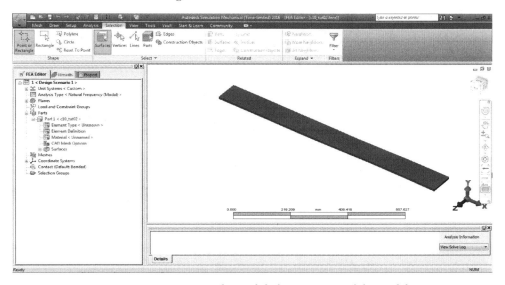

Figure 10-15 The modified orientation of the model

Setting the Units for analysis

Now you need to change the default Model unit system to Metric (mm).

1. Expand the **Unit Systems** node available in the **Tree View**. Next, set the units of the current session to Metric (mm) by double-clicking on the **Display Units < Metric mmks >** option of the expanded **Unit Systems** node in the **Tree View**.

Generating Mesh on the Model

Once the model has been imported in the Autodesk Simulation Mechanical, you need to generate mesh on it. In this tutorial, you will generate the mesh with default mesh settings.

1. Choose the **Generate 3D Mesh** tool from the **Mesh** panel of the **Mesh** tab in the **Ribbon**; the **Meshing Progress** window is displayed and the process of generating mesh with the default settings starts. Once the mesh has been generated, the **View Mesh Results** window is displayed.

2. Choose the **No** button from the window. The model after generating the mesh is displayed, refer to Figure 10-16.

Figure 10-16 *Model displayed after generating mesh*

Assigning Material

Now, you need to assign material to the meshed model.

1. Select the **Material <Unnamed>** option from the expanded **Part 1** node of the **Tree View** and then right-click.

2. Choose the **Edit Material** option from the shortcut menu displayed; the **Element Material Selection** dialog box is displayed.

3. Expand the **Steel** material family node by clicking on the **+** sign available on the left of the **Steel** node; the categories available in the **Steel** material family are displayed.

4. Expand the **Stainless** category; the materials available in the expanded category are displayed. Select the **Stainless Steel (AISI 202)** material from the list in the **Steel** material family; all the properties of the selected material are displayed on the right in the dialog box.

5. Choose the **OK** button from the dialog box; all the properties of the selected material are assigned to the model.

Applying Fixed Constraint

Now, you will apply fixed constraints at one end of the model.

1. Click on the **Selection** tab of the **Ribbon** and then choose the **Point or Rectangle** tool from the **Shape** panel of the **Selection** tab if not already chosen. Next, choose the **Surfaces** tool from the **Select** panel of the **Selection** tab.

2. Rotate the model such that its back planar face is visible in the graphic area, refer to Figure 10-17.

3. Select the back planar face of the model, refer to Figure 10-17.

4. Choose the **General Constraint** tool in the **Constraints** panel of the **Setup** tab; the **Creating 1 Surface General Constraint Objects** dialog box is displayed.

5. Choose the **Fixed** button in the dialog box; the fixed constraint is applied to the selected face of the model. Next, choose the **OK** button. Figure 10-18 shows the cantilever beam after applying the fixed constraint.

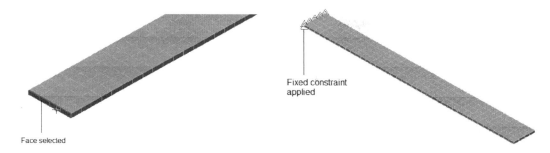

Figure 10-17 *Face selected for applying fixed constraint*

Figure 10-18 *Model after applying fixed constraint*

Setting Analysis Parameters

Now, you need to specify the analysis parameters.

1. Click on the **Setup** tab of the **Ribbon**. Next, choose the **Parameters** tool in the **Model Setup** panel; the **Analysis Parameters - Natural Frequency (Modal)** dialog box is displayed, as shown in Figure 10-19.

2. Make sure the number of frequencies to be calculated is set to 5 in the **Number of frequencies/ modes to calculate** edit box of the dialog box.

3. Make sure the lower cut off and upper cut off frequencies are set to 0 in respective edit boxes.

 Note
*The **Lower cut-off frequency** and **Upper cut-off frequency** edit boxes of the dialog box are used to specify the lower and upper cut off frequency range, respectively. Any frequency lower or higher than the specified frequency range is filtered out from the result. By default, the value 0 is set in these edit boxes. As a result, no lower or higher frequency is filtered out from the result.*

4. After accepting the default parameters, refer to Figure 10-19. Choose the **OK** button.

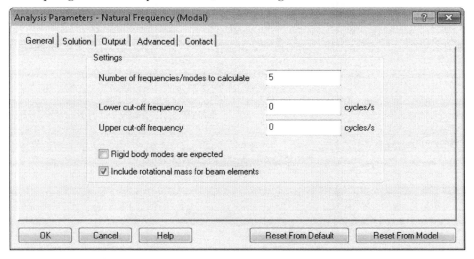

Figure 10-19 *The **Analysis Parameters - Natural Frequency (Modal)** dialog box*

Running Analysis

After specifying all the required input parameters such as loads, constraints, and material; you can run the analysis and review the results.

1. Choose the **Analysis** tab in the **Ribbon**. Next, choose the **Run Simulation** tool in the **Analysis** panel; the process of structural analysis is started.

2. Once the process of analysis is done, the **Result** environment is invoked with the frequency result of the first mode, refer to Figure 10-20.

3. Click on the **Next** button of the **Load Case Options** panel of the **Results Contours** tab in the **Ribbon**; the frequency result for second mode is displayed, refer to Figure 10-21.

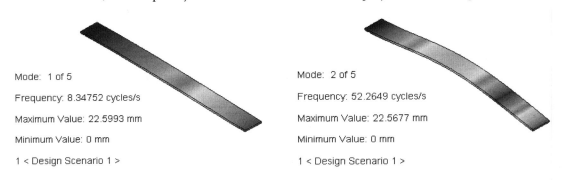

Mode: 1 of 5

Frequency: 8.34752 cycles/s

Maximum Value: 22.5993 mm

Minimum Value: 0 mm

1 < Design Scenario 1 >

Mode: 2 of 5

Frequency: 52.2649 cycles/s

Maximum Value: 22.5677 mm

Minimum Value: 0 mm

1 < Design Scenario 1 >

Figure 10-20   *The frequency result for mode 1 of 5 constraint*   *Figure 10-21*   *The frequency result for mode 2 of 5 constraint*

4. Similarly, you can review the frequency for different modes by clicking the **Next** button ▶ of the **Load Case Options** panel. Note that to review the frequency of the previous mode, choose the **Previous** button ◀ in the **Load Case Options** panel.

Animating the Mode Shapes

Now, you need to animate the frequency mode shapes.

1. Choose the **Start** button 🖳 from the **Captures** panel of the **Results Contours** tab in the **Ribbon**; the animation gets started in the graphics area.

2. Once the animated results have been reviewed, stop the animation by choosing the **Stop** button 🖳 from the **Captures** panel of the **Results Contours** tab in the **Ribbon**.

Saving the Model

1. Choose the **Save** button from the **Quick Access Toolbar** to save the model.

2. Choose **Close** from the **Application Menu** to close the file.

Tutorial 3   Harmonic/Frequency Response Analysis

In this tutorial, you will open the Tutorial 2 model of this chapter and calculate the frequency response of the model at different frequencies 100, 200, 300, and the natural frequencies to the harmonic load 200N at the end of the beam, refer to Figure 10-22. The damping ratio to be used is 0.01.   **(Expected time: 30 min)**

Figure 10-22 The model for Tutorial 3

The following steps are required to complete this tutorial:

a. Open the Tutorial 2 model of Chapter 10.
b. Save the *.fem* file with a new name.
c. Identify nodes for applying harmonic load.
d. Specify analysis type.
e. Specify the parameters.
f. Run analysis.
g. Animate the result.
h. Save and close the model.

Opening the Tutorial 2 Model of this Chapter
You need to open the Tutorial 2 model for calculating the frequency response at different frequencies.

1. Start Autodesk Simulation Mechanical and then choose the **Open** button from the **Quick Access Toolbar**; the **Open** dialog box is displayed.

2. Browse to *C:\Autodesk Simulation Mechanical\c10\Tut02*.

3. Select the *c10_tut02.fem* file and then choose the **Open** button from the dialog box; the selected *.fem* file is opened in the Autodesk Simulation Mechanical.

Saving the .fem File with New Name
Now, save the model as *c10_tut03* file.

1. Save the model with the name *c10_tut03* in *Tut03* folder of chapter 10. Note that you need to create a *Tut03* folder under the *c10* folder.

Identifying the Nodes for Applying the Harmonic Load
Now, you need to identify the nodes for applying the harmonically varying load.

1.   Choose the **Results** tab from the **Tree View** to invoke the result environment.

     Now, you need to turn on the visibility of the node numbers to identify the node number on which the harmonic load is to be applied.

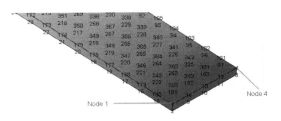

2.   Choose the **Results Options** tab in the **Ribbon**. Next, choose the **Show Numbers** tool of the **View** panel; a flyout is displayed.

Figure 10-23 Partial view of the model with node numbers displayed

3.   Select the **Node Numbers** check box; all the node numbers are displayed in the graphics area. Figure 10-23 shows the partial view of the model with the visibility of the node numbers turned on.

4.   It is evident from Figures 10-22 and 10-23 that the harmonic load needs to be applied to node numbers 1 and 4. Note that in your case node numbers may be changed.

5.   Switch to the FEA Editor environment by clicking on the **FEA Editor** tab in the **Tree View**.

Specifying the Analysis Type

Now, you need to specify the analysis type for determining the frequency response.

1.   Choose the **Analysis** tab in the **Ribbon**. Next, choose the **Type** tool in the **Change** panel; a drop-down is displayed, refer to Figure 10-24.
2.   Choose **Linear > Frequency Response** from the drop-down, refer to Figure 10-24; the **Autodesk Simulation Mechanical** message window is displayed, as shown in Figure 10-25.

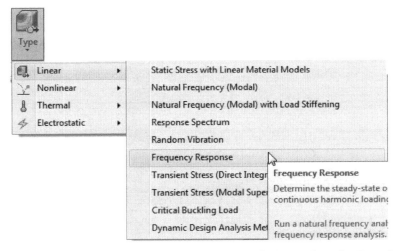

*Figure 10-24 Selecting the **Frequency Response** analysis type from the **Type** drop-down*

3.   Choose the **Yes** button from this message window to create a new design scenario rather than changing the analysis type for the previous one.

*Figure 10-25 The **Autodesk Simulation Mechanical** message window*

Specifying Parameters

Now, you need to specify different parameters to calculate the frequency response at different frequencies and at the harmonic load.

1. Choose the **Setup** tab in the **Ribbon** and then choose the **Parameters** tool; the **Frequency Response Analysis Input** dialog box is displayed, as shown in Figure 10-26.

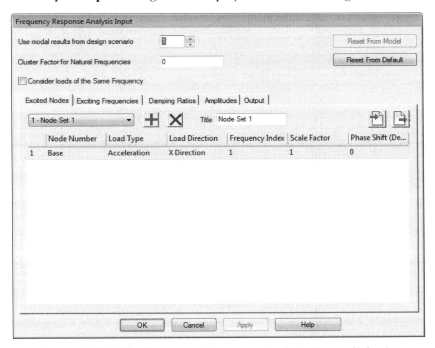

*Figure 10-26 The **Frequency Response Analysis Input** dialog box*

2. Make sure that **1** is entered in the **Use modal results from Design Scenario** edit box of the dialog box.

   The **Use modal results from Design Scenario** edit box is used to specify the existing design scenario whose results are to be used in the current design scenario. In this tutorial, you will use the result of Tutorial 2, which is design scenario 1.

3. Choose the **Exciting Frequencies** tab and select the **Include Natural Frequencies** check box to include the natural frequencies result of the design scenario 1.

4. Choose the **Excited Nodes** tab. Next, select the Base from the Node Number column and enter **1** in the edit box.

5. Next, click in the Load Type column in the row; flyout is displayed. Next, select the **Force Input option** from the flyout.

6. Next, click in the Load Direction column in the row; flyout is displayed. Next, select the **Y Direction** option from the flyout.

7. Enter **1** in the edit box in the Scale Factor column. Next, choose the **New** button; the **Excited Nodes Set ID** window is displayed.

8. Enter **2** in this window and then choose the **OK** button.

9. Select the Base in second row from the Node Number column and then enter **4** in the edit box. Next, click in the Load Type column in the row, a flyout is displayed.

10. Select the **Force Input** and **Y Direction** options in the Load Direction column and enter **1** in the edit box in Scale Factor column.

11. Choose the **Exciting Frequencies** tab and enter **100**, **200**, and **300** in the first three rows of the **Frequency (HZ)** column. You can insert the rows by using the **Insert Row** button, refer to Figure 10-27.

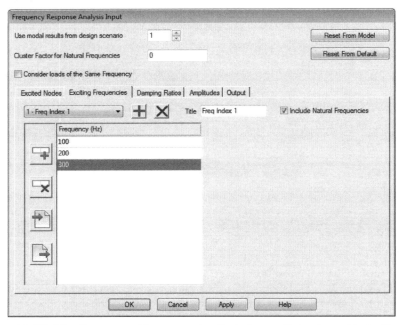

Figure 10-27 The *Frequency Response Analysis Input* dialog box with the *Exciting Frequencies* tab chosen

12. Choose the **Damping Ratios** tab and enter the frequency and damping ratio in their respective columns, as shown in Figure 10-28.

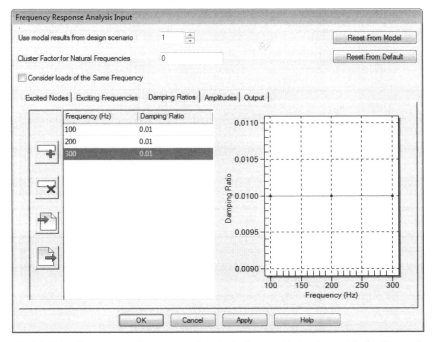

Figure 10-28 *The **Frequency Response Analysis Input** dialog box with the **Damping Ratio** tab chosen*

13. Choose the **Amplitudes** tab and fill the **Frequency (HZ)**, **Acceleration (g)**, and **Force (N)** columns, as shown in Figure 10-29.

14. After specifying all the input parameters, choose the **OK** button.

Running Analysis

After specifying all the required input parameters, you can run the analysis and review the results.

1. Choose the **Analysis** tab in the **Ribbon** and then choose the **Run Simulation** tool from the **Analysis** panel; the process of structural analysis is started.

2. Once the analysis is complete, the **Result** environment is invoked with the displacement magnitude result of the default load case, refer to Figure 10-30. Note that total 8 load cases have been created due to the five natural frequencies of the design scenario 1 and the three new frequencies are specified in the input parameters. figure 10-30 through 10-37 show the displacement magnitude at different frequencies and harmonic load specified.

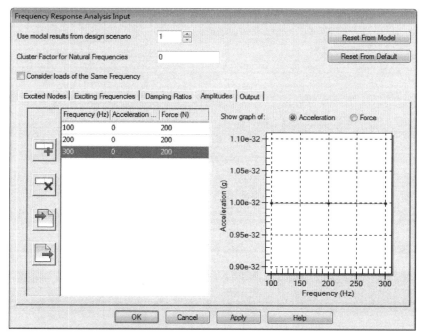

Figure 10-29 *The* **Frequency Response Analysis Input** *dialog box with the* **Amplitudes** *tab chosen*

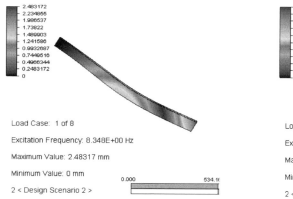

Figure 10-30 *The displacement magnitude result for load case 1 of 8*

Figure 10-31 *The displacement magnitude result for load case 2 of 8*

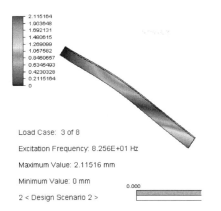

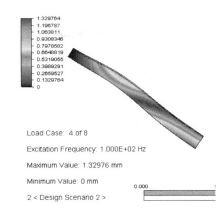

Figure 10-32 *The displacement magnitude result for load case 3 of 8*

Figure 10-33 *The displacement magnitude result for load case 4 of 8*

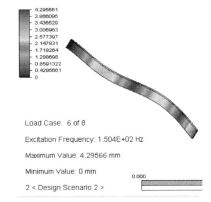

Figure 10-34 *The displacement magnitude result for load case 5 of 8*

Figure 10-35 *The displacement magnitude result for load case 6 of 8*

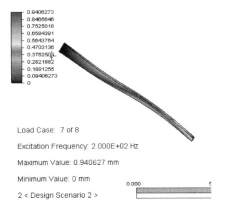

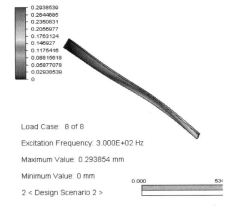

Figure 10-36 *The displacement magnitude result for load case 7 of 8*

Figure 10-37 *The displacement magnitude result for load case 8 of 8*

Note

*To switch from one load case to another, you need to choose the **Next** button ▶ from the **Load Case Options** panel. On choosing the **Previous** button ◀ , you can review the results for the previous load case.*

Animating the Result of the Current Load Case

Now, you need to animate the frequency mode shapes.

1. Choose the **Start** button ▧ from the **Captures** panel of the **Results Contours** tab in the **Ribbon**; the animation for the current load case starts in the graphics area.

2. Once the animated results are reviewed, stop the animation by choosing the **Stop** button ▧ from the **Captures** panel of the **Results Contours** tab in the **Ribbon**. Similarly, you can review the results for the other load cases.

Saving the Model

Now, you need to save the model.

1. Choose the **Save** button from the **Quick Access Toolbar** to save the model.

2. Choose **Close** from the **Application Menu** to close the file.

Tutorial 4   Transient Analysis

In this tutorial, you will open the Tutorial 3 model of this chapter and perform the transient stress analysis on it. Calculate the response of the model to the arbitrarily time varying load 1000N, refer to Figure 10-38. The damping ratio to be used is 0.01.

(Expected time: 30 min)

Figure 10-38 The model for Tutorial 4

The following steps are required to complete this tutorial:

a. Open the Tutorial 3 model of the chapter.
b. Save the *.fem* file with a new name.
c. Specify the analysis type.
d. Specify arbitrary time varying load.
e. Specify parameters.
f. Run analysis.
g. Create a presentation.

h. Generate the report.
i. Save and close the model.

Opening the Tutorial 3 Model of this Chapter

1. Start Autodesk Simulation Mechanical and then open the *c10_tut03.fem* model.

Saving the .fem File with a new Name

You need to save the model with the name *c10_tut04*.

1. Save the model with the name *c10_tut04* in *Tut04* folder of chapter 10.

Specifying the Analysis Type

Now, you need to specify the analysis type to Transient Stress (Direct Integration).

1. Choose the **Analysis** tab from the **Ribbon**. Next, choose the **Type** tool from the **Change** panel; a drop-down is displayed, refer to Figure 10-39.

2. Choose **Linear > Transient Stress (Direct Integration)** from the drop-down, refer to Figure 10-39; the **Autodesk Simulation Mechanical** message window is displayed.

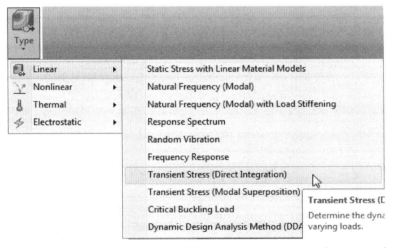

*Figure 10-39 The **Transient Stress (Direct Integration)** analysis type selected*

3. Choose the **Yes** button from this message window to create a new design scenario rather than changing the analysis type for the previous one.

Specifying the Arbitrary Time Varying Load

Now, you need to specify the arbitrary time varying load on the top nodes of the free end of the beam.

1. Change the current orientation of the model, as shown in Figure 10-40, by using the ViewCube and now you need to specify the arbitrary time varying load on the top nodes of the free end of the beam, refer to Figure 10-41.

2. Click on the **Selection** tab in the **Ribbon** and then choose the **Point or Rectangle** tool from the **Shape** panel of the **Selection** tab. Next, choose the **Surfaces** tool from the **Select** panel of the **Selection** tab.

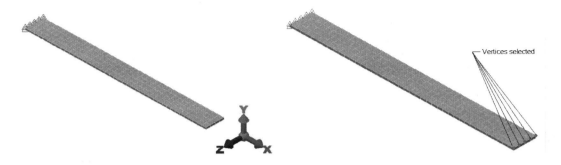

Figure 10-40 *The modified orientation of the model*

Figure 10-41 *The vertices selected*

3. Click on the **Setup** tab in the **Ribbon** and then choose the **Force** tool from the **Loads** panel; the **Creating Surface Force** dialog box is displayed, refer to Figure 10-42.

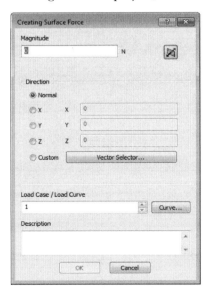

Figure 10-42 *The **Creating Surface Force** dialog box*

 Note
The total magnitude of the force applied on to the model is 1000 N and the number of nodes selected is 5 therefore the magnitude of the force per node is equal to 200 N.

4. Choose the **Curve** button from the dialog box; the **Multiplier Table Editor** dialog box is displayed, refer to Figure 10-43.

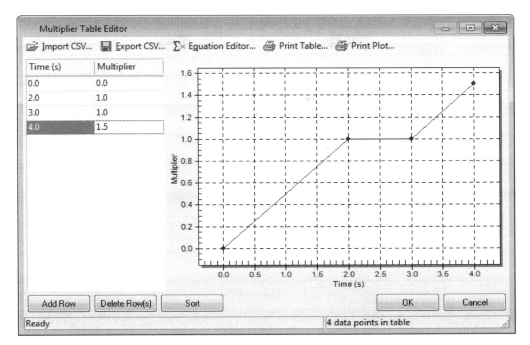

Figure 10-43 The **Multiplier Table Editor** *dialog box*

5. Choose the **Add Row** button thrice to add three additional rows to the multiplier table, refer to Figure 10-43.

6. In the second row, enter **2** in the **Time (s)** column and **1** in the **Multiplier** column, refer to Figure 10-43.

7. In the third row, enter **3** in the **Time (s)** column and **1** in the **Multiplier** column, refer to Figure 10-43.

8. In the fourth row, enter **4** in the **Time (s)** column and **1.5** in the **Multiplier** column, refer to Figure 10-43. Choose the **OK** button from the **Multiplier Table Editor** dialog box.

9. Click on the **Selection** tab in the **Ribbon** and then choose the **Point or Rectangle** tool from the **Shape** panel of the **Selection** tab. Next, choose the **Vertices** tool from the **Select** panel of the **Selection** tab; the **Creating Surface Force** dialog box is changed into the **Creating Nodal Force** dialog box.

10. Select the top five vertices of the free end of the beam by pressing the CTRL key, refer to Figure 10-41.

11. Enter **-200** in the **Magnitude** edit box of the dialog box. Next, select the **Y** radio button in the **Direction** area of the dialog box. Note that the negative and positive magnitude values indicate the direction of the force applied.

12. Next, choose the **OK** button from the
    Creating 5 Nodal Force Objects dialog box;
    the force of magnitude 200 N is applied
    and the arrows representing the force are
    displayed in the graphics area, as shown in
    Figure 10-44.

Specifying the Parameters

Now, you need to specify input parameters
such as number of time steps and time step
size.

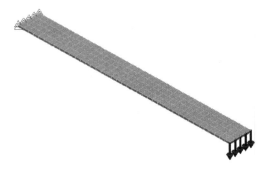

Figure 10-44 The model after applying force

1. Choose the **Setup** tab from the **Ribbon**
   and then choose the **Parameters** tool; the **Analysis Parameters - Transient Stress (Direct
   Integration)** dialog box is displayed.

2. Enter **80** in the **Number of time steps** edit box and **0.05** in the **Time-step size** edit box of
   the dialog box, refer to Figure 10-45.

3. Enter **0.01** in the **Alpha** edit box and **0.01** in the **Beta** edit box of the dialog box, refer to
   Figure 10-45.

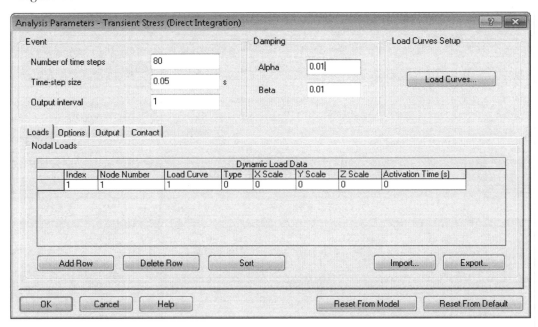

*Figure 10-45 The **Analysis Parameters - Transient Stress (Direct Integration)** dialog box*

4. Choose the **OK** button from the dialog box.

Running the Analysis

After specifying all the required input parameters, you can run the analysis and review the results.

1. Choose the **Analysis** tab in the **Ribbon**. Next, choose the **Run Simulation** tool from the **Analysis** panel; the process of structural analysis is started.

2. Once the process of analysis is done, the **Result** environment is invoked with the displacement magnitude result for the default time step, refer to Figure 10-46. Note that the result for the total 80 time steps are displayed. Figure 10-46 shows the displacement result for time step 1 on time 0.05 sec and Figure 10-47 shows the results for the second time step at 0.1 sec time.

Note

*To switch from the result of one time step to another, you need to choose the **Next** button* ▶ *located in the **Load Case Options** panel. On choosing the **Previous** button* ◀, *you can review the results for the previous time step.*

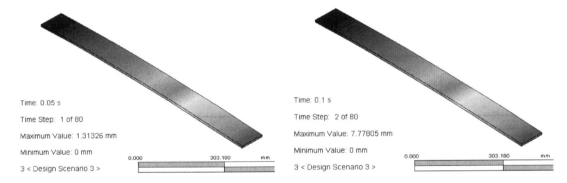

Figure 10-46 The displacement result for time step 1

Figure 10-47 The displacement result for time step 2

Animating the Result

Now, you need to animate the frequency mode shapes.

1. Choose the **Start** button 🖼 from the **Captures** panel of the **Results Contours** tab in the **Ribbon**; the animation is started in the graphics area and different result values for different time steps are displayed.

2. Once the animated results are reviewed, stop the animation by choosing the **Stop** button 🖼 from the **Captures** panel of the **Results Contours** tab in the **Ribbon**.

Creating the Presentation of Different Results

After reviewing the results, you need to create a presentation for different results. Note that the presentation for the displacement is created by default.

1. Expand the **Saved Presentations** node of the **Tree View** by clicking on the +sign available on its left, if not expanded by default.

2. Select **Stress** from the expanded **Saved Presentations** node of the **Tree View** and then right-click; a shortcut menu is displayed.

3. Choose the **Activate** option; the stress presentation is activated and added under the **Presentations** node of the **Tree View**. Also, the stress result is displayed in the graphics window.

4. Similarly, create the strain presentation.

 Note

*The presentations created in the result environment and displayed under the **Presentations** node of the **Tree View** are also added in the report.*

Generating the Report

Once the required results are reviewed and different presentations are created in the result environment, you can generate the report.

1. Choose the **Report** tab available at the top of the **Tree View**; the report environment is invoked and a report of different results and input parameters is generated.

   Note that on the left in the report environment, the default contents added in the report are displayed in the **Tree View**. You can click on the content whose result you want to display in the display area. You can control the display of contents in the report by using the **Configure** tool in the **Setup** panel of the **Report** tab.

2. Choose the **PDF** tool from the **Save As** panel of the **Report** tab in the **Ribbon**; the **Save As** dialog box is displayed.

3. Browse to the location where you want to save the report file and then choose the **Save** button; the report is saved in pdf file format.

Saving the Model

Now, you need to save the file.

1. Choose the **Save** button from the **Quick Access Toolbar** to save the model.

2. Choose **Close** from the **Application Menu** to close the file.

Tutorial 5   Nonlinear Analysis

In this tutorial, you will open Tutorial 4 of this chapter and perform a nonlinear analysis to calculate the response of the beam to a load of 80000 N, refer to Figure 10-48. Note that the load applied on the beam is a heavy load and therefore the resulting stresses are greater than the yield strength of the material. As a result, the material becomes plastic and nonlinear.

(Expected time: 30 min)

Figure 10-48 *The model for Tutorial 5*

The following steps are required to complete this tutorial:

a. Open the Tutorial 4 model of this Chapter.
b. Save the *.fem* file with a new name.
c. Specify the analysis type.
d. Specify loading and boundary conditions.
e. Specify the element definition.
f. Specify the other input parameters.
g. Run the analysis.
h. Animate the result.
i. Create the presentation.
j. Generate the report.
k. Save and close the model.

Opening Tutorial 4 Model
You need to open the Tutorial 4 model file for analysis.

1. Start Autodesk Simulation Mechanical and then open the *c10_tut04.fem* model.

Saving the .fem File with a New Name
You need to save the model with the *c10_tut05* file name.

1. Save the model with the name *c10_tut05* in the *Tut05* folder of chapter 10.

Specifying the Analysis Type
Now, you need to specify the analysis type to Transient Stress (Direct Integration).

1. Choose the **Analysis** tab in the **Ribbon**. Next, choose the **Type** tool in the **Change** panel; a drop-down is displayed, refer to Figure 10-49.

2. Choose **Nonlinear > MES with Nonlinear Material Models**, refer to Figure 10-49; the **Autodesk Simulation Mechanical** message window is displayed.

3. Choose the **Yes** button from the message window; another message window may be displayed. Choose **Yes** to create a new design scenario rather than changing the analysis type for the previous one.

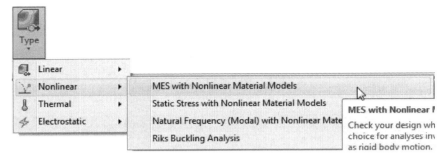

Figure 10-49 *The **MES with Nonlinear Material Models** analysis type being selected*

Specifying Loading and Boundary Conditions

Now, you need to specify load, boundary conditions, and material properties. In this tutorial, you will use the previously defined properties.

1. Change the current orientation of the model to the one shown in Figure 10-50, by using the ViewCube.

    It is evident from Figure 10-50 that one end of the beam is fixed and the load is applied on its other end. Now, you need to edit the loading magnitude to 80000N.

2. Expand the **Load and Constraint Groups** node of the **Design Scenario 4** in the **Tree View** by clicking on the +sign available on its left.

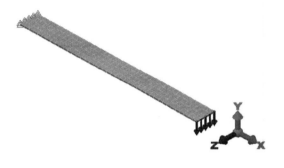

Figure 10-50 *Modified orientation of the model*

3. Select **Nodal Forces** from the expanded **Load and Constraint Groups** node of the **Tree View** and then right-click; a shortcut menu is displayed.

4. Choose the **Edit** option; the **Creating 5 Nodal Force Objects** dialog box is displayed.

5. Enter **-16000** in the **Magnitude** edit box of the dialog box.

Note
The total magnitude of the force applied on to the model is 80000 N and the number of nodes are 5 therefore the magnitude of the force per node is equal to 16000 N.

6. Next, choose the **Curve** button from the dialog box; the **Multiplier Table Editor** dialog box is displayed, refer to Figure 10-51.

7. Add additional rows to the multiplier table by using the **Add Row** button of the dialog box and then enter the time and multiplier values in the table, refer to Figure 10-51.

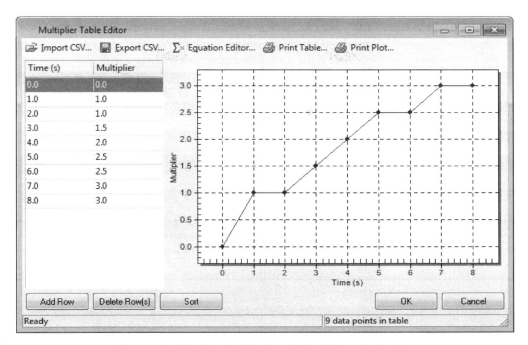

*Figure 10-51 The **Multiplier Table Editor** dialog box*

8. Choose the **OK** button from the **Multiplier Table Editor** dialog box. Next, choose the **OK** button from the **Creating 5 Nodal Force Objects** dialog box; the total force of magnitude 80000 N is applied.

Specifying the Element Definition

Now, you need to specify the element definition.

1. Expand the **Parts** node of the **Design Scenario 4** in the **Tree View** if not expanded already.

2. Select **Part1** from the expanded **Parts** node and then right-click; a shortcut menu is displayed.

3. Choose the **Edit > Element Definition** from the shortcut menu; the **Element Definition - Brick/Tetrahedral** dialog box is displayed, refer to Figure 10-52.

4. Make sure that the **Large Displacement** option is selected in the **Analysis Type** drop-down list of the dialog box, refer to Figure 10-52.

5. Choose the **OK** button from the dialog box.

Specifying Other Input Parameters

Now, you need to specify the other input parameters required for analysis.

1. Choose the **Setup** tab in the **Ribbon** and then choose the **Parameters** tool; the **Analysis Parameters - MES with Nonlinear Material Models** dialog box is displayed , refer to Figure 10-53.

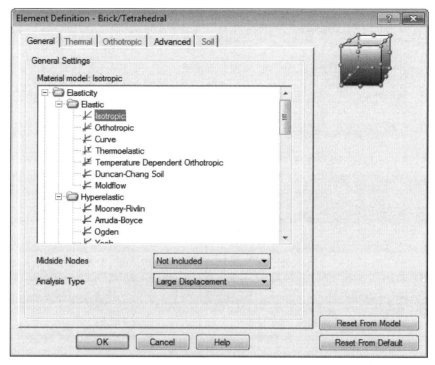

*Figure 10-52 The **Element Definition - Brick/Tetrahedral** dialog box*

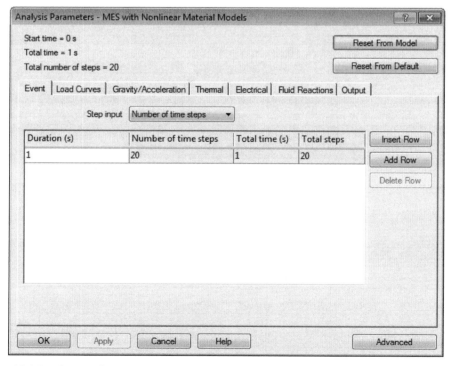

*Figure 10-53 The **Analysis Parameters - MES with Nonlinear Material Models** dialog box*

2. Make sure the **Number of time steps** option is selected in the **Step input** drop-down list of the **Event** tab in the dialog box.

3. Enter **8** in the **Duration (s)** column and **80** in the **Number of time steps** column of the dialog box, refer to Figure 10-53.

4. Choose the **OK** button from the dialog box.

Running Analysis

After specifying all the required input parameters, you can run the analysis and review the results.

1. Choose the **Analysis** tab in the **Ribbon**. Next, choose the **Run Simulation** tool in the **Analysis** panel; the process of structural analysis is started.

2. Once the process of analysis is done, the result environment is invoked with the displacement magnitude result for the default time step, refer to Figure 10-54. Note that the result for the total 80 time steps has been created. Figure 10-54 shows the displacement result for time step 1 on time 0.1 sec and Figure 10-55 shows the results for the second time step at 0.2 sec.

Note

*To switch from the result of one time step to another, you need to choose the **Next** button ▶ of the **Load Case Options** panel. On choosing the **Previous** button ◀ , you can review the results for the previous time step.*

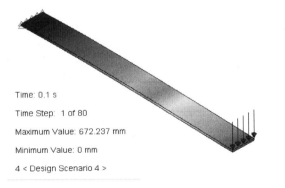

Time: 0.1 s

Time Step: 1 of 80

Maximum Value: 672.237 mm

Minimum Value: 0 mm

4 < Design Scenario 4 >

Time: 0.2 s

Time Step: 2 of 80

Maximum Value: 1027.49 mm

Minimum Value: 0 mm

4 < Design Scenario 4 >

Figure 10-54 The displacement result for time step 1

Figure 10-55 The displacement result for time step 2

Animating the Result

Now, you need to animate the frequency mode shapes.

1. Choose the **Start** button 🖼 from the **Captures** panel of the **Results Contours** tab in the **Ribbon**; the animation starts in the graphics area and different result values for different time steps are displayed.

2. Once the animated results are reviewed, stop the animation by choosing the **Stop** button 🔳 from the **Captures** panel of the **Results Contours** tab in the **Ribbon**.

Creating a Presentation

After reviewing the results, you need to create presentations for different results. Note that the presentation for the displacement is created by default.

1. Expand the **Saved Presentations** node of the **Tree View** by clicking on the +sign available on its left if not expanded by default.

2. Select **Stress** from the expanded **Saved Presentations** node of the **Tree View** and then right-click; a shortcut menu is displayed.

3. Choose the **Activate** option; the stress presentation is activated and added under the **Presentations** node of the **Tree View**. Also, the stress result is displayed in the graphics window.

4. Similarly, create other required presentations.

 Note
*The presentations created in the result environment and displayed under the **Presentations** node of the **Tree View** are also added to the report.*

Generating the Report

Once the required results are reviewed and different presentations are created in the Result environment, you can generate the report.

1. Choose the **Report** tab available at the top of the **Tree View**; the report environment is invoked and a report of different results and input parameters is generated.

   Note that on the left in the report environment, the default contents added to the report are displayed in the **Tree View**. You can select the content whose result you want to display in the display area. You can control the display of contents in the report by using the **Configure** tool of the **Setup** panel of the **Report** tab.

2. Choose the **PDF** tool from the **Save As** panel of the **Report** tab in the **Ribbon**; the **Save As** dialog box is displayed.

3. Browse to the location where you want to save the report file and then choose the **Save** button; the report is saved in the pdf file format.

Saving the Model

Now, you need to save the model file.

1. Choose the **Save** button from the **Quick Access Toolbar** to save the model.

2. Choose **Close** from the **Application Menu** to close the file.

Self-Evaluation Test

Answer the following questions and then compare them to those given at the end of this chapter:

1. In dynamic analysis, the load or field conditions vary with time. (T/F)

2. When the intensity of the stress exceeds the elastic limit, the member loses its elastic property. (T/F)

3. In Autodesk Simulation, you can filter out the results of some of the frequencies by defining the lower and upper cut off. (T/F)

4. The _____ analysis is used to calculate the vibration characteristics of a structure.

5. The _____ analysis is used to calculate the response of a structure to arbitrary time varying loads.

Review Questions

Answer the following questions:

1. If the load is removed after the elastic limit is exceeded, the member will not retain its original shape. (T/F)

2. When the natural frequency of a system is very close to the operating conditions or to the excitation frequency, the component becomes resonant and can fail. (T/F)

3. The _____ analysis is used mainly for analyzing rotating machinery.

4. The _____ analysis is used to calculate the response of a structure to harmonically time varying loads.

5. The _____ button of the **Load Case Options** panel in the result environment is used to review the results of next time step or load case.

EXERCISES

Exercise 1

In this exercise, you will perform a simple Model analysis on a cantilever beam shown in Figure 10-56, and then find the first three natural frequencies and mode shapes.

(Expected time: 30 min)

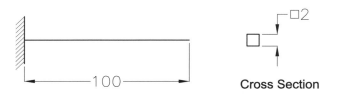

Dimensions are in Inches

Figure 10-56 *The model for Exercise 1*

Exercise 2

In this exercise, you will perform the transient stress analysis on a cantilever beam shown in Figure 10-57 and calculate the response of the model to arbitrary time varying load 100 N. The damping ratio to be used is 0.01.   (**Expected time: 30 min**)

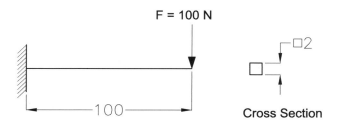

Dimensions are in Inches

Figure 10-57 *The model for Exercise 2*

Answers to Self-Evaluation Test
1. T, **2.** T, **3.** T, **4.** Model, **5.** Transient

Index

Other Publications by CADCIM Technologies

The following is the list of some of the publications by CADCIM Technologies. Please visit *www.cadcim.com* for the complete listing.

AutoCAD Textbooks
- AutoCAD 2016: A Problem-Solving Approach, 3D and Advanced, 22nd Edition
- AutoCAD 2016: A Problem-Solving Approach, Basic and Intermediate, 22nd Edition
- AutoCAD 2015: A Problem-Solving Approach, 3D and Advanced, 21st Edition
- AutoCAD 2014: A Problem-Solving Approach

AutoCAD Plant 3D Textbooks
- AutoCAD Plant 3D 2016 for Designers, 3rd Edition
- AutoCAD Plant 3D 2015 for Designers
- AutoCAD Plant 3D 2014 for Designers

Autodesk Inventor Textbooks
- Autodesk Inventor 2016 for Designers, 16th Edition
- Autodesk Inventor 2015 for Designers, 15th Edition
- Autodesk Inventor 2014 for Designers

AutoCAD MEP Textbooks
- AutoCAD MEP 2016 for Designers, 3rd Edition
- AutoCAD MEP 2015 for Designers
- AutoCAD MEP 2014 for Designers

Solid Edge Textbooks
- Solid Edge ST7 for Designers, 12th Edition
- Solid Edge ST6 for Designers
- Solid Edge ST5 for Designers

NX Textbooks
- NX 9.0 for Designers, 8th Edition
- NX 8.5 for Designers
- NX 8 for Designers

SolidWorks Textbooks
- SOLIDWORKS 2015 for Designers, 13th Edition
- SolidWorks 2014 for Designers
- SolidWorks 2013 for Designers

CATIA Textbooks
- CATIA V5-6R2014 for Designers, 12th Edition
- CATIA V5-6R2013 for Designers

Creo Parametric and Pro/ENGINEER Textbooks
- PTC Creo Parametric 3.0 for Designers, 3rd Edition
- Creo Parametric 2.0 for Designers
- Creo Parametric 1.0 for Designers

Autodesk Simulation Mechanical Textbooks
- Autodesk Simulation Mechanical 2015 for Designers
- Autodesk Simulation Mechanical 2014 for Designers

ANSYS Textbooks
- ANSYS Workbench 14.0: A Tutorial Approach
- ANSYS 11.0 for Designers

Creo Direct Textbook
- Creo Direct 2.0 and Beyond for Designers

Autodesk Alias Textbooks
- Learning Autodesk Alias Design 2016, 5th Edition
- Learning Autodesk Alias Design 2015, 4th Edition
- Learning Autodesk Alias Design 2012

AutoCAD LT Textbooks
- AutoCAD LT 2015 for Designers, 10th Edition
- AutoCAD LT 2014 for Designers

EdgeCAM Textbooks
- EdgeCAM 11.0 for Manufacturers
- EdgeCAM 10.0 for Manufacturers

AutoCAD Electrical Textbooks
- AutoCAD Electrical 2015 for Electrical Control Designers, 6th Edition
- AutoCAD Electrical 2014 for Electrical Control Designers
- AutoCAD Electrical 2013 for Electrical Control Designers

Autodesk Revit Architecture Textbooks
- Autodesk Revit Architecture 2016 for Architects and Designers, 12th Edition
- Autodesk Revit Architecture 2015 for Architects and Designers, 11th Edition

Autodesk Revit Structure Textbooks
- Exploring Autodesk Revit Structure 2016, 6th Edition
- Exploring Autodesk Revit Structure 2015, 5th Edition

AutoCAD Civil 3D Textbooks
- Exploring AutoCAD Civil 3D 2016, 6th Edition
- Exploring AutoCAD Civil 3D 2015, 5th Edition

AutoCAD Map 3D Textbooks
- Exploring AutoCAD Map 3D 2016, 6th Edition
- Exploring AutoCAD Map 3D 2015, 5th Edition
- Exploring AutoCAD Map 3D 2014

3ds Max Design Textbooks
- Autodesk 3ds Max Design 2015: A Tutorial Approach, 15th Edition
- Autodesk 3ds Max Design 2014: A Tutorial Approach
- Autodesk 3ds Max Design 2013: A Tutorial Approach

3ds Max Textbooks
- Autodesk 3ds Max 2016: A Comprehensive Guide, 16th Edition
- Autodesk 3ds Max 2016 for Begginers: A Tutorial Approach, 16th Edition
- Autodesk 3ds Max 2015: A Comprehensive Guide, 15th Edition

Autodesk Maya Textbooks
- Autodesk Maya 2016: A Comprehensive Guide, 8th Edition
- Autodesk Maya 2015: A Comprehensive Guide, 7th Edition
- Character Animation: A Tutorial Approach

ZBrush Textbook
- Pixologic ZBrush 4R6: A Comprehensive Guide

Fusion Textbooks
- Blackmagic Design Fusion 7.0 Studio: A Tutorial Approach
- The eyeon Fusion 6.3: A Tutorial Approach

Flash Textbooks
- Adobe Flash Professional CC: A Tutorial Approach
- Adobe Flash Professional CS6: A Tutorial Approach

AutoCAD Textbooks Authored by Prof. Sham Tickoo and Published by Autodesk Press
- AutoCAD: A Problem-Solving Approach: 2013 and Beyond
- AutoCAD 2012: A Problem-Solving Approach
- AutoCAD 2011: A Problem-Solving Approach

Textbooks Authored by CADCIM Technologies and Published by Other Publishers

3D Studio MAX and VIZ Textbooks
- Learning 3DS Max: A Tutorial Approach, Release 4
  Goodheart-Wilcox Publishers (USA)
- Learning 3D Studio VIZ: A Tutorial Approach
  Goodheart-Wilcox Publishers (USA)

CADCIM Technologies Textbooks Translated in Other Languages

SolidWorks Textbooks
- SolidWorks 2008 for Designers (Serbian Edition)
  Mikro Knjiga Publishing Company, Serbia
- SolidWorks 2006 for Designers (Russian Edition)
  Piter Publishing Press, Russia
- SolidWorks 2006 for Designers (Serbian Edition)
  Mikro Knjiga Publishing Company, Serbia

NX Textbooks
- NX 6 for Designers (Korean Edition)
  Onsolutions, South Korea
- NX 5 for Designers (Korean Edition)
  Onsolutions, South Korea

Pro/ENGINEER Textbooks
- Pro/ENGINEER Wildfire 4.0 for Designers (Korean Edition)
  HongReung Science Publishing Company, South Korea
- Pro/ENGINEER Wildfire 3.0 for Designers (Korean Edition)
  HongReung Science Publishing Company, South Korea

AutoCAD Textbooks
- AutoCAD 2006 (Russian Edition)
  Piter Publishing Press, Russia
- AutoCAD 2005 (Russian Edition)
  Piter Publishing Press, Russia
- AutoCAD 2000 Fondamenti (Italian Edition)

Coming Soon from CADCIM Technologies
- CATIA V5-6R2015 for Designers
- NX 10.0 for Designers, 9th Edition
- NX Nastran 9.0 for Designers
- Exploring Primavera P6 V8.1
- Exploring Bentley STAAD.Pro V8i

Online Training Program Offered by CADCIM Technologies
CADCIM Technologies provides effective and affordable virtual online training on animation, architecture, and GIS software, computer programming languages, and Computer Aided Design, Manufacturing and Engineering (CAD/CAM/CAE) software packages. The training will be delivered 'live' via Internet at any time, any place, and at any pace to individuals, students of colleges, universities, and CAD/CAM/CAE training centers. For more information, please visit the following link: *http://www.cadcim.com*

29131874R00240

Made in the USA
Middletown, DE
08 February 2016